Promoting Innovation in New Ventures and Small- and Medium-Sized Enterprises

Series on Technology Management

Series Editor: J. Tidd (*University of Sussex, UK*) ISSN 0219-9823

The Technology Management Series is dedicated to the advancement of academic research and management practice in the field of technology and innovation management. The series features titles which adopt an interdisciplinary, multifunctional approach to the management of technology and innovation, and includes work which seeks to integrate the management of technological, market and organisational innovation. All titles are based on original empirical research, and includes research monographs and multiauthor edited works. The focus throughout is on the management of technology and innovation at the level of the organisation or firm, rather than on the analysis of sectoral trends or national policy.

Published

Vol. 28 *Promoting Innovation in New Ventures and Small- and Medium-Sized Enterprises*
edited by Joseph Tidd (University of Sussex, UK)

Vol. 27 *The Role of Creativity in the Management of Innovation: State of the Art and Future Research Outlook*
edited by Alexander Brem (University of Southern Denmark, Denmark), Rogelio Puente-Diaz (University Anáhuac, Mexico) & Marine Agogué (HEC Montréal, Canada)

Vol. 26 *Total Value Development: How to Drive Service Innovation*
by Frank M. Hull (Cass Business School and Fordham University Graduate School of Business-Executive Education, USA) & Christopher David Storey (University of Sussex, UK)

Vol. 25 *Small Firms as Innovators: From Innovation to Sustainable Growth*
by Helena Forsman (University of Tampere, Finland)

Vol. 24 *The Knowledge Enterprise: Innovation Lessons from Industry Leaders (Second Edition)*
by E. Huizenga (University of Amsterdam, The Netherlands)

Vol. 23 *Open Innovation Research, Management and Practice*
edited by J. Tidd (University of Sussex, UK)

Vol. 22 *Discontinuous Innovation: Learning to Manage the Unexpected*
by P. Augsdörfer (Technische Hochschule Ingolstadt, Germany), J. Bessant (University of Exeter, UK), K. Möslein (Universität Erlangen-Nürnberg, Germany), B. von Stamm (Innovation Leadership Forum, UK) & F. Piller (RWTH Aachen University, Germany)

More information on this series can also be found at http://www.worldscientific.com/series/stm

(Continued at the end of the book)

SERIES ON TECHNOLOGY MANAGEMENT – VOL. 28

Promoting Innovation in New Ventures and Small- and Medium-Sized Enterprises

Editor

Joe Tidd

Science Policy Research Unit (SPRU)
University of Sussex

World Scientific

EW JERSEY • LONDON • SINGAPORE • BEIJING • SHANGHAI • HONG KONG • TAIPEI • CHENNAI • TOKYO

Published by

World Scientific Publishing Europe Ltd.
57 Shelton Street, Covent Garden, London WC2H 9HE
Head office: 5 Toh Tuck Link, Singapore 596224
USA office: 27 Warren Street, Suite 401-402, Hackensack, NJ 07601

Library of Congress Cataloging-in-Publication Data
Names: Tidd, Joseph, 1960– editor.
Title: Promoting innovation in new ventures and small- and medium-sized enterprises / [edited by] Joseph Tidd, Science Policy Research Unit (SPRU), University of Sussex, UK.
Description: New Jersey : World Scientific Publishing Company, [2017] | Series: Series on technology management ; vol. 28 | Includes index.
Identifiers: LCCN 2017010488 | ISBN 9781786343475 (hardcover)
Subjects: LCSH: New business enterprises--Management. | Small business--Management. | Technological innovations--Management. | Organizational change. | Knowledge management.
Classification: LCC HD62.5 .P736 2017 | DDC 658.4/063--dc23
LC record available at https://lccn.loc.gov/2017010488

British Library Cataloguing-in-Publication Data
A catalogue record for this book is available from the British Library.

Desk Editor: Karimah Samsudin

Typeset by Stallion Press
Email: enquiries@stallionpress.com

Printed in Singapore

CONTENTS

INTRODUCTION

Joe Tidd

Science Policy Research Unit (SPRU)
University of Sussex, U.K.

Ever since the pioneering work of early scholars such as Joseph Schumpeter and Peter Drucker, the fields of Innovation and Entrepreneurship have evolved to become two separate and distinct disciplines. The early work of Schumpeter, so-called Schumpeter 1, focussed on the contribution of entrepreneurial start-ups and smaller firms, whereas the later work, Schumpeter 2, emphasized the role of formal research and development and industrial innovation in larger firms. However, we believe that the study and practice of each field has suffered as a result: entrepreneurship has become preoccupied with individual entrepreneurs and small business creation, and innovation dominated by corporate R&D and new product development (Tidd, 2014).

In this book, we aim to identify themes which can reunite the study and practice of entrepreneurship and innovation by examining a potentially bridging phenomenon. Our focus here is on high-growth, innovative SMEs, rather than the more commonplace lifestyle businesses, and the interactions between SMEs and larger organizations, private and public. This book is organized around three overlapping themes: SME innovation performance, practices, and networks.

Performance

Too often, entrepreneurs, the self-employed, and SMEs are grouped together as a single group for policy purposes, whereas research confirms that these have very different characteristics, motives, and outcomes. For example, the Global Entrepreneurship Monitor (GEM) tracks entrepreneurship at the national level, and in addition to cultural, demographic, and educational factors, includes indicators of infrastructure and national context. However, the focus of the GEM is on start-up activity, rather than subsequent success or growth, so it also fails to differentiate small businesses from successful entrepreneurial activity.

In terms of innovation, the performance of SMEs is easily exaggerated. Early studies based on innovation counts consistently indicated that when adjusted for size, smaller firms created more new products than the larger counterparts. However, methodological shortcomings appear to undermine this clear message. When the divisions and subsidiaries of larger organizations are removed from such samples, and the innovations weighted according to their technological merit and commercial value, the relationship between firm size and innovation is reversed: larger firms create proportionally more significant innovations than SMEs.

Therefore, we need to move beyond such generalizations, and develop a finer-grained taxonomy to reveal the relationships between innovation and performance in SMEs. The relationship between innovation and performance is different in SMEs than in larger firms, and varies significantly by sector. Overall, formal R&D by SMEs is only weakly associated with profitability, and is not correlated with success or growth, but broader innovation management investment and practices can have a positive influence on SME performance, including incremental organizational and process innovations (Forsman, 2015). However, different sectors and technological and market characteristics will demand different innovation strategies and management practices. For example, in service SMEs, integration of suppliers and partners is critical to innovation. In software ventures, the level of R&D expenditure, use of external technology, and management of intellectual property are associated with success, whereas for biotechnology start-ups, success is more influenced by the commercial experience of the founder, quality of scientific staff (measured by citation counts), and location within a significant concentration of similar firms. Such sector-specific studies confirm that the environment in which small firms operate significantly influences both the opportunity for innovation, in a technological and market sense, and the most appropriate strategy and practices for innovation (Bessant and Tidd, 2017).

Practices

In much of the entrepreneurship literature, capabilities are conceived in the narrow sense of individual attributes, but attempts to measure more general entrepreneurial traits have been less than successful. For example, the General Enterprise Tendency test assesses five types of trait — need for achievement, drive and ambition, risk-taking, autonomy and creativity, and potential for innovation. This instrument has proven effective at identifying potential owner-mangers, but is unable to distinguish these from successful entrepreneurs and ventures.

Therefore, it may be more productive to consider innovation management practices in the collective and aggregate sense more commonly adopted in the strategy and innovation literatures on dynamic capabilities and resource-based view of the firm. Capabilities can be conceptualized as collective, consisting of more than individual intellectual ability or level of formal education, and include social, tacit and experiential knowledge. Expertise, knowledge base, and practices all contribute to new venture creation and success. In this way, we can begin to explore the integration of multiple founders, from entrepreneurship studies, with cross-functional teams, from the field of innovation management and new product development (Tidd and Thuriaux-Alemán, 2016).

At each of the different stages of developing a venture, there are specific significant challenges to overcome to make a successful transition to the next stage, so-called "critical junctures". For example, the initial opportunity recognition, at the interface of the research and opportunity framing phases, requires the ability to connect a specific technology or know-how to a commercial application, and is based on a rather rare combination of skill, experience, aptitude, insight, and circumstances. A key issue here is the ability to synthesize scientific knowledge and market insights, which increases with the entrepreneur's social capital — linkages, partnerships, and other network interactions. This requires a delicate balance between differentiation and integration of capabilities, and strong ties within disciplines to develop depth, but weaker ties across functions to promote innovation.

Networks

The ability to identify, develop, and exploit different networks over time appears to be central to the successful transition and growth of a new venture. Innovative SMEs are likely to have diverse and extensive linkages with a variety of external sources of innovation, and in general, there is a positive association between the level of external technical and professional inputs and the performance of an SME. The sources of innovation and precise types of relationship vary by sector, but links with contract research organizations, suppliers, customers, and universities are consistently rated as being highly significant, and constitute the 'social capital' of the firm. However, such relationships are not without cost, and the management and exploitation of these linkages can be difficult for an SME, and can overwhelm the limited technical and managerial resources of SMEs. As a result, in some cases, the cost of collaboration may outweigh the benefits and, in the specific case of collaboration between SMEs and universities, there is an inherent mismatch between the short-term, near-market focus of most SMEs and the long-term, basic research interests of universities.

The size and location of SMEs also has an influence on performance. Geographic closeness increases the likelihood of informal linkages and encourages the mobility of skilled labour across firms. Larger SMEs are associated with a greater spatial reach of innovation-related linkages, and with the introduction of more novel product or process innovations for international markets. In contrast, smaller SMEs are more embedded in local networks, and are more likely to be engaged in incremental innovations for the domestic market. However, networks are more than just ways of assembling and deploying knowledge to promote innovation. They can also have "emergent properties", which can deliver a wide range of benefits beyond the collective knowledge efficiency, including sharing risks, accessing new markets and technologies and strategic positioning and power. Therefore, the position an SME occupies in a network is a matter of great strategic importance, including the identification and development of innovative business models (Tidd and Bessant, 2017).

References

Bessant, J. and J. Tidd (2017). *Entrepreneurship.* Wiley, 2017; Bessant J. and J. Tidd *Innovation and Entrepreneurship.* Third edition, Wiley, 2015.

Forsman, H. (2015). *Small Firms as Innovators: From Innovation to Sustainable Growth.* Imperial College Press, London.

Tidd, J. (2014). Conjoint innovation: Building a bridge between innovation and entrepreneurship. *International Journal of Innovation Management*, 18(1), 1450001.

Tidd J. and B. Thuriaux-Alemán (2016). Innovation management practices: Cross-sectorial adoption, variation and effectiveness, *R&D Management*, 46(3), 1024–1043.

Tidd J. and J. Bessant (2017). *Managing Innovation: Integrating Technological, Market and Organizational Change.* Sixth edition, Wiley.

SME Innovation and Performance

Chapter 1

INNOVATION AND GROWTH IN THE CITY REGION: MICROECONOMIC EVIDENCE OF ASYMMETRIES

David Devins, George Lodorfos,
Ioannis Kostopoulos and Don Webber

Leeds Beckett University
United Kindgom

Using a micro-economic perspective, this study explores the type, scale and impact of innovation activities that small and medium-sized enterprises (SMEs) undertake with focus on the sub-national level. Based on a representative sample of 250 independent SMEs, based in the Leeds City Region (LCR) and using quantitative methods, this study provides evidence of the types of innovation that are most likely to result in achieving policy aims associated with for example the development of international markets or growth of employment opportunities. The findings show that there is a need for aligning business support for innovation with human capital (education/training) interventions. Furthermore, the findings provide a strong evidence base, at the local level, to inform the development of innovation policy and illustrate the connections between different types of innovation SMEs' growth at regional level.

Keywords: Innovation; growth; region; microeconomics.

Introduction

Small and Medium Enterprises (SMEs) are regarded as the preferred vehicle for the regeneration and development of economies (Gavron *et al.*, 1998; Carr *et al.*, 2000; Beaver and Carr, 2002; Stanworth and Gray, 1991) and for the enhancement of a country's competitiveness (Dobbs and Hamilton, 2007). Specifically,

This paper was first published in *International Journal of Innovation Management*, Vol. 20, No. 2 (February 2016), 1650032 (25 pages). © Imperial College Press. DOI: 10.1142/S1363919616500328 they are considered essential for the establishment of a solid industrial base and innovation that facilities industrialisation (Dickson and Weaver, 2008; Smallbone *et al.*, 2010; Arando *et al.*, 2009; Harris and Gibson, 2006; Sauser, 2005; Tonge *et al.*, 2000; Robbins *et al.*, 2000). Notably, they play a vital role in the economy and society not just because performance affects a large degree of the country's economic growth but due to the large number of people they employ and jobs created (Aziz and Mahmood, 2011; Ahmad and Xavier, 2012). In addition to SMEs, innovation is frequently cited as a critical element of the growth of individual firms and the economy in general (Hasan and Tucci, 2010).

Even though it is agreed that there is correlation between firm size and innovation, many studies examine the issue of innovation relating to company size with inconclusive findings (Acs and Audretsch, 1987; 1990; 1991; Cohen and Klepper, 1996; Damanpour, 2010; Freeman and Soete, 1997; Kamien and Schwartz, 1975; Pavitt *et al.*, 1987; Rothwell and Zegveld, 1986; Tsvetkova *et al.*, 2014; Wagner and Hansen, 2005). For instance, Acs and Audretsch's (1987) findings suggest the average innovation rate for small firms was 43% higher than large firms in 1982, however this does not imply the answer to the question of 'Which firm size is more innovative?' is the 'small firm'. Despite the relative innovative activities between large- and small-medium enterprises, as De Jong and Marsili (2006) argued, there is research evidence to suggest new small firms are continually entering the market with new ideas, products and processes.

Furthermore, Griliches (1986) argued that larger firms can afford to invest in long-term research with high potential returns but smaller firms cannot — either for financial reasons, or because of the high technological complexity of these projects. This is contrary to other studies, which have contested that because of their nature, SMEs — particularly those operating in manufacturing, electronics, engineering and general high technology industries — are able to undertake radical innovation more easily than large firms. The same studies argue that introducing pioneering products is essential for the sustainability of SMEs (Kanter, 1985; Simon *et al.*, 2002). Similarly, empirical findings confirm small firms cannot wait for the payoff from their innovative activities, and focus on short-term incremental (rather than radical) innovations (Kafouros, 2008; Pavitt *et al.*, 1987; Piergiovanni *et al.*, 1997). Moreover, several studies suggest the opportunity for small firm innovation tends to be highest when the industry is in the early stages of the product life-cycle (e.g., Pavitt and Wald, 1971).

Markedly, a number of British small, medium-size manufacturing enterprises (SMMEs) have survived and thrived through the release of innovative new products (Laforet and Tann, 2006). Thus, the contention is that small firms who innovate successfully increase their chances of survival and growth (Cefis and Marsili, 2003). In the UK there is evidence to suggest policy makers see an important link between innovation and business, and economic growth. The recognition of the importance of innovation in SMEs has led to the development of the National Systems of Innovation in several countries, as well as the formation of innovation clusters and regional reservoirs of knowledge (Filippetti and Archibugi, 2011).

From all the above it is evident that not only is innovation related to firm size, but also that different types of innovation affect companies in different ways. The ability to determine the scale of innovation activities and the characteristics of innovative firms is a prerequisite for the pursuit and analysis of policies aimed at fostering innovation at a regional level. The standard measures of innovation activity such as Research and Development activity or R&D intensity or other common measures such as patents provide little indication of the nature of innovation. Policy development requires further advances in the analysis of information, particularly in terms of the relationship between different types of innovation and growth without excluding the need for survival. Furthermore, previous research was more focused on technology intensive or dependent industries and sectors, with less recognition on the importance of innovation in less R&D intensive industries such as services, low-technology manufacturing and other sectors of the economy that contribute significantly to employment and economic growth.

Shelton and Percival's (2003) findings suggest innovation priorities vary widely by sector, with the focus spread across products, technology and services yet innovation is a key driver of organic growth for all companies regardless of sector or geography. Similarly, Anderson *et al.* (2014) argue innovation has become an increasingly important determinant of organisational performance, success and long-term survival.

Additionally, previous studies have focused on technological, product and process (TPP) innovations. Since then, there has been a growing sense that much of innovation in service sectors of the economy is not adequately captured in the TPP concept. As a result, the OECD (2005) argue the concept of innovation needs to be expanded to include marketing and organisational innovation, and take account of different forms of product and market innovation, in order to provide a more complete analytical framework and one better able to capture the changes that affect firm performance. Moreover, the majority of previous studies took a national or even international perspective.

The relationship between economic growth and innovation on one hand, and firm size on the other, is evident and well recognised. The relationship between different types of innovation and growth has been a more controversial issue. Aiming to fill these gaps in the pertinent literature, this study is set within a public policy context defined by localism, and an aspiration to reflect business interests in the development and implementation of local policies (Gov, 2010) to establish the connections between different forms of innovation and the anticipated growth of SMEs at City Region level.

Literature Review

Innovation and growth in SMEs

Research on innovation spans a number of disciplines with economic approaches alone adopting several alternative theoretical perspectives each of which offers useful and significant insights. In order to provide the theoretical basis for this study, we reviewed previous studies on the types of innovation and literature that investigated the link between innovation, SMEs and growth. Previous studies cover a range of products, services, processes and organisations and encompass a broad range of innovation types including technological (Pratali, 2003), and non-technological such as, strategic (Turock, 2001), organisational (Salavou *et al.*, 2004), marketing (Johne and Davies, 2000), process (Khazanchi *et al.*, 2007) and/or administrative (Tarry, 2000).

On this basis, Schumpeter (1934; 1939) identified several types of innovation classified as technological and non-technological innovations, including the introduction of new products, the introduction of new methods of production and the opening of new markets. Product innovation, which may include new product offerings or improvements in existing products, appears to be the most common form of innovation. Also, service innovation has become increasingly important and can be described as new developments in those activities undertaken to deliver the core product and make it more attractive to consumers. Gadrey *et al.* (1995) defined service innovations as innovations in processes and innovations in organisation for existing service products. Service innovations can therefore be described as new developments in activities undertaken to deliver core service products for various reasons, for instance to make those core service products more attractive to consumers. Organisational innovation refers to the implementation of new organisational methods not previously used in the firm, while a marketing innovation is the implementation of a new marketing method (OECD/ESTAT, 2005: 172–180; Abernathy and Utterback, 1978; Trott, 1998).

The technological view on innovation has been criticised as biased towards innovation in manufacturing and not able to fully capture innovation in services (Hipp and Grupp, 2005; Hipp *et al.*, 2000 cited in Schmidt and Rammer, 2007). Furthermore, firms often adopt and re-organise business routines, internal organisation, external relations and marketing (Baranano, 2003; Boer and During, 2001). Innovation management scholars stress the importance and emphasise the crucial role of integrating technological and non-technological innovations for successfully transferring new ideas and new business opportunities into market success (Cozzarin and Percival, 2006; Tidd *et al.*, 2001; Griffin and Hauser, 2001).

Similarly, Slater and Narver (1995) argued the most likely way in which service companies innovate is by developing new services or reformulating existing ones, plus creating new distribution channels and discovering new approaches for management. These are essentially marketing innovations as they represent ways in which companies can develop new methods of marketing to potential or existing customers. *Marketing* innovations should not be confused with *market* innovations that, according to Johne and Davies (2000), are concerned with activities such as the entry into new markets. Another type, process innovation, involves creating or improving methods of production, service or administrative operations (Khazanchi *et al.*, 2007) as well as, developments in the processes, systems and re-engineering activities undertaken to develop new products.

The innovation process has been defined as radical or incremental innovation, imitated or completely new, improved or revolutionary, technological or organisational types and many other kinds. (Edosomwan, 1989; Tohidi and Jabbari, 2011; Tohidi, 2011; Evangelista *et al.*, 2001; Tohidi *et al.*, 2010). In Schumpeter's view 'radical' innovations create major disruptive changes, whereas 'incremental' innovations continuously advance the process of change.

The idea that innovation is a linear and sequential process proceeding through specific steps (see Kline and Rosenberg's model 1986), has been replaced by a systemic approach to innovation. The systemic approach has recognised, therefore, that in addition to the usual innovation inputs, external sources of innovation have to be considered in order to explain in particular small firms' innovation (De Propris, 2002). Oke *et al.* (2007) have distinguished between innovation performance and business performance based on the premise that a good innovation performance does not necessarily translate into a good business performance.

Previous studies have used different measures to assess the innovation performance of firms. Generally, outcome level measures of innovation have been based on financial and non-financial metrics (Avlonitis *et al.*, 2001; Cooper and Kleinschmidt, 1995). Edquist and McKelvey (2000) and Lundvall (1992) argue the innovation process should rather be considered as a circular and complex system embracing interactive elements. In addition, a lot of the literature focuses on learning regions (Simmie, 1997; Maskell *et al.*, 1998; Boekema *et al.*, 2000; Nyhan and Attwell, 2000; Lorenzen, 2001), in order to bind together the concepts of innovation and locality.

Some of the early work on innovation and SMEs tends to adopt the view that innovation involves making fundamental or radical changes comprising the transformation of a new idea into a marketable product of process (Freeman, 1971). Conversely, Porter (2000), views innovation as an attempt for companies to defend their existing competitive position as well as seek new competitive advantages. A recent survey by Shelton and Percival (2013) found that innovation is a driver for rapid and profitable revenue growth and recognised by the executives interviewed as being integral to sustain the long-term future as well as the competitive position of their business.

Summarising, a number of studies have documented the positive impact of innovation on SMEs' rate of growth (Becchetti and Trovato, 2002; Lu and Beamish, 2001; Yasuda, 2005). In particular, innovation can create new, uncontested product markets, as well as increased consumers' willingness to pay by adding or improving product features (Cho and Pucik, 2005; Cohen and Klepper, 1996). Moreover, innovative firms are better able to benefit from spillovers, and are less sensitive to adverse macroeconomic shocks (Geroski *et al.*, 1993). Therefore, a firm may take a reactive approach and innovate to avoid losing market share to an innovative competitor or it may take a proactive approach to gain a strategic market position relative to its competitors by developing new products or services.

Most of the previous studies have analysed the impact of innovation on SMEs' growth and there is 'agreement' in the literature that innovation in different settings plays an important role in the survival, growth and competitiveness of firms but the previous research findings are inconclusive with regards to the type and scale of innovation activities at local levels and in particular the impact on SMEs growth. However a number of scholars, including Schumpeter (1934), assumed similar effects on market structures and firm performance which raises the question about the role played by various types of innovation.

Hypotheses development

It is widely recognised that growth is vital to all businesses and in particular to SMEs (Golovko and Valentini, 2011). There is substantial evidence suggesting SMEs in a variety of sectors engage in innovative activities, and these activities are likely to be an important determinant of performance and growth (See Becchetti and Trovato, 2002; Cooper and Kleinschmidt, 1986; Rothwell, 1991; Lawton-Smith *et al.*, 1991; SBRC, 1992; Joyce *et al.*, 1994; Keeble, 1992; Barber *et al.*, 1989; Moore, 1993; Herbert, 1988; Yasuda, 2005). Love and Roper (2013) argue those SMEs with a track record of innovation are more likely to export successfully, and more likely to generate growth from exporting than non-innovating firms. They also suggest for firms of all sizes that there is a strong positive relationship between innovation, exporting and productivity and/or growth but evidence on the inter-relation between innovation and exporting in SMEs specifically, is limited.

Previous studies suggest that SMEs' innovative activities are more likely to involve product innovation than process innovation. In addition, their innovative activities are focused more on producing products for niche markets rather than mass markets, and generate both radical and incremental innovations likely to be associated with growth in output, turnover and employment (Brouwer *et al.*, 1993; Pratten, 1991; Storey, 1993; Keeble, 1992; Cannon, 1985; Keeble and Walker, 1993; Oakey, 1990; Lee and Oakes, 1995; Hartman *et al.*, 1994; Henderson, 2002; Goh and Ridgway, 1994). Moreover, the literature provides considerable evidence of the direct contribution of workforce, managerial and marketing skills to innovation, and link between innovation and growth in employment, but again the literature on SMEs specifically is limited (See Leiponen, 2005; Freel, 2005; Vivarelli, 2012). The majority of empirical studies have found a positive relationship between product innovation and employment growth (König *et al.*, 1995; Van Reenen, 1997; Greenan and Guellec, 2000; Hall *et al.*, 2008; Peters *et al.*, 2013; Harrison *et al.*, 2014).

However, the scope and size of performance enhancing effects of innovation at the level of the innovating firm depends on the type of innovative activity and degree to which innovation outputs (new products, new processes) are successfully implemented and succeed the market (see Stoneman, 1983; Katsoulacos, 1984; Peters, 2008). Empirical evidence on the employment effects of process innovations is less clear. Van Reenen (1997), Entorf and Pohlmeier (1990) and Hall *et al.* (2008) found the effect of process innovation on employment to be small or statistically insignificant. On the contrary, other studies found a significant positive effect of process innovation on employment growth (Lachenmaier and Rottmann, 2011; König *et al.*, 1995; Smolny, 2002; Greenan and Guellec, 2000).

On the basis of the above discussion and aiming to contribute to the debate about the link between innovation and SMEs' growth and in particular to provide empirical evidence as to what role different types of innovation play on SMEs' growth, the following hypotheses have been developed:

H1: *Firms are more likely to anticipate growth in sales when they undertake (a) product, (b) process, (c) marketing and (d) organisational innovation.*

H2: *Firms are more likely to anticipate growth in profitability when they undertake (a) product, (b) process, (c) marketing and (d) organisational innovation.*

H3: *Firms are more likely to anticipate growth in employment when they undertake (a) product, (b) process, (c) marketing and (d) organisational innovation.*

H4: *Firms are more likely to anticipate growth in exports when they undertake (a) product, (b) process, (c) marketing and (d) organisational innovation.*

Methodology

The Leeds City Region Small- and Medium-Sized Enterprise (SME) Innovation Survey is a telephone-based survey of 250 independent organisations based in the Leeds City Region (LCR). This unique survey is designed and conducted at the enterprise as opposed to establishment level since the survey sought to identify the innovation dynamics of local firms and did not want to include sites that were part of larger groups based outside the LCR. The survey has two principal aims: (i) to test out a methodology to empirically explore firm-based innovation activity and performance amongst SMEs in the LCR and (ii) to collect data to inform the development and targeting of local services to support firm-based innovation in the area. For the purposes of this survey, SMEs are defined as those employing between 10 to 250 employees. The exclusion of micro enterprises (employing under 10) from the study's sampling frame is purposeful and influenced by several factors, most importantly that micro enterprises tend to be qualitatively different from 'larger' SMEs and increasingly seen as a distinct analytical category in their own right.

A stratified random sampling with both size and sector was deployed. The survey's sample was selected in order to be representative of independent SMEs in the LCR in the commercial and charitable sectors of the economy. The profile of the population was established through the Office for National Statistics based on data from Inter-Departmental Business Register (IDBR) statistics. Quotas were set for the numbers of interviews to be achieved related to size and sector to minimise the impact of sample bias, which may have occurred due to differential rates of response among enterprises of different sizes or sectors. Given the policy emphasis on the manufacturing sector and its pre-eminence in innovation discourse, a relatively large quota was set for this sector to ensure the sample base was sufficient to provide a foundation for sub-sector analysis. The contact details for employers were sourced from a commercial database of establishments and interviews were conducted with the owner or senior business representative of the SME. The characteristics of the sample are indicated in Table 1 below.

Table 1. Sample characteristics.

	n	%
Size		
10–49	158	63
50–249	91	36
Sector		
Primary and manufacturing	9	4
Construction	74	30
Wholesale/Retail	30	12
Transport/storage	20	8
Food/accommodation	9	4
Information/communication	19	8
Finance/Real Estate	16	6
Professional services	26	10
Administrative and public	47	19
Single Multi Site		
Single site	182	73
Multi–site (HQ in Leeds)	68	27
Age		
Less than 5 years	28	12
6–10 years	40	16
11–20 years	64	26
20–50 years	79	32
50+ years	33	14

Note: Base: All respondents (non-weighted).

Although obtaining data from a single key informant increases the risk of common method variance (Podsakoff *et al.*, 2012), in the current survey the characteristics of the variables, the informants and the firms offer reassurance on the validity and reliability of the data. Specifically, in the study the questions relate to (a) objective constructs, (b) salient events and (c) phenomena internal to the firm. Also, the informants (d) are working in higher hierarchical positions, (e) have a long tenure and the firms are (f) relatively small and (g) relatively old. All these indicate the reliability and validity of the data were ensured by the data collection process (Homburg *et al.*, 2012).

The survey questionnaire was informed by the OECD guidelines associated with firm-based innovation (OECD/ESTAT, 2005). It contained sections that sought to ensure only independent firms contributed to the research and collected data on the characteristics of independent SMEs and a variety of forms of innovation and SME performance.

Four classifications are used to indicate innovative activity that had taken place in the SMEs during the last 12 months:

- Product: the introduction of a good or service that is new with respect to its characteristics or intended uses.
- Process: the implementation of a new production or delivery method.
- Marketing: aimed at better addressing customer needs or opening up new markets with an objective of increasing sales.
- Organisational: the implementation of a new organisational method in the firm's business practices or workplace organisation.

Four measures of anticipated growth were used:

- Sales (turnover)
- Profitability
- Number employed and
- International sales

The analysis draws on the questions associated with innovation activity in the past 12 months and the entrepreneurs' expectations of growth in the next 12 months. The analysis in this paper explores the relationship between types of innovation and likelihood of SME growth and high growth (defined as anticipated growth in sales or employment by 20% or more). The questionnaire also included sections on performance management and networking that do not feature in the following analysis.

Analysis

Method

Mantel–Haenszel common odds ratios are widely used in academic studies for two main reasons: they provide an estimate (with confidence interval) of the association between two binary variables and they have a convenient interpretation. The particular method has been used several times in recent similar studies (Vuong *et al.*, 2013; Hechavarria *et al.*, 2012). Odds ratios represent the ratio of the probability that an event will happen relative to the probability the event will not happen; in our case we link the probability that a manager will expect a type of growth to the probability that the firm has introduced an innovation. If the probabilities of the events occurring are p_1 for expected future growth and p_2 for past innovation, then the odds ratio is:

$$\frac{p_1/(1-p_1)}{p_2/(1-p_2)}. \qquad (1)$$

An odds ratio equal to 1 indicates the expectation of growth is equally likely to occur irrespective of whether the firm has undertaken innovation activities, and hence the type of growth and innovation are likely to be independent and unrelated. An odds ratio greater than 1 indicates managers are more likely to expect growth if they did introduce an innovation. The Mantel–Haenszel common odds ratio estimate is asymptotically normally distributed around the common odds ratio assumption of unity. To test whether the estimates of the odd ratios have occurred by chance Pearson Chi^2 test was employed.

Results

Table 2, presents a cross-tabulation (counts) corresponding to whether firms introduced new products over the last 12 months and whether the manager considers the business will grow in terms of sales (turnover) in the next 12 months. Out of 245 managers who responded to both questions, 46.1% (113/245) indicated they had introduced new products to their organisations in the previous 12 months and expected sales growth in the next 12 months; this corresponds to 14.6% (36/245) of the sample who stated the opposite. Also, there are firms who introduced new products to their organisation in the previous 12 months but do not expect sales growth in the next 12 months (6.9%; 17/245) as well as firms

Table 2. Cross-tabulation of new product innovation with anticipated sales growth.

		In the last 12 months, has your organisation introduced any new products?		Total
		Yes	No	
In the next 12 months, do you think that the business will grow in terms of . . .? Sales (turnover)	Yes	113	79	192
	No	17	36	53
Total		130	115	245

who did not introduce new products in the previous 12 months but expected growth of sales in the next 12 months (32.2%; 79/245).[1]

Application of the Mantel–Haenszel common odds ratio procedure noted above to the cross-tabulation in Table 2 provides the results presented in Table 3. The estimated odds ratio of 3.029, which is statistically significant at the 1% level, indicates the following: the odds of a firm expecting sales growth in the next 12 months is 3.029 times higher if the firm had introduced new products in the last 12 months than if the firm had not introduced new products in the last 12 months. This estimation procedure is replicated for all measures of innovation and all measures of growth across the entire sample in Table 4.

Table 4 is categorised according to the four groups of innovations and four types of growth. The results indicate that growth in sales is associated with all

Table 3. Mantel–Haenszel common odds ratio estimate.

Estimate		3.029
ln (Estimate)		1.108
Standard error of in (Estimate)		0.329
Asymptotic significance (two-sided)		0.001
Asymptotic 95% confidence interval	Lower bound	1.590
	Upper bound	5.770

[1]In order to be able to estimate odds ratios all four cells require a positive number, when this does not occur, as is the case in a couple of occasions in Table 5 below, it is not possible to compute a useful odds ratio and therefore we highlight this instance with an 'n/a'.

Table 4. All firms.

	Increase sales (q10a)	Increase profits (q10c)	Increase employment (q10e)	Increase exports (q10g)
Product innovations				
Introduced new product codes (extensions)	3.029***	1.994**	1.349	2.312***
Introduced products new to business	1.350	0.990	0.616	0.941
Introduced products new to market	1.374	1.961	1.755	2.181*
Process innovation				
Introduced new ways to improve efficiency	2.288**	2.143**	1.623	1.964*
Marketing innovations				
Radical changes in packaging	4.727	5.276	1.155	2.370*
Radical changes in placement	5.265**	1.873	1.554	2.298**
Radical changes in promotion	2.952***	2.039**	1.738**	1.432
Radical changes in pricing	3.637***	1.825*	1.839**	0.829
Radical changes in product design	3.375***	2.726***	2.430***	2.018**
Organisational innovations				
Policies and procedures	2.403**	1.268	1.705**	1.283
Outsourcing	2.766*	1.460	1.948*	1.402
Initiatives to encourage job satisfaction	3.033***	1.739*	2.134***	1.362
Investment in new technology	4.235***	3.066***	3.409***	1.846**
Investment in training and development	3.740***	2.517***	3.319***	2.003*

Note: ***, ** and * represent statistical significance at the 1%, 5% and 10% levels, respectively.

four types of innovation, since all relevant estimates are greater than 1 and most of them are statistically significant. Hence, hypotheses H1a, H1b, H1c and H1d are accepted. For product innovations new to the business or new to the market, as well as for radical changes in packaging the estimates, although greater than 1, were not found significant ($p > 0.1$). Similarly, all four types of innovation influence the possibility of growth in terms of profit increase, as most of the pertinent estimates are statistically significant. The estimates for some of the subcategories were not found significant ($p > 0.1$), but since the influence on profit growth is strong for at least one subcategory from each type of innovation, hypotheses H2a, H2b, H2c and H2d can be accepted.

Regarding the third type of growth included in the study, product and process innovations are not strongly associated with increases in employment ($p > 0.1$).

The innovation-type mostly associated with employment growth is organisational innovation, such as investment in new technology or in training and development. Firms with installed marketing innovations are also more likely to be expecting an expansion in their number of workers, but less so if these marketing innovations are related to radical changes in packaging or placement ($p > 0.1$). On this basis, hypotheses H3a and H3b are rejected, whereas hypotheses H3c and H3d are accepted.

Finally, all four types of innovation are associated with expectations of increases in exports. Even though the estimates for some of the subcategories were not found significant ($p > 0.1$), the impact on expected export growth is strong for at least one subcategory from each type of innovation. Hence, hypotheses H4a, H4b, H4c and H4d are accepted. Of great importance here is that radical marketing innovation is significantly associated with expectations of increases in exports. This may reflect business activity stimulated by improved information associated with export markets and distribution channels to reach these markets. Radical changes in product design in response to receiving information from overseas markets may also be the reason behind why results have identified that radical changes in product design over the last 12 months are associated with expectations of increases in exports.

From a policy perspective, the results of the study indicate for those companies wishing to stimulate greater sales growth it may be opportune to encourage innovation in radical changes in placement, as odds of a firm expecting sales growth in the next 12 months is 5.265 times higher if the firm had introduced these changes in the last 12 months relative to those firms who had not introduced radical changes in placement in the last 12 months. Also, the results indicate those policy makers wishing to stimulate increases in profits should encourage investment in new technologies, as odds of a firm expecting profit growth in the next 12 months is 3.066 times higher if the firm had introduced these changes in the last 12 months relative to those firms who had not invested in new technologies over the last 12 months.

Similarly, those policy makers wishing to stimulate increases in employment should encourage investment in new technologies or training and development, as odds of a firm expecting employment growth in the next 12 months is 3.409 and 3.319 times higher, respectively, if the firm had introduced these changes in the last 12 months relative to those firms who had not. Furthermore, those policy makers wishing to stimulate increases in exports should encourage radical changes in packaging and placement, as emphasised above, or product innovation (new products to the market), as odds of a firm expecting employment growth in the next 12 months are at least two times higher, respectively, if the firm had introduced these changes in the last 12 months relative to those firms who had not.

High growth firms

There is much interest and attention in high growth firms, either to identify who these currently are or who they are likely to be in the near future. The results presented above were re-estimated using the dichotomous growth indicator related to whether the manager predicted growth in sales or employment over the next 12 months to be greater than 20%. The corresponding results are presented in Table 5 and several points are worth highlighting:

First, it becomes clear that fewer and slightly different innovations are associated with expectation of very high levels of growth. For instance, evidence in Table 5 suggests a much stronger relationship between the introduction of new products to the market and growth expectations and product innovation is not significantly associated with expectations of high growth unless they are new products to the market.

Second, firms that undertook organisational innovations over the last 12 months remain much more likely to expect high sales growth, but the innovations associated with growth in other areas are less clear. For instance, radical changes in product design, which are probably associated with the introduction of new products to the market, are the only product, process or marketing innovation associated with expectations of high profit growth. It may be unsurprising that outsourcing is the innovation-type with the strongest association with expectations for high profit growth, although this may have a strong product life-cycle component and vary across sectors.

Third, policy makers seeking to encourage innovation in order to stimulate high rates of employment growth appear to have four innovation areas to encourage: one is to encourage new products to the market which again is probably closely associated with radical changes in product design; another is outsourcing, which may save costs that could then be redirected to other areas of the business with a shortage of workers. A further area to encourage innovation in order to stimulate larger employment growth expectations is linked to investments in new technologies. Extant literature emphasises this is a complex issue as technological change can have a skilled-bias, unskilled-bias, sector-bias or a combination of all of three, thereby suggesting an intricate knowledge of the firm is required for the most effective policy recommendation.

Table 5. Comparable results for high growth firms.

	High sales growth (>20%)	High profit growth (>20%)	High employment growth (>20%)	High exports growth (>20%)
Product innovations				
Introduced new product codes (extensions)	1.500	0.860	0.874	1.768
Introduced products new to business	0.765	0.625	0.708	n/a
Introduced products new to market	2.424**	3.611*	3.481*	2.667
Process innovation				
Introduced new ways to improve efficiency	1.528	1.242	1.812	2.135
Marketing innovations				
Radical changes in packaging	1.789	0.887	2.143	2.300
Radical changes in placement	1.121	1.449	1.157	1.777
Radical changes in promotion	1.558	1.335	0.714	3.879**
Radical changes in pricing	1.867*	1.747	1.226	2.876*
Radical changes in product design	1.629	2.195**	2.000*	1.460
Organisational innovations				
Policies and procedures	3.127***	1.903*	0.889	5.238**
Outsourcing	5.109***	4.042***	2.196*	3.419**
Specific initiatives to encourage job satisfaction	3.592***	1.497	1.369	2.494
New technology	4.763***	2.167*	3.437***	2.794
Training and development	2.130*	1.103	1.558	2.122

Note: ***, ** and * represent statistical significance at the 1%, 5% and 10% levels, respectively. '*na*' indicates that values are not available due to a lack of observations in cells i.e., there are no high export growth firms that reported no organisational innovations.

Conclusions and Discussions

Innovation as a concept and its empirical measurement are open to multiple understandings and interpretations. The literature is wide ranging and diverse, adopting a range of perspectives and presenting results associated with different typologies of innovation at the micro and macro level. The starting point for this research was the indication arising from previous work, that there is a connection between innovation and SMEs' growth but there was not enough previous research that was looking into the connections between different forms of innovation and the anticipated growth within different regions.

The contribution of the current study lies in that it established a methodology that provides an evidence base to inform the development of innovation policy for the SME 'market' at the local level and established the connections between different forms of innovation and the anticipated growth (using four measures) of SMEs. The survey methodology provides an empirical basis for discussion on the nature of innovation and its contribution to different areas of SME growth. In addition, the focus on the LCR allows us to provide a focused and detailed assessment of the association between types of innovation and types of growth that could not be done accurately at a national level without an extremely large representative dataset.

This study also provides robust and timely intelligence to inform operational support for SMEs' innovation and strategic development of local innovation policy. The research may be used for a number of purposes including the targeting of further intelligence gathering associated with, for example, the barriers to innovation faced by local SMEs, raising awareness of existing services, identifying gaps in supply or the potential development of new services to support different types of innovation. It has the potential to contribute to the development of local innovation strategy, informed by the practice of local SMEs as opposed to the requirements of programmes developed by, for example, Government Departments, LEPs, Further and Higher Educational Institutions and other public sector organisations who often struggle to engage a representative sample of SMEs in the design of programmes and policies. The authors plan to adopt a reflexive process to explore research impact with local stakeholders that will help assess the utility of the methodology and its findings, and continue to shape further research in this area.

The findings confirm the central role innovation plays in shaping the growth of SMEs employing between 10 and 250 people in Leeds, and are consistent with prior research suggesting firms engaged in multiple forms of innovation are more likely to grow (and grow faster) than those that do not (Von Tunzelmann, 1995). It provides evidence of the types of innovation most likely to result in achieving policy aims associated with, for example, the development of international markets (product development and investment in training) or growth of employment opportunities (investment in new technology and training and development), highlighting the need to align business support for innovation with human capital (education/training) interventions at the local level. In the case of 'high growth' firms, innovations associated with development of policies and procedures are most likely to lead to anticipated growth in export sales, those associated with outsourcing most likely to lead to growth in domestic sales and profit growth and innovations associated with investments in new technology most likely to lead to high employment growth. The findings of this study illustrates a clear link between innovation, growth and economic performance, something which highlights why innovation policy is so important to the regions. Finally, the synergies identified between SME innovations and exporting suggest the value of co-ordinated support mechanisms and conducive policies at regional level.

Limitations and Suggestions for Future Research

This study, although offering useful conclusions, is tempered by some limitations. Firstly, it has adapted and applied the conceptual framework of the OSLO Manual (OECD/ESTAT, 2005) and sought to apply some common meanings to common forms of innovation — product, process, marketing and organisation. This does not provide a holistic framework and we recognise further conceptual and empirical development may be required. For instance, a key limitation of the current analysis is the absence of discussion associated with open innovation and networking which is recognised as a key feature of SME innovation (Rothwell, 1992). Similarly, no formal models of industrial concentration and industrial geography have been used. Future research should follow this approach and enrich our study's findings.

In addition, the present study is focused on one region — specifically the LCR. Another suggestion for future researchers would be to carry out comparable research focused on other city regions. This would prove useful in assessing the external validity of our results, in enhancing their generalisability, and in drawing out important city-region-specific idiosyncrasies. Finally, often the innovating firms are very adept at working across the interface of local and global knowledge transfers networks. This raises questions that future research could investigate over the relative importance of regional versus international knowledge spill overs for the innovative.

References

Abernathy, WJ and J Utterback (1978). Pattern of industrial innovation. In *Readings in the Management of Innovation*, ML Tushman and WL Moore (eds.), New York, NY: HarperCollins.

Acs, ZJ and DB Audretsch (1987). Innovation, market structure, and firm size. *The Review of Economics and Statistics*, 69(4), 567–574.

Acs, ZJ and DB Audretsch (1991). R&D, firm size, and innovative activity. In *Innovation and Technological Change: An International Comparison*, ZJ Acs and DB Audretsch (eds.), New York: Harvester Wheatsheaf.

Ahmad, ZS and RS Xavier (2012). Entrepreneurial environments and growth: Evidence from Malaysia GEM data. *Journal of Chinese Entrepreneurship*, 4(1), 50–269.

Anderson, N, K Potočnik and J Zhou (2014). Innovation and creativity in organizations: A state-of-the-science review, prospective commentary, and guiding framework. *Journal of Management*, 40(5), 1297–1333.

Arando, S, I Peña and I Verheul (2009). Market entry of firms with different legal forms: An empirical test of the influence of institutional factors. *International Entrepreneurship and Management Journal*, 5(1), 77–95.

Avlonitis, GJ, PG Papastathopoulou and SP Gounaris (2001). An empirically-based typology of product innovativeness for new financial services: Success and failure scenarios. *Journal of Product Innovation Management*, 18(5), 324–342.

Aziz, AS and R Mahmood (2011). The relationship between business model and performance of manufacturing small and medium enterprises in Malaysia. *African Journal of Business Management*, 5(22), 8918–8932.

Baranano, AM (2003). The non-technological side of technological innovation: State-of-the-art and guidelines for further empirical research. *International Journal of Entrepreneurship and Innovation Management* 3, 107–125.

Barber, J, S Metcalfe and M Porteous (eds.) (1989). *Barriers to Growth in Small Firms*. London: Routledge.

Beaver, G and P Carr (2002). The enterprise culture: Understanding a misunderstood concept. *Journal of Strategic Change*, 11(2), 105–113.

Becchetti, L and G Trovato (2002). The determinants of growth for small and medium sized firms: The role of the availability of external finance. *Small Business Economics*, 19(4) 291–306.

Boekema, F, K Morgan, S Bakkers and R Rutten (2000). *Knowledge, Innovation and Economic Growth*, Cheltenham: Edward Elgar.

Boer, H and WE During (2001). Innovation, what innovation? A comparison between product, process and organisational innovation. *International Journal of Technology Management*, 22, 83–107.

Brouwer, E, A Kleinknecht and J Reijnen (1993). Employment growth and innovation at the firm level: An Empirical Study. *Journal of Evolutionary Economics* 3, 153–159.

Cannon, T (1985). Innovation, creativity and small firm organisation. *International Small Business Journal*, 4, 33–67.

Carr, M, A Chen and J Tate (2000). Globalization and home-based workers. *Feminist Economics*, 6(3), 123–142.

Cefis, E and O Marsili (2003). Survivor: the role of innovation in firm's survival, No.03-18. WPTKoopans Institute, USE, Utrecht University in Laforet, S. (2008) Size, strategic, and market orientation affects on innovation. *Journal of Business Research* 61(7), 753–764.

Cho, H and V Pucik (2005). Relationship between innovativeness, quality, growth, profitability, and market value. *Strategic Management Journal*, 26(6), 555–575.

Cohen, WM and S Klepper (1996). Firm size and the nature of innovation within industries: The case of product and process R&D. *The Review of Economics and Statistics*, 78(2), 232–243.

Cooper, RG and E Kleinschmidt (1986). An investigation into the new product process: Steps, deficiencies and impact. *Journal of Product Innovation Management*, 3(2), 71–85.

Cooper, RG and EJ Kleinschmidt (1995). Benchmarking the firm's critical success factors in new product development. *Journal of Product Innovation Management*, 12(5), 374–391.

Cozzarin, B and J Percival (2006). Complementarities between organizational strategies and innovation. *Economics of Innovation and New Technology*, 15, 195–217.

Damanpour, F (2010). An integration of research findings of effects of firm size and market competition on product and process innovations. *British Journal of Management*, 21(4), 996–1010.

De Jong, PJJ and O Marsili (2006). The fruit flies of innovations: A taxonomy of innovative small firms. *Research Policy*, 35, 213–229.

De Propris, L (2002). Types of innovation and inter-firm co-operation. *Entrepreneurship and Regional Development*, 14, 337–353.

Dickson, PH and KM Weaver (2008). The role of the institutional environment in determining firm orientation towards entrepreneurial behaviour. *International Entrepreneurship Management Journal*, 4(4), 467–483.

Dobbs, M and R Hamilton (2007). Small business growth: Recent evidence and new directions. *International Journal of Entrepreneurial Behaviour and Research*, 13(5), 296–322.

Edosomwan, JA (1989). *Integrating Innovation and Technology Management*, John Wiley & Sons: New York.

Edquist, C and M McKelvey (2000). *Systems of Innovation: Growth, Competitiveness and Employment*, Cheltenham: Edward Elgar.

Evangelista, R, S Imammarino, V Mastrostefano and A Silvani (2001). Measuring the regional dimension of innovation: Lessons from the Italian innovation survey. *Technovation*, 21, 733–755.

Filippetti, A and D Archibugi (2011). Innovation in times of crisis: National Systems of Innovation, structure, and demand. *Research Policy*, 40(2), 179–192.

Freel, MS (2005). Patterns of innovation and skills in small firms. *Technovation*, 25, 123–134.

Freeman, C (1971). The role of small firms in innovation in the united kingdom since 1945: Report to the Bolton Committee of inquiry on small firms. Research Report No. 6. HMSO, London.

Freeman, C and L Soete (1997). *The Economics of Industrial Innovation*, Cambridge, MA: The MIT Press.

Freeman, J, GR Carroll and MT Hannan (1983). The liability of newness: Age dependence in organizational death rates. *American Sociological Review*, 48(5), 692–710.

Gadrey, J, F Gallouj and O Weinstein (1995). New modes of innovation: how services benefit Industry. *International Journal of Service Industry Management*, 6(3), 4–16.

Gavron, R, M Cowling, G Holtham and A Westall (1998). *The Entrepreneurial Society, London, Institute for Public Policy Research.*

Geroski, P, S Machin and J Van Reenen (1993). The profitability of innovating firms. *Rand Journal of Economics*, 24(2), 198–211.

Goh, PL and K Ridgway (1994). Implementation of TQM in SMEs: A survey of progress and attitudes, operation and performance. In *1st International Conf. the European Operation Management Association*, University of Cambridge, June, pp. 567–573.

Golovko E and G Valentini (2011). Exploring the complementarity between innovation and export for SMEs' growth. *Journal of International Business Studies*, 42, 362–380.

Gov, HM (2010). Local Growth. Realising Every Place's Potential. Available at http://www.bis.gov.uk/assets/biscore/economic-development/docs/l/cm7961-local-growth-white-paper.pdf (accessed on July 2013).

Greenan, N and D Guellec (2000). Technological innovation and employment reallocation. *Labour*, 14, 547–590.

Griffin, A and JR Hauser (2001). Integrating R&D and marketing: a review and analysis of the literature. *Journal of Product Innovation Management*, 13, 191–215.

Griliches, Z (1986). Productivity, R&D, and basic research at the firm level in the 1970s. *American Economic Review*, 76(1), 141–154.

Hall, BH, F Lotti and J Mairesse (2008). Employment, innovation, and productivity: Evidence from Italian microdata. *Industrial and Corporate Change*, 17, 813–839.

Harris, ML and SG Gibson (2006). Determining the common problems of early growth of small businesses in Eastern North Carolina. *SAM Advanced Management Journal*, 71(2), 39–45.

Harrison, R, J Jaumandreu, J Mairesse and B Peters (2014). Does innovation stimulate employment? A firm-level analysis using comparable micro-data from four European countries. *International Journal of Industrial Organization*, 35, 29–43.

Hartman, A, E Burk, C Tower and T Sebora (1994). Information sources and their relationship to organisational innovation in small businesses. *Journal of Small Business Management*, 32, 36–46.

Hasan, I and CL Tucci (2010). The innovation–economic growth nexus: Global evidence. *Research Policy*, 39(10), 1264–1276.

Hechavarria, DM, M Renko and CH Matthews (2012). The nascent entrepreneurship hub: Goals, entrepreneurial self-efficacy and start-up outcomes. *Small Business Economics*, 39(3), 685–701.

Henderson, J (2002). Building the rural economy with high-growth entrepreneurs. *Economic Review*, 87(3), 45–70.

Herbert, C (1988). Inter-firm information flows and innovation in small firms. In *11th National Small Firms Policy and Research Conf.*

Heseltine, M (2012). No stone unturned in pursuit of growth. Available at https://www.gov.uk/government/uploads/system/uploads/attachment_data/file/34648/12-1213-no-stone-unturned-in-pursuit-of-growth.pdf (accessed on July 2013).

Hipp, C and H Grupp (2005). Innovation in the service sector: The demand of service-specific innovation measurement concepts and typology. *Research Policy*, 34, 517–535.

Hipp, C, B Tether and I Miles (2000). The incidence and effects of innovation in services: Evidence from Germany. *International Journal of Innovation Management*, 4, 417–453.

Homburg, C, M Klarmann, M Reimann and O Schilke (2012). What drives key informant accuracy? *Journal of Marketing Research*, 49(4), 594–608.

Johne, A and R Davies (2000). Innovation in medium-sized insurance companies: How marketing adds value. *International Journal of Bank Marketing*, 18, 6–14.

Joyce, P, C Seaman and A Woods (1994). The economic growth implications of control and innovation strategies in small businesses. London Central Training and Enterprise Council.

Kafouros, IM (2008). Economic Returns to Industrial Research. *Journal of Business Research*, 61, 868–876.

Kamien, MI and NL Schwartz (1975). Market structure and innovation: A survey. *Journal of Economic Literature*, 13(1), 1–37.

Kanter, RM (1985). Supporting innovation and venture development in established companies. *Journal of Business Venturing*, 1(1), 47–60.

Katsoulacos, YS (1984). Product innovation and employment, *European Economic Review*, 26, 83–108.

Keeble, D (1992). Small firm creation, innovation and growth and the urban–rural shift. Working Paper 52 Technovation Vol. 18 No. 1 No. 24, Small Business Research Centre, University of Cambridge, September.

Keeble, D and S Walker (1993). *New Firm Formation and Small Business Growth in the United Kingdom: Spatial and Temporal Variations and Determinants*. Research Series No. 15, Cambridge, Department of Geography and Small Business Research Centre, University of Cambridge.

Keeble, D, P Tyler, G Broom and J Lewis (1992). *Business Success in the Countryside, the Performance of Rural Enterprise*. London: HMSO, Department of the Environment.

Khazanchi, S, MW Lewis and K Boyer (2007). Innovation-supportive culture: The impact of organizational values on process innovation. *Journal of Operations Management*, 25(4), 871–884.

Kline, SJ and N Rosenberg (1986). An overview of innovation. In *The Positive Sum Strategy*, R Landau and N Rosenberg (eds.), pp. 275–305 (Washington DC: National Academy Press).

König, H, G Licht and H Buscher (1995). Employment, investment and innovation at the firm level, In *The OECD Jobs Study – Investment, Productivity and Employment*, OECD (ed.), pp. 67–81, Paris.

Lachenmaier, S and H Rottmann (2011). Effects of innovation on employment: A dynamic panel analysis. *International Journal of Industrial Organization*, 29, 210–220.

Laforet, S and J Tann (2006). Innovative characteristics of small manufacturing firms. *Journal of Small Business and Enterprise Development*, 13(3), 363–380.

Lawton-Smith, H, K Dickson and S Lloyd Smith (1991). There are two sides to every story: Innovation and collaboration within networks of large and small firms. *Research Policy*, 20, 457–468.

Lee, GL and IK Oakes (1995). Impact of process innovation on supply chain relationships, some smaller firm perspectives. *Proc. European Conf. Management of Technology, Aston*, University, Birmingham, pp. 413–426.

Leiponen, A (2005). Skills and innovation. *International Journal of Industrial Organization*, 23(5–6), 303–323.

Lorenzen, M (2001). Localized learning and policy: Academic advice on enhancing regional competitiveness through learning. *European Planning Studies*, 9, 163–185.

Love, J and S Roper (2013). SME innovation, exporting and growth. ERC White Paper No.5. Available at http://www.enterpriseresearch.ac.uk/wp-content/uploads/2013/12/ERC-White-Paper-No_5-Innovation-final.pdf (accessed on 7 May 2015).

Lu, J and P Beamish (2001). Internationalization and performance of SMEs. *Strategic Management Journal*, 22(6/7), 565–586.

Lundvall, BA (1992). *National Systems of Innovation: Towards a Theory of Innovation and Interactive Learning*. London: Pinter.

Lundvall, BA and S Borrás (1999). *The Globalising Learning Economy: Implications for Innovation Policy*. Office for Official Publications of the European Communities, Luxembourg.

Maskell, P, H Eskelinen, I Hannibalsson, A Malmberg and E Vatne (1998). *Competitiveness, Localised Learning and Regional Development. Specialization and Prosperity in Small Open Economies*. Routledge, London.

Moore, B (1993). Financial constraints to the growth and development of small high technology firms. Small Business Research Centre, Cambridge University (mimeo).

Nyhan, B and G Attwell (2000). Towards the learning region: Education and regional innovation in the European Union and the United States. Luxembourg: Office for Official Publications of the European Communities.

Oakey, R (1990). Problems of sustaining growth in established small high-tech firms. In *Symp. Growth and Development of Small High-Technology Business*, Cranfield.

OECD and Eurostat (2005). *Oslo Manual - Proposed Guidelines for Collecting and Interpreting Technological Innovation Data*, 3rd Edition. Paris.

Oke, A, G Burke and A Myers (2007). Innovation types and performance in growing UK SMEs. *International Journal of Operations and Production Management*, 27(7), 735–753.

Pavitt, K and S Wald (1971). *The Conditions for Success in Technological Innovation*, Paris: OECD.

Pavitt, K, M Robson and J Townsend (1987). The size distribution of innovating firms in the UK: 1945–1983. *Journal of Industrial Economics*, 55, 291–316.

Peters, B (2008). *Innovation and Firm Performance. An Empirical Investigation for German Firms, ZEW Economic Studies*, 38. Heidelberg: Physica.

Peters, B, R Riley and I Siedschlag (2013). The influence of technological and non-technological innovation on employment growth in european service firms. Service gap Discussion Paper No. 40, Mannheim.

Piergiovanni, R, E Santarelli and M Vivarelli (1997). From which source do small firms derive their innovative inputs? Some evidence from Italian industry. *Review of Industrial Organization*, 12, 243–258.

Podsakoff, PM, SB MacKenzie and NP Podsakoff (2012). Sources of method bias in social science research and recommendations on how to control it. *Annual Review of Psychology*, 63, 539–569.

Porter, M (2000). Location, competition and local economic development: Local clusters in a global economy. *Economic Development Quarterly*, 14(1), 15–34.

Pratali, P (2003). Strategic management of technological innovations in the small to medium enterprise. *European Journal of Innovation Management*, 6, 18–31.

Pratten, C (1991). *The Competitiveness of Small Firms*. London: Cambridge University Press.

Robbins, DK, LJ Pantuosco, DF Parker and BK Fuller (2000). An empirical assessment of the contribution of small business employment to US state economic performance. *Small Business Economics*, 15(4), 293–302.

Rothwell, R (1991). External networking and innovation in small and medium-sized manufacturing firms in Europe. *Technovation*, 11(2), 93–112.

Rothwell, R (1992). Successful industrial innovation: critical success factors for the 1990s. *R&D Management*, 23(2), 221–240.

Rothwell, R and W Zegveld (1986). *Innovation and the Small and Medium Sized Firm*. London: Francis Pinter.

Salavou, H, G Baltas and S Lioukas (2004). Organisational innovation in SMEs: The importance of strategic orientation and competitive structure. *European Journal of Marketing*, 38, 1091–1112.

Sauser, WI Jr (2005). Starting your own business? Prepare for success. *SAM Management in Practice*, 3(1), 1–4.

SBRC (Small Business Research Centre) (1992). The state of British enterprise: Growth, innovation and competitive advantage in small and medium sized firms. SBRC, University of Cambridge.

Schmidt, T and C Rammer (2007). Non-technological and technological innovation: Strange Bedfellows? Discussion Paper No. 07–052. Available at http://ftp.zew.de/pub/zew-docs/dp/dp07052.pdf (accessed on 7 May 2015).

Schumpeter, J (1934). *The Theory of Economic Development*. Cambridge, Mass: Harvard University Press.

Schumpeter, J (1939). *Business Cycles: A Theoretical, Historical, and Statistical Analysis of the Capitalist Process*, Vol. 2. New York: McGraw-Hill.

Shelton, R and D Percival (2013). Breakthrough innovation and growth-PWC Available at http://www.pwc.ch/user_content/editor/files/publ_adv/pwc_breakthrough_innovation_and_growth_e.pdf (accessed on 7 May 2015).

Simmie, J (1997). *Innovation, Networks and Learning Regions*? Regional Studies Association, Regional Policy and Development Series.

Simon, M, B Elango, S Houghton and S Savelli (2002). The successful product pioneer: Maintaining commitment while adapting to change. *Journal of Small Business Management*, 40(3), 187–203.

Slater, SF and JC Narver (1995). Market orientation and the learning organisation. *Journal of Marketing*, 59, 63–74.

Smallbone, D, F Welter, A Voytovich and I Egorov (2010). Government and entrepreneurship in transition economies: The case of small firms in business services in Ukraine. *Service Industries Journal*, 30(5), 655–670.

Smolny, W (2002). Employment adjustment at the firm level. A theoretical model and an empirical investigation for West German manufacturing firms. *Labour*, 16, 65–88.
Stanworth, J and C Gray (1991). *Bolton 20 Years On: The Small Firm in the 1990s*. London: Paul Chapman.
Stoneman, P (1983). *The Economic Analysis of Technological Change*, New York: Oxford University Press.
Storey, D (1993). Should we abandon support to startup businesses? In *Small Firms, Recession and Recovery*, F Chittenden, M Robertson and D Watkins (eds.), Paul Chapman, London: Institute for Small Business Affairs.
Tarry, SE (2000). Innovation in the administration of public airports. Grant Report March 2000, The PricewaterhouseCoopers Endowment for the Business of Government, Arlington (2000) in Halpern N. (2010) Marketing innovation: Sources, capabilities and consequences at airports in Europe's peripheral areas. *Journal of Air Transport Management*, 16(2), 52–58.
Tidd, JJ, K Bessant and Pavitt (2001). *Managing Innovation: Integrating Technological, Market and Organizational Change*, 2nd Edition. Chichester: Wiley.
Tohidi, H (2011). Review the benefits of using value engineering in information technology project management. *Procedia-Computer Science Journal*, 3(1), 917–924.
Tohidi, H and MM Jabbari (2011). The main requirements to implement an electronic city. *Procedia-Computer Science Journal*, Elsevier, USA.
Tohidi, H, AA Afshar and A Jafari (2010). Using balanced scored card in educational organizations. *Procedia - Social and Behavioral Sciences*, Elsevier, May, USA.
Tonge, R, P Larsen and M Roberts (2000). Information systems investment within high-growth medium-sized enterprises. *Management Decision*, 38(7), 489–502.
Trott, P (1998). *Innovation Management and New Product Development*, London: Financial Times Publishing.
Tsvetkova, A, JC Thill and D Strumsky (2014). Metropolitan innovation, firm size, and business survival in a high-tech industry. *Small Business Economics*, 43(3), 661–676.
Turock, A (2001). Strategic innovations. *Executive Excellence*, 19, 9–10.
Van Reenen, J (1997). Employment and technological innovation: Evidence from U.K. manufacturing firms. *Journal of Labor Economics*, 15, 255–284.
Vivarelli, M (2012). Innovation, employment and skills in advanced and developing countries: A survey of the literature. IZA Discussion Papers 6291.
Von Tunzelmann, N (1995). Technology and industrial progress: The foundations of economic growth. Cheltenham: Edward Elgar.
Vuong, QH, NK Napier and TD Tran (2013). A categorical data analysis on relationships between culture, creativity and business stage: The case of Vietnam. *International Journal of Transitions and Innovation Systems*, 3(1), 4–24.
Wagner, E and E Hansen (2005). Innovation in large versus small companies: insights from the US wood products industry. *Management Decision*, 42(6), 837–850.
Yasuda, T (2005). Firm growth, size, age and behaviour in Japanese manufacturing. *Small Business Economics*, 24(1), 1–15.

Chapter 2

INNOVATION AND ORGANISATIONAL SIZE IN IRISH SMES: AN EMPIRICAL STUDY

Rodney McAdam
School of Business Organisation & Management
University of Ulster, UK

Renee S. Reid
The Centre for Family Enterprise
Glasgow Caledonian University

David A. Gibson
School of Management
Queen's University Belfast, UK

The aim of this paper is to conduct an empirical study to explore the influence of size category on innovation incorporation in Irish SMEs. Changes in markets and in large organisation strategies have resulted in the need for SMEs to re-examine and modify their competitive strategies to increase innovation. Although there is some literature on innovation incorporation in SMEs, there is a lack of direct studies on this issue, with a reliance on related but indirect studies such as Reengineering and New product development in SMEs. Furthermore, there is a paucity of studies and data on the innovation incorporation in different organisational size categories within SMEs.

This study uses a grounded SME innovation model developed from earlier studies as the basis of a questionnaire survey of 2086 SMEs in Northern Ireland. The results indicate that SME organisational size categories has a significant effect on innovation incorporation in relation to leadership, people and culture, TQM/CI, product and process and Knowledge and information management.

Keywords: Innovation; SMEs (Irish); empirical study; size; sector; survey; incorporation; organisational size; conceptual model.

This paper was first published in *International Journal of Innovation Management*, Vol. 8, No. 2 (June 2004), pp. 147–165.

Introduction

SMEs are becoming increasingly involved in global competitive markets, either as part of supply chains or due to expansion and growth (Cagliano *et al.*, 2001). Niche markets, once the preserve of SMEs is being aggressively targeted by larger organisations, which have become more agile and responsive to more refined market segmentation. This pressure is particularly strong in more peripheral regions where indigenous SMEs have traditionally relied more on local markets and now find themselves ill-equipped to face market challenges of the need for growth and exports in a highly competitive environment. In addition to local government help, SMEs in these regions must "re-examine and modify their competitive strategies by fully incorporating innovation within their people processes and products (Ghobadian and Gallear, 1997). Although there is some literature on innovation incorporation within organisational levels and activities (e.g., Vossen, 1999), there is a lack of direct studies on this issue, with a reliance on related but indirect studies such as Reengineering and New product development in SMEs. Furthermore, there is a paucity of studies and data on the innovation incorporation in different organisational size categories within SMEs.

Thus a number of key questions remain, for example, how can innovation be expanded beyond that of just a technical taxonomy to address a much wider view of innovation suitable for SMEs (Cagliano *et al.*, 2001)? Moreover, can a wider view of innovation be evaluated within SMEs? There is also a need to avoid, where possible, any study being affected by different regional policies and infrastructures for innovation (Vossen, 1999).

Thus, any study, which can clarify these issues, will have a twofold benefit. Firstly, SMEs and local government agencies will be able to make more informed decisions in regard to targeting their scarce resources for incorporating innovation and hence increase competitiveness. Secondly, there will be a contribution to knowledge, from which models, methodologies and further studies can be developed.

The aim of this paper is to conduct an empirical study to explore the influence of size category on innovation incorporation in Irish SMEs. The study is within a single region, namely Northern Ireland, to minimise the effects of regional variation and to concentrate on a peripheral region in need of indigenous SME growth. The paper briefly reviews the literature, describes the innovation model derived from previous studies, the research methodology and the results and discussion.

Definition of Innovation within SMEs

Zairi (1994) states "what makes innovation challenging is the fact that it is very difficult to agree on a common definition, and also to decide which firms are the most innovative and how to quantify innovation activity". Peters and Waterman (1982) state "innovative companies are especially adroit at continually responding to change of any sort in their environments and are characterised by creative people developing new products and services". Both of these definitions make reference to the terms change and creativity — they imply that innovation is the harnessing of creative ability within individuals and the workforce in response to change.

Tushman and Nadler (1986) refer to product and process innovation. They contend that "Innovation is the creation of any product, service or process, which is new to a business ... the vast majority of successful innovations are based on the cumulative effect of incremental change in ideas or methods". It is a process of continuous renewal involving the whole company and is an essential part of Business Strategy and every day practice (DTI, 1998). The innovation considered is defined as a technology or practice being used for the first time by members of an organisation, whether or not other organisations have already used it previously (Nord and Tucker, 1987). Thus, innovation based studies in organisations must move beyond a solely technical definition of innovation and include new business practices (Slappendel, 1996). Therefore, reflecting the above discussion, effective business innovation will be defined as the harnessing of creative ability within people and processes in an organisation, in response to customer and market led change.

Innovation in SMEs

Building on this organisational definition, innovation is discussed in relation to the general characteristics of SMEs, in order to identify the capabilities, which support innovation in small firms.

Innovation in relation to leadership, people and culture

The literature defines leadership as being of central importance to all organisations seeking to develop innovation throughout the organisation (Tidd *et al.*, 2001). In SMEs developing innovation, leadership is especially vital for success for a number of reasons. Sadler-Smith *et al.* (1998) study of development of SME employees shows that leadership training and development is often *ad hoc* resulting in adequate leadership skills when faced with business challenges such as the need to innovate. In most SMEs, the owner-manager or management has a larger influence or effect when compared with large organisations (Hale *et al.*, 1996). Thus the leader's vision and drive must be focused on innovation if it is to be successfully incorporated in the organisation. Davenport and Bibby (1999) speak of the "entrepreneurial dynamism" which leaders in SMEs can instil in the behaviours of others in the organisation. This effect instills a dynamic innovative culture within the organisation. Furthermore, SMEs can make rapid decisions when the leadership is dynamic and focussed on innovation (Voss, 1998). Ghobadian and Gallear (1997) and Culkin and Smith (2000) stress that SMEs with leaders close to the action can act as the main innovative catalysts for change.

Ghobadian and Gallear (1997) state that "SMEs are more likely to be people-orientated than system orientated". Throughout the innovation literature the central role of people and the ensuing culture this creates are continually identified as being the key to innovation (e.g., Voss, 1998; Schmidt, 1990). In this area, SMEs must weigh up their basic lack of people resources against their increased flexibility and response. Furthermore, the SME leaders must go beyond autocratic control and facilitate empowerment amongst the SME workforce (Davenport and Bibby, 1999) and achieve effective two-way communications (Ghobadian and Gallear, 1996). Innovative ideas from employee's ideas must be backed by planning and resource allocation and appropriate and timely reward systems (Ghobadian and Gallear, 1997).

Innovation in relation to processes, knowledge and continuous improvement

Scarce resources of people, materials, finance and other types must be carefully managed in organisations which seek to develop (Freel, 2000; Alstrup, 2000; Tidd *et al.*, 2001). SMEs have a generic lack of resources and overall resource strategies and action plans. Gunasekaran *et al.* (1996) states that "overall productivity and quality improvement strategies are lacking in SMEs". Thus rapid decision making by inspired leaders can fail to incorporate innovative practice due to a basic lack of fundamental resources. Mosey *et al.*'s (2002) research into innovation in SMEs showed that what they called "low growth incremental improvers" made poor use of knowledge and information, especially external customer and market information, in comparison to innovative high growth SMEs. Effective knowledge and information management includes new knowledge construction, knowledge embodiment, knowledge dissemination and knowledge use/benefit as shown in Demerest's model (1997), which is based on the concepts of tacit and explicit knowledge conversion as developed by Nonaka and Takeuchi (1995). Thus, knowledge and information management has the potential to be a catalyst for innovation within organisations. Voss (1998) considers close customer proximity and hence, knowledge of the customer to be a key factor in innovation in SMEs. This knowledge will inform SMEs about potential product or service changes in their immediate market. When this knowledge is linked to that of rapid decision making then SMEs can make beneficial innovative decisions in a relatively short space of time (Wiele and Brown, 1999).

In innovative organisational change, Gunasekaran *et al.* (1996) and Bessant and Caffyn (1997) consider continuous improvement to be a fundamental step along the innovation pathway. Wiele and Brown (1998), in their study of Australian SMEs, consider continuous improvement and Total Quality Management to building blocks for wider innovative change within SMEs. Continuous improvement can thus lead to radical innovation where reengineering is attractive to SMEs who want to develop innovation, in that cross-functional activity occurs naturally within SMEs (Raymond *et al.*, 1998). This process approach can enable SMEs to improve supplier compatibility with large organisations, where, for example, 50 percent of sales in manufacturing is generated by SME suppliers (Ghobadian and Gallear, 1997). Appiah-Adu and Singh (1998) suggest this more radical approach can lead to innovation in new product development: "business practitioners and executives of SMEs should combine innovativeness and product quality because such a combination is likely to result in higher levels of new product success rates". Soderquist *et al.* (1997) also sees Reenineering's link with key customers as being supportive of radical innovation within SMEs. This view is supported by Mosey *et al.*'s research where 90 percent of companies incorporating radical innovations were growing, compared with 45 percent of incremental improvers.

Innovation in relation to customers and the market

Appiah-Adu and Singh (1998) see the need for new product development leading to market differentiation as a major opportunity for innovative SMEs in competitive markets. Their research involving 500 SMEs concluded that "there is a strong positive link between the extent of adoption of innovation orientation and the degree of customer orientation in SMEs". In discussing the relative strengths and weaknesses of SMEs in regard to innovation Vossen (1999) and Gadenne (1999) identifies the ability to react rapidly, or even to anticipate, market changes as a key element of innovation. For most organisations market changes are reflected in the actions of their main customers (Tidd *et al.*, 2001). SMEs must use customer-based knowledge to develop innovative products and services through a customer-pull approach (Temtime and Solomon, 2002; Gunasekaran *et al.*, 1997) as distinct from the job-shop ethos. They must become closer to the customer and ensure they are making innovative use of privileged customer information. Soderquist *et al.* (1997) points out that SMEs have little choice but to foster closer links with a very few key customers due to a lack of resources and increased market sector competition.

Innovation and SME Size Categories

SMEs are frequently categorised as organisations below a given number of employees (usually 250 or 500 employees). However, there is lack of homogeneity within SMEs in this category beyond that of being different to large organisations. Although size classifications within SMEs are not the sole differentiating factor, they represent a key factor and initial starting point for studying innovation incorporation in SMEs, which is not restricted to large-SME comparison studies.

Ghobadian and Gallear (1997) indicate that size classifications of SMEs "vary markedly". For example, the European Observatory, produce a range of sizes which attempt to reflect that over 99 percent of EU and UK businesses are SMEs, representing more than 60 percent of the turnover and 65 percent of employment. Welsh and White (1981) encourage size classification studies within SMEs by stating that "a small business is not a little large business", pointing out that differences exist in many areas. Furthermore, Cagliano *et al.*'s (2001) study on world class innovative practices in a large-scale study of SMEs indicates substantial size effects within the 250-employee range. Thus innovation models for SMEs must be developed for SMEs, rather than accepting scaled down versions of large organisation innovation models (McAdam, 2000). Studies have shown that SME size classification affects behaviour, structure, decision-making, process development and leadership style, communications and change implementation, for example TQM (see Ghobadian and Gallear's (1997) study of 5,279 UK SMEs with supporting follow up case studies). These findings were supported by Husband and Mandal's (1999) study of the literature and conceptual model building for SMEs. Both sets of authors highlight the lack of studies with size classifications within SMEs. Therefore, size classification studies are essential in initially investigating innovation incorporation in SMEs, where innovation is seen as affecting the entire business.

A range of authors (e.g., Cagliano *et al.*, 2001; Ghobadian and Gallear, 1996; Bessant and Caffyn, 1997) using both large-scale surveys and case-based methodologies to investigate change in SMEs have used a number of different size classifications within both the 500 employee and 250 employee limits) — helpfully summarised in table form by Martin (1999). As indicated by Cagliano *et al.* (2001), the size classifications are best set in survey work by firstly conducting exploratory studies of the data set across a range of size classifications, and secondly, by matching these classifications to those used in the literature, where possible, in order to increase the opportunities for comparative studies. The current data set and previous studies (Mosey *et al.*, 2002) have shown that an upper limit of 250 employees for SMEs is appropriate (this limit is applied in most UK SME studies) given the structure of the organisations, the markets and the method of government grant apportionment. Therefore, the size classifications chosen were, 1 to 25 employees; 26 to 50 employees; 51 to 200 employees and 200 to 250 employees. In this classification, it is suggested that 1–25 employees is the entrepreneurs span of control and 50 plus employees call for additional organisational arrangements. In summary, it should be noted that in many ways firm size can mask many contextual factors such as sector (Nooteboom, 1994). This issue is further addressed in the recommendations for further research.

Previous Research and Model Building

Previous exploratory EU funded research was undertaken (McAdam *et al.*, 2000) to determine the keys issues involved in incorporating a wider definition of innovation, as derived from the literature, in SMEs. The research methodology was a combined quantitative and qualitative study. In the qualitative approach, the research survey involved a study of 30 SMEs, all with less than 250 employees to investigate how innovation can be incorporated in SMEs. For this initial research, the CENTRIM Innovation model was used to measure innovation. This model was initially developed for large organisations by the University of Brighton, with the wording adapted to be more suitable to SMEs. SMEs were selected which had shown a high level of growth and which had a commitment to ongoing change. The hope was that many of these companies would exhibit innovative practice. Information regarding company selection was sourced from government funding agencies.

In total, 30 companies were surveyed. The exploratory research programme involved the collection of two types of data — qualitative data was obtained from the interviews with the Managing Directors and Management Teams. The analysis of these interviews was produced by the University research team (McAdam *et al.*, 2000). Secondly, focus groups representing a "diagonal Slice" through the organisation were asked to complete a questionnaire based on the model. On average 27 questionnaires per company were collected — 810 in total. These questionnaires were analysed using SPSS. In addition to feedback from the University research team, each business received a report on the questionnaire analysis.

Furthermore, Action Learning Sets were formed involving three clusters of the 30 organisations. These Sets were monitored over a two-year period to see how innovation had been longitudinally incorporated within the SMEs.

Based on this earlier research, a conceptual model for innovation within SMEs in NI has been developed to "reflect the factors and influences forming the very fabric of an SME", as stated by Husband and Mandal (1999) in constructing a conceptual model based on existing research. That is, the derivation of the conceptual model has been grounded in the author's existing work with SMEs over a longitudinal period. Shaw (1999) also used this approach based on SME qualitative case analysis, in developing a conceptual model. The current model, as derived from this research, is shown in Fig. 1.

A survey questionnaire was developed from this model and the exploratory research to give a fuller and more specific representation of innovation

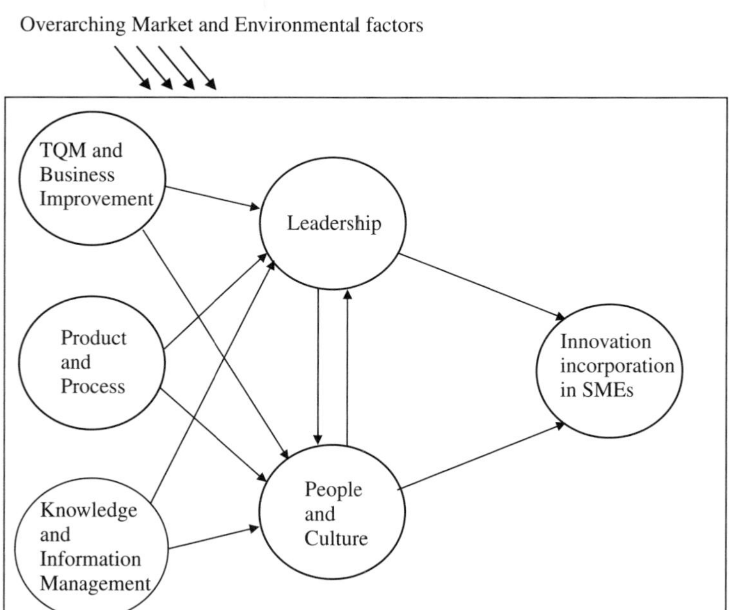

Fig. 1. The innovation model for SMEs.

incorporation in SMEs, as distinct from large organisation models being extrapolated to SME contexts. The current data set will be used to test and further refine this model (Fig. 1).

The Need for Further Research

The discussion on the literature with regard to SMEs and innovation has revealed a number of points. First, most of the innovation studies on SMEs focus on national systems of innovation, rather than the processes whereby innovation is incorporated in SMEs. Second, innovation studies on SMEs are lacking in size classification data below the 250 or 500 employee mark and often are limited to SME — large organisation comparisons. Third, during a grounded model building phase an SME specific innovation model has been developed. This model will require model testing. Finally, it has been found that the effects of SME size classification is a key starting point for examining the current data set in preparation for model testing.

Research Methodology

To address the aims and research questions of the paper it was decided that a large-scale postal survey approach should be adopted. This method was chosen as it is suited to model testing (Remenyi *et al.*, 1999) and determining the effects of size categories within SMEs (Cagliano *et al.*, 2001). Questions were derived for each section of the grounded model postulated from the exploratory research. Therefore, each question had been derived from a triangulation of case data, previous questionnaires and semi structured interviews in line with the suggestion of Remenyi *et al.* (1999); a survey was carried out based on a questionnaire. Furthermore, the questionnaire was piloted in 10 SMEs and with 10 academics working in related disciplines, each with extensive SME experience. These pilot tests led to improvements in the wording throughout the questionnaire and removal and addition of some questions. The final version of the questionnaire contained 84 closed questions relating to each of the areas shown in Fig. 1, namely Leadership, People and Culture, Total Quality/Continuous Improvement, Product and Process, and Knowledge and Information. Issues relating to innovative practices and activities were examined within the context of the market environment. Physical business characteristics such as turnover, size, the number of employees and business sector were also noted. The questionnaires were designed using Formic (design and capture software) and SPSS for handling and analysis.

All SMEs in Northern Ireland (NI) with 250 employees or less, as determined from the Fair Employment register were surveyed (do we need to say this registration is compulsory in NI). This amounted to 2086 SMEs. There was a usable percentage return rate of 19 percent. Given the absolute size of the sample, no attempt was made to increase the response rate as this size of response was considered to be representative of the main population as indicated by Remenyi *et al.* (1999). Checks were made to confirm that the response rate was representative of the population in terms of industry grouping.

Results and Discussion

The focus of this paper is to explore the innovative characteristics (within the previously identified areas) of companies within the context of the size of the company. Results follow the format of the questionnaire, using the reported practices within each of the following sections (Leadership, People and Culture, Total Quality/Continuous Improvement, Product and Process, and Knowledge and Information). Within each section respondents were offered five options by way of a response, ranging from "Strongly Disagree" to "Strongly Agree" with a middle option indicating they neither agreed nor disagreed with the statement. For ease of interpretation each of the responses were recoded into a numeric format reflecting the five choice options ranging from "−2 = 'Strongly Disagree'" to "2 = 'Strongly Agree'". Mean scores were produced for each question within the five sections.

A Pearson's correlation coefficient was then used to measure whether a linear association existed between company-size and the answers to each question in the various sections previously outlined. For brevity only those reported as statistically significant are reported. Company size was recoded into four categories, depending on the number of employees: 1–25, 26–50, 51–200, 200–250. Table 1 shows the percentage breakdown of useable returns relating to company size.

Section 1 — leadership

This section contained thirteen questions relating to issues of leadership within the organisation. Table 2 contains mean scores (based on the underlying variables ranging from −2 to 2)[1] and the level of statistical significance (* = 0.05, ** = 0.01 and *** = 0.001) of the correlation coefficient of being able to reject

[1]Thus, an overall mean score of 0 signifies "neither disagree nor agree", while a negative (positive) value is in the "disagree" (agree) range.

Table 1. Percentage breakdown of returns.

Employment size	No. of respondents	Percent
1–25	49	12.4
26–50	139	35.2
51–200	153	38.7
200–250	54	13.7
Total	395	100

Table 2. Mean value of responses to questions on leadership in the organisation.

Company size	Q1*	Q3**	Q4**	Q5***	Q7***	Q8**	Q13*
1–25	−1.65	−0.71	−0.43	−0.19	0.03	−0.30	0.91
26–50	−1.83	−1.05	−1.01	−0.50	−0.18	−0.31	1.01
51–200	−1.63	−0.48	−0.28	0.06	−0.28	−0.32	0.68
200+	−1.15	−0.40	−0.32	0.35	−0.88	−1.07	0.76
Overall	−1.65	−0.72	−0.59	−0.14	−0.28	−0.42	0.84

Level of statistical significance: * = 0.05, ** = 0.01 and *** = 0.001.

the null hypothesis that there is *no* linear association between the answers to each question and the size of the company.

Question 1 in the section dealing with Leadership in the Organisation asked respondents to rank the extent to which they agreed/disagreed with the statement that "the senior management makes a point of being seen around the organisation". Overall the mean response to this question (with a score of −1.65) indicates that firms were somewhere between "strongly disagree" and "disagree". Companies 200+ were less likely to disagree compared to companies employing between 1–200. Overall, the null hypothesis that smaller companies disagreed more strongly is accepted at the 5 percent significance level.

There are also strong (i.e., significant) associations between company size and the answers respondents gave to the questions on whether "management fosters creative thinking and innovation in the company" (Q3), "our top managers like to try new ways of doing things" (Q4) and "management spend adequate time planning change" (Q5). With regard to the latter, large companies employing over 200 tended to agree with the statement while the smallest (employing 1–50) on average disagreed. In contrast, when asked to rank whether "the rate of change in the organisation is quite fast" (Q7), perhaps not surprisingly smaller companies tended to agree (or be fairly neutral in their response) while the largest companies employing over 200 disagreed. Thus, there is a strong linear association with size for Q7 but this time in the opposite direction to that indicated in Qs 3–5. Interestingly, larger companies were also more likely to disagree with the statement "the organisation is working to a clear business plan".

Finally, in this first section, there was a significant relationship between firm size and the answers to the question "the management team doesn't seem to have the time to think constructively/creatively about the future": the smallest firms were more likely to agree, presumably reflecting their lower access to resources.

Overall, the findings for this first section indicate that leadership tends to be seen as less visual, creative, innovative or proactive within the smaller company (up to 50). This is consistent with the findings of Cagliano *et al.* (2001) who found that SMEs in this bracket are more operationally focused and dependant on "contingent and accidental factors". In the companies with 200–250 employees, the most important leadership issue was perceived as the failure to work within a clear business plan. Cagliano *et al.* (2001) indicates that in this size bracket, there is more of a focus on strategic development. However, the current studies also showed a lack of clear strategic planning among the SMEs.

Section 2 — people and culture

This section contained seventeen questions relating to issues of organisational culture and change, communications and performance (including target setting). Again, only those questions where there was a significant difference with respect to the size of the company are considered.

Question 1 asked respondents whether "there is a strong team spirit at all levels of the organisation". Overall (see Table 3 last row) respondents tended to disagree with this statement, but particularly the smaller the company concerned. With respect to communications (Q3), larger companies (those employing 200+) generally agreed that two-way communication happened at all levels of the organisation, but smaller companies disagreed. In terms of what is expected of staff (Q5), on average respondents disagreed with the statement that "everyone knows what is expected of them", especially those belonging to smaller enterprises (particularly those employing between 26 and 50). Larger companies tended to agree (or at least not disagree) that "the structure of the organisation facilitates change" (Q6), while smaller companies were less convinced that this was the case. In terms of Q7 (whether the organisation is not bureaucratic), small firms

Table 3. Mean value of responses to questions on people and culture.

Company size	Q1*	Q3*	Q5**	Q6*	Q7***	Q10*
1–25	−0.50	−0.53	−0.65	−0.17	−0.47	−0.88
26–50	−0.56	−0.74	−0.92	−0.42	−0.96	−1.03
51–200	−0.35	−0.23	−0.38	−0.17	−0.52	−0.43
200+	−0.06	0.29	−0.28	0.19	0.24	0.13
Overall	−0.39	−0.35	−0.60	−0.21	−0.53	−0.60

Company size	Q11**	Q12*	Q13**	Q15**	Q16***	Q17***
1–25	−1.08	0.48	−0.84	−1.00	−0.43	−0.65
26–50	−0.81	−0.01	−1.08	−1.13	−0.28	−0.43
51–200	−0.66	−0.23	−0.90	−0.84	−0.15	−0.13
200+	−0.38	−0.05	−0.19	−0.30	−0.13	0.13
Overall	−0.72	−0.03	−0.85	−0.89	−0.22	−0.21

Level of statistical significance: * = 0.05, ** = 0.01 and *** = 0.001.

were significantly more negative on this issue and thus tended to disagree. Related to this (Q10: "there is a feeling of openness in this organisation"), smaller companies were much more likely to disagree that the firm was characterised by "openness", while larger firms were generally neutral on this issue. Similarly, small companies were likely to disagree with the statement (Q11) that "the whole management team is competent", while larger firms also had an overall negative response to this but with a score close to 0 (indicating a neutral position).

In terms of remuneration packages, Q12 asked if "bonuses are paid according to the organisation's performance", and the smallest companies (employing 1–25 staff) were more likely to agree with this while larger firms were relatively neutral in their responses. When asked if "overall, employees have access to all the resources needed to get the job done" (Q13), smaller firms were much more likely to disagree than were larger firms, reflecting greater resource constraints in smaller businesses. Q15 asked respondents to comment on whether the organisation was an enjoyable place to work, with smaller companies providing a significantly more negative response in reply. Similarly, smaller companies were more likely to disagree with the statement that "everyone in the company has a good grasp of how the organisation is performing" (Q16) and whether "employees get useful feedback about their work". With respect to the latter, respondents working in companies employing over 200 were much more positive in their replies.

Overall in this section, there is evidence that smaller firms face more problems with communications, resources and in providing a working environment that makes their workforce more content. In a related study Smith *et al.* (2002) found that the smaller firm's focus on operations led to a lack of people development and feelings of insecurity in relation to jobs and promotion. These arguments run counter to the suggestions that the smaller firms "family" atmosphere overcomes such problems (Gadenne, 2000).

Section 3 — total quality and continuous improvement (TQ/CI)

The questions in this section were clearly taken to be more appropriate for manufacturing operations. Moreover, only 2 of the 13 questions asked resulted in a significant difference across companies with respect to size.

Q12 (Table 4) asked whether a number of financial benefits had been achieved from the TQ/CI programme. Overall, responses were generally positive although tending towards being neutral; however, the largest companies employing over 200 were more inclined to disagree that there had been financial benefits. Similarly, when asked if other business benefits had been achieved (Q13), smaller companies were positive (verging between neutral and agreeing) but the largest companies were more negative.

Section 4 — product and process

This section concentrated on new product and process innovations and improvements, and therefore concentrated on the technological capabilities of companies.

Table 4. Mean value of responses to questions on TQ/CI.

Company size	Q12**	Q13*
1–25	0.43	0.36
26–50	0.39	0.36
51–200	0.23	0.20
200+	−0.26	−0.15
Overall	0.26	0.23

Level of statistical significance: * = 0.05, ** = 0.01 and *** = 0.001.

Table 5. Mean value of responses to questions on product and process.

Company size	Q1**	Q7***	Q8**	Q11***	Q12**
1–25	0.46	0.74	0.53	−0.06	0.53
26–50	0.50	0.30	0.28	−0.50	−0.03
51–200	0.73	0.08	0.26	−0.99	−0.02
200+	0.81	−0.06	−0.22	−1.00	−0.21
Overall	0.63	0.23	0.25	−0.66	−0.02

Level of statistical significance: * = 0.05, ** = 0.01 and *** = 0.001.

Q1 asked respondents to consider if "we are committed to making our existing products and services obsolete by introducing new ones". Overall, there was agreement with the statement, with stronger commitment the larger the company. As to whether existing products/services had a high level of technology built into them (Q7), the smallest companies (employing 1–25 workers) generally responded that this was the case, with positive agreement declining with firm size with the largest companies tended on average to slightly disagree that technology standards were high.

As to the sources of technology, Q8 stated that "customers are regularly involved in the development of new products and services". Again, small companies tended to agree while the largest were significantly more likely to disagree. Q11 asked if "technology is important in this organisation"; generally, companies tended to disagree with this statement, but much less so for smaller companies *vis-à-vis* larger companies. Finally, when asked if the products and services produced use better technology than competitors (Q12), companies employing 1–25 workers tended to agree, while larger companies were more likely to disagree (the more so the larger the company). This confirms that small firms tended to use technology to establish niches through exploiting new ideas and through having to be more up-to-date in order to survive.

Overall this section suggests that small companies are more technologically innovative and that larger companies within the sample recognise the need to replace older, more obsolete products and services in order to maintain their competitive stance in the markets in which they operate. These findings challenge the traditional SME barrier of lack of resources (Gadenne, 1999) and shows that innovative approaches can overcome this obstacle. Thus, these results, in agreement with those of Cagliano *et al.* (2001), suggest that the smaller SMEs can compensate for their lack of strategic planning by being inherently flexible and close to the customer and thus addressing product and process issues in new product development.

Section 5 — knowledge and information

This section contained fourteen questions relating to issues on the transfer and management of knowledge and information within the organisation. Again, only those questions where there was a significant difference with respect to the size of the company are considered.

Question 1 asked respondents whether "everyone is in possession of the information/knowledge necessary to do their job". Smaller companies disagreed more strongly with this statement, (in particular, the 26–50 employee group). Regarding the effective management of knowledge and information, question 2 asked respondents whether they felt "information/knowledge is effectively managed and used throughout the organisation". Once again the (26–50) disagreed more strongly with this statement, whilst the 200+ company group reported a neutral to positive response regarding the use and management of information/knowledge. Regarding the capture and use of information and knowledge (Q3) the responses were very similar, with companies in the 26–50 category viewing this negatively whilst the larger the company the more positive their view on this statement became. Question 4 followed the capture and use of knowledge asking, "responsibilities for information/knowledge capture are clearly defined". Again, the larger the company the stronger the agreement. This theme was followed in question 5 "knowledge that employees hold in their heads is managed and captured effectively". The larger companies unsurprisingly agreed more strongly with this statement although disagreement was less pronounced within the 26–50

Table 6. Mean value of responses to questions on knowledge *and* information in the organisation.

Company size	Q1**	Q2***	Q3***	Q4***	Q5***	Q8**
1–25	−1.05	−0.59	−0.38	−0.15	0.04	−0.04
26–50	−1.22	−0.82	−0.63	−0.18	−0.07	−0.66
51–200	−0.98	−0.38	−0.28	0.05	0.22	−0.11
200+	−0.39	0.07	0.21	0.39	0.47	0.22
Overall	−0.99	−0.49	−0.34	0.00	0.14	−0.23

Level of statistical significance: * = 0.05, ** = 0.01 and *** = 0.001.

employee category and was almost neutral/positive in the 1–25 employee category. Finally, companies responded to the statement "lessons learned from daily experiences and projects are captured and disseminated". Consistent with previous responses in this section larger companies reacted more positively to this statement although the 51–200 employee category took a more negative to neutral stance. Again the 26–50 employee group disagreed more strongly with this statement.

Overall, as intuitively expected, the larger the company the more likely the management and use of knowledge and information was supported by clear structures both through the systems created and with clear guidelines regarding personal responsibility. These findings challenge the idea that SME networks can overcome the barrier of lack of knowledge systems in SMEs (Gadenne, 1999). The findings are consistent with those of Mosey *et al.* (2002) who show that the larger SMEs can more readily use knowledge and innovation in a systematic manner to increase innovation within the organisations.

Conclusions

The aim of this paper is to conduct an empirical study of innovation incorporation in Irish SMEs, primarily from an organisational size perspective. The study has included a brief analysis of relevant literature; the construction of a conceptual model with the help of previous exploratory research and a large scale survey based on the model of SMEs in the Northern Ireland region.

A study of the literature reveals that there is a lot of information on national systems of innovation, however, there is a paucity of studies on how SMEs can incorporate the principles and practices of innovation. Furthermore, there are few studies that use size classifications within SMEs. Such an approach treats SMEs as a homogenous grouping and restricts studies to that of SME: large organisation comparisons and contrasts. Based on the literature and the earlier exploratory research a conceptual model of innovation in SMEs has been developed and then applied in this paper. The key areas within the model were Leadership, People and Culture, Total Quality/Continuous Improvement, Product and Process, and Knowledge and Information and the overriding influence of the Market and Customers.

The first stage of model testing, investigating the effects of SME size in relation to innovation, has shown significant effects in a number of areas.

Regarding issues of leadership this survey supports the findings of Cagliano *et al.* (2001) that smaller SMEs are more operationally focussed and depend on more contingent and accidental factors, whereas the larger SME (200–250) focussed more on issues relating to strategic development.

In relation to people and culture within the organisation smaller SMEs experience more difficulties with communication, resources and providing a working environment that make the workforce more content. This supports the findings of Smith *et al.* (2002) suggesting that the small firm's focus on operational issues can have a detrimental effect on people development.

Section 3 addressed issues of Total Quality and Continuous Improvement. This section (although perceived as perhaps more appropriate to the manufacturing sector) elicited generally overall positive responses in terms of the positive benefits of the introduction of TQ/CI. Surprisingly the smaller SMEs reported that the positive effects of introducing TQ/CI in terms of financial and business rewards were greater than for the larger SME.

The final section examined issues relating to the management of Knowledge and Information. As expected the larger SME had in place structures to support the storage and movement of information as well as clear divisions for the responsibility of managing that knowledge. This supports the findings of Mosey *et al.* (2002) who suggest that the larger organisation manages knowledge and information more systematically.

The findings of this survey highlight the differences in terms of how 'innovativeness' is incorporated into SMEs from small (1–25) to larger (200–250) companies. For policy makers and advisers who are responsible for the design of interventions to increase the innovative nature and competitive edge of the SME sector, it is vital that they take into account the range of practices and perceptions held by SMEs (according to size). This study has identified several important differences in terms of practices according to firm size that could potentially lessen the effectiveness of programmes or interventions introduced. One of the purposes of this survey was to examine the appropriateness of defining SMEs as one group with homogeneous practices. The survey sheds doubt on this approach, suggesting that a variety of interventions are necessary if advisors/policy makers intend to improve either innovative behaviour or competiveness across all SMEs.

In terms of how innovation is incorporated into SMEs through the utilisation of the conceptual model, it has been demonstrated that size effects the practices used. The effects of SME size classification is a key starting point for examining the current data set in preparation for future model testing.

Thus, there is a challenge to entrepreneurs to incorporate a wider definition of innovation within their organisations, rather than solely promoting technical innovation. Efforts to achieve incorporation of such innovation must take account of the effect of organisational size and the need for different innovation implementation strategies at different stages of the firm's growth. Equally, there is a challenge to Government support agencies to devise methods of helping these SMEs by tailoring their efforts to address the different size categories of SMEs. Finally, there is a word of caution for researchers in this area. As shown by Nooteboom (1994), firm size can be a proxy for, or can mask, the effects of other organisational factors. One such factor is organisational sector. It is suggested that an area for further research would be to repeat the study controlling for organisational sector.

References

Alstrup, L (2000). Coaching continuous improvement in small enterprises. *Integrated Manufacturing Systems*, 11, 165–170.

Appiah-Adu, K and S Singh (1998). Customer orientation and performance: A study of SMEs. *Management Decision*, 26, 385–394.

Bessant, J and S Caffyn (1997). High-involvement innovation through continuous improvement. *International Journal of Technology Management*, 14, 14–22.

Cagliano, R, K Blackmon and C Voss (2001). Small firms under the MICROSCOPE: International differences in production/operations management practices and performance. *Integrated Manufacturing Systems*, 12, 469–482.

Culkin, N and D Smith (2000). An emotional business: A guide to understanding the motivations of small business decision takers. *Qualitative Market Research: An International Journal*, 3, 145–157.

Davenport, S and D Bibby (1999). Rethinking a national innovation system: The small country as "SME". *Technology Analysis and Strategic Management*, 11, 431–462.

Demarest, M (1997). Understanding knowledge management. *Journal of Long Range Planning*, 30, 374–384.

DTI (1998). *White Paper on Innovation and Competitiveness*. London.

Freel, M (2000). Barriers to product innovation in small manufacturing firms. *International Small Business Journal*, 18, 60–81.

Gadenne, D (1999). Critical success factors for small business: An inter-industry comparison. *International Small Business Journal*, 17, 36–51.

Ghobadian, A and D Gallear (1996). Total quality management in SMEs. *OMEGA*, 24, 83–93.

Ghobadian, A and D Gallear (1997). TQM and organisation size. *International Journal of Operations and Production Management*, 17, 21–163.

Gunasekaran, A, P Okko, T Martikainen and P Yli-Olli (1996). Improving productivity and quality in small and medium enterprises: Cases and analysis. *International Small Business Journal*, 15, 59–72.

Hale, A and P Cragg (1996). Business process reengineering in the small firm: A case study. *INFOR*, 34, 15–29.

Husband, S and P Mandal (1999). A conceptual model for quality integrated management in small and medium size enterprises. *International Journal of Quality and Reliability Management*, 16, 699–713.

McAdam, R (2000). The implementation of reengineering in SMEs: A grounded study. *International Small Business Journal*, 18, 29–45.

McAdam, R, P Stevenson and G Armstrong (2000). Innovative change management in SME's: Beyond continuous improvement. *Journal of Logistics Information Management*, 13, 138–149.

Martin, L (1999). Looking for the right stuff, human capital formation in small firms. PhD Thesis, University of Warwick.

Mosey, S, J Clare and D Woodcock (2002). Innovation decision making in British manufacturing SMEs. *Integrated Manufacturing Systems*, 13, 176–183.

Nonaka, I and K Takeuchi (1995). *The Knowledge Creating Company: How Japanese Companies Create the Dynamics of Innovation.* Oxford: Oxford University Press.

Nord, W and S Tucker (1987). *Implementing Routines and Radical Innovations.* Lexington, MA.

Nooteboom, B (1994). Innovation and diffusion in small firms: Theory and evidence. *Small Business Economics*, 6, 327–347.

Peters, T and R Waterman (1982). *In Search of Excellence.* NY: Harper and Row.

Raymond, L, F Bergeron and S Rivard (1998). Determinants of business process reengineering success in small and large enterprises: An empirical study in the Canadian context. *Journal of Small Business Management*, 36, 72–85.

Remenyi, D, B Williams, A Money and E Swartz (1999). *Research in Business and Management.* NY: Sage.

Sadler-Smith, E, A Sargeant and A Dawson (1998). Higher level skills training and SMEs. *International Small Business Journal*, 16, 84–94.

Schmidt, L (1990). The innovative attitude of small and medium-sized enterprises. *Journal of Small Business Management*, 28, 68–80.

Shaw, E (1999). A guide to the qualitative research process: Evidence from a small firm study. *Qualitative Market Research: An International Journal*, 2, 59–70.

Slappendel, C (1996). Perspectives in innovations in organisations. *Organisation Studies*, 17, 107–129.

Smith, A, G Boocock, J Loan-Clarke and J Whittaker (2002). IIP and SMEs: Awareness, benefits and barriers. *Personnel Review*, 31, 62–85.

Soderquist, K, J Chanaron and J Motwani (1997). Managing innovation in French small and medium-sized enterprises: An empirical study. *Benchmarking for Quality Management and Technology*, 4, 1–8.

Temtime, Z and G Solomon (2002). Total quality management and the planning behaviour of SMEs in developing economies. *The TQM Magazine*, 14, 181–191.

Tidd, J, J Bessant and K Pavitt (2001). *Managing Innovation: Integrating Technological, Market and Organisational Change.* UK: Wiley.

Tushman, M and D Nadler (1986). Organising for Innovation. *California Management Review*, 28, 9–18.

Voss, C (1998). Made in Europe: Small companies. *Business Strategy Review*, 9, 1–19.

Vossen, R (1999). Relative strengths and weaknesses of small firms in innovation. *International Small Business Journal*, 16, 88–94.

Welsh, J and J White (1981). A small business is not a little big business. *Harvard Business Review*, July–August, 18–32.

Wiele, T and A Brown (1998). Venturing down the TQM path for SME's. *International Small Business Journal*, 16, 50–69.

Zairi, M (1994). Innovation or Innovativeness, results of a benchmarking study. *TQM Magazine*, 5, 10–16.

Chapter 3

EFFECTS OF PRODUCT INNOVATION AND ORGANISATIONAL CAPABILITIES ON COMPETITIVE ADVANTAGE: EVIDENCE FROM UK SMALL AND MEDIUM MANUFACTURING ENTERPRISES

Ana Cristina O. Siqueira and Andy D. Cosh

University of Cambridge Judge Business School

We investigate the extent to which product innovation moderates the relationship between capabilities and competitive advantage among small and medium enterprises (SMEs). Using resource-based and capabilities theories, we examine capabilities as organisational routines, focusing on job rotation and multi-skilling. We examine competitive advantage by using logistic regression to assess the probability of top performance in productivity relative to most other firms in the same industry. Considering the path-dependence in developing capabilities and innovation, we use a longitudinal sample of 300 UK manufacturing SMEs in traditional and high-technology industries to evaluate the effects of innovating and using capabilities continuously over time. The results suggest that firms using job rotation or multi-skilling and introducing product innovations consistently from 2002 through 2004 are more likely to be top performers in 2004. The findings support a theoretical model according to which the association between capabilities and competitive advantage is moderated by innovation.

Keywords: Product innovation; capability; competitive advantage; SME; UK.

Introduction

Competitive advantage and performance often depend on firms' internal resources, and on their ability to coordinate and combine their resources and capabilities (Barney, 1991; Grant, 1991). Several studies have examined how firm

This paper was first published in *International Journal of Innovation Management*, Vol. 12, No. 2 (June 2008), pp. 113–137.

performance is connected with innovation (Cainelli *et al.*, 2004, 2006; Cefis and Ciccarelli, 2005; Cefis and Marsili, 2005; Wolff and Pett, 2006). Despite the importance of these studies, they lack a focus on evaluating performance from a competitive advantage standpoint. There is a scarcity of research on SMEs that examines the role of firm capabilities and innovation in facilitating competitive advantage.

Many studies of firm innovation have used theories from the industrial organisation tradition, providing an examination of how innovation varies according to firm size and industry structure (Galende and Suárez, 1999; Galende, 2006). Despite the relevance of the resource-based theory as a perspective to examine the role of internal resources, recent empirical studies on the relationship between innovation and firm performance (Cainelli *et al.*, 2004, 2006; Cefis and Ciccarelli, 2005; Cefis and Marsili, 2005) lack an explicit resource-based approach examining the importance of firm capabilities. Thus, little is known about the connections between organisational capabilities and competitive advantage in the context of innovation.

This paper helps bridge this gap in the literature by examining the importance of firm capabilities and innovation for explaining SME competitive advantage. Specifically, this study focuses on the questions of (1) the extent to which organisational capabilities are associated with competitive advantage, and (2) the extent to which this association is moderated by innovation.

This research contributes to the innovation, capabilities, and small business literatures in three ways. First, this paper uses an analytical approach to empirically assess competitive advantage from a probabilistic standpoint. This approach identifies significant patterns in the data by evaluating the probability that firms will be top performers in productivity relative to most other firms in the same industry. Second, this paper offers new empirical evidence focused on SMEs. Despite notable exceptions (Cosh *et al.*, 1999), many studies on how firm performance is connected with innovation (e.g., Cainelli *et al.*, 2004, 2006; Cefis and Ciccarelli, 2005; Cefis and Marsili, 2005) provide empirical findings that may not necessarily apply to SMEs. Although a central interest in the field of small business management is to identify how smaller enterprises can be managed in ways that will lead to continuing success, there is a dearth of research that assesses how firm capabilities and innovation may facilitate SME competitive advantage. Third, this study provides a unique empirical assessment of the resource-based theory and the process/capability school to explain competitive advantage in the context of innovation. This assessment is consistent with Callon *et al.*'s (1999) suggestion, which highlights the importance of building on a sound empirical base for research dealing with innovation to formulate valuable contributions to the theory.

Theoretical Considerations

In the business economics and management literatures, principal approaches to the study of firms' innovative activities have been industrial organisation, agency theory, transaction costs economics, evolutionary theory, and the resource-based approach (Galende, 2006). The approaches more concerned with the relationship between innovation and factors internal to the firm, are the evolutionary and resource-based theories. Evolutionary theory considers technological development to be an interdependent process. Accordingly, empirical research (e.g., Cainelli *et al.*, 2006; Cosh *et al.*, 1999) has used evolutionary theory to study the interdependence between innovation and firm performance. Although evolutionary theory provides valuable contributions for analysing innovation from a perspective internal to the firm, it is less concerned with studying how a firm can optimise its behaviour in activities related to innovation (Galende, 2006). Compared to industrial organisation, agency theory, transaction costs economics, and evolutionary theory, the resource-based theory is particularly suited to the analysis of the implications of firm capabilities for competitive advantage in the context of innovation.

The resource-based theory contends that performance differentials among firms depend to a large extent on the possession of idiosyncratic resources and capabilities (Conner, 1991; Grant, 1991). Resources and capabilities refer to different constructs. We use the following definitions: *Resources are defined as "anything which could be thought of as a strength or weakness of a given firm"* (Wernerfelt, 1984, p. 172). Resources comprise tangible and intangible assets such as capital, machinery, technological competences, and the employment of skilled personnel (Wernerfelt, 1984). *Capabilities are defined as organisational routines* (Grant, 1991). Capabilities entail patterns of coordination between people and between people and other resources (Grant, 1991). The literature drawing on the resource-based view encompasses various definitions of capabilities. For instance, Sanchez (2004) describes capabilities as intangible assets that determine the uses of tangible and intangible resources. Zahra *et al.* (2006) show nine existing definitions of dynamic capabilities and propose a distinction between dynamic and substantive capabilities. In their view, dynamic capabilities represent the firm's ability to change substantive capabilities, and substantive capabilities are the firm's ability to solve a problem or to produce a tangible or intangible result. This study builds on Grant's (1991) original description of capabilities, because his text is a founding source of the resource-based theory. This definition is congruent with later works, such as Sanchez's (2004) notion of capabilities and Zahra *et al.*'s (2006) concept of substantive capabilities.

A firm can achieve a competitive advantage by using resources that are valuable, rare, difficult to substitute, and difficult to imitate (Barney, 1991). Competitive advantage is viewed within the resource-based theory as an outcome associated with a value-creating strategy that is rare among competitors (Barney, 1991). The notion of competitive advantage implies relative performance, in the sense of performing better than a benchmark, such as a group of competitors or all other firms in the industry. In this study, *competitive advantage is defined as being among the top-performing firms relative to most other firms in the same industry*.

The resource-based perspective embraces two principal schools of thought (Schulze, 1994). First, the structural/content school focuses on organisational features that can produce Ricardian rents, in the sense that rents derive from features that are scarce. Scholars in this school maintain that the source of a firm's competitive advantage depends on possessing key resources that are unavailable to others in the industry, and on preventing others from acquiring the same resources (Lane *et al.*, 2006). Second, the process/capability school centers on organisational capabilities that can produce efficiency rents. According to this school, competitive advantage depends on how resources are used within the firm to accomplish tasks and to develop organisational capabilities (Lane *et al.*, 2006).

This study draws on the process/capability school. Research by Grant (1991) and Teece *et al.* (1997) are principal references within this school (Schulze, 1994). The process/capability school is similar to the structural/content school in the sense that both assume the importance of having resources that are valuable, rare, difficult to imitate, and difficult to substitute for competitive advantage (Schulze, 1994). But the process/capability school focuses on how firm performance and competitive advantage depend on the way in which organisational resources and capabilities are managed (Grant, 1991; Schulze, 1994). It is important to avoid tautology by noting that firm capabilities are *not* defined as those routines that necessarily create effectiveness and superior firm performance (Zahra *et al.*, 2006). Decision-makers in the firm, or the owner and a few other managers in small enterprises, have a critical role in observing market and environmental conditions and adjusting the firm's use of capabilities (Zahra *et al.*, 2006). Thus, performance and competitive advantage are influenced by the way capabilities are managed in the firm (Zahra *et al.*, 2006). This study addresses the connections between firm capabilities, innovation, and competitive advantage in the next section.

Innovation, organisational capabilities, and competitive advantage

As stated before, capabilities are defined as organisational routines, which entail the patterns of coordination between people and between people and other resources (Grant, 1991). According to this definition, firms build on their internal resources such as technological competences and skilled personnel to create organisational capabilities or routines, which constitute repeatable procedures that firms develop to produce a tangible or intangible outcome. Organisational routines involve recurrent interaction patterns and provide a key repository of knowledge in the organisation (Becker, 2004). Routines not only store productive knowledge about what resources to use and how to transform them, but they are also able to store tacit knowledge and can be held by individuals or by the firm (Becker, 2004). Organisational capabilities or routines are seen according to the resource-based theory as a major factor influencing firm performance and competitive advantage (Becker, 2005).

This research focuses on specific capabilities, namely, job rotation and multi-skilling. These routines particularly suit the goals of this study for two main reasons. First, they center on processes internal to the organisation. Second, they represent not only a means for firms to store productive knowledge, but also a mechanism through which firms can distribute internally knowledge about different products and operational processes.

There are reasons to expect that the use of job rotation and multi-skilling is associated with firm performance. The use of these routines can bring a set of advantages (Hoyt and Matuszek, 2001). Multi-skilling refers to cross-training of workers in more than one area in such a way that they can perform different functions. The use of multi-skilling makes it easier for firms to reassign workers to different processes. Moreover, regardless of the absence of specific employees, multi-skilled workers can help firms complete a task within a fixed time frame.

One mechanism to implement multi-skilling is the rotation of employees. The use of job rotation can bring many benefits in addition to improving employee skill diversity. Job rotation facilitates the development of informal communication channels within the organisation. Moreover, rotation helps to improve socialisation by disseminating knowledge of other departments' responsibilities. By introducing variety, both multi-skilling and job rotation can enhance employee morale and efficiency. Indeed, Campion *et al.* (1994) found that employee salary and promotion, positive affect, and perception of skill acquisition, along with employee tenure and performance were associated with job rotation.

Above all, the use of multi-skilling capabilities is useful to distribute productive knowledge within the organisation, so that different employees are aware of the firm's know-how and are prepared to perform various routines. Multi-skilling capabilities can help firms enhance their performance. Indeed, Eriksson and Ortega (2006) found that rotation schemes are associated with firm growth rates suggesting a positive relationship between the use of job rotation and firm performance. In this way, organisations that use multi-skilling capabilities may have greater chances of achieving a competitive advantage than organisation not using these capabilities.

H1: *Firms that use multi-skilling capabilities will have a greater probability of being top-performing enterprises than firms not using these capabilities.*

Prior empirical studies found evidence of an association between employee's skill diversity and firm productivity. For instance, Kim and Park (2003) examined 206 Korean manufacturing firms and concluded that multi-skill formation was positively related to labour productivity growth. Dyck and Halpern (1999) examined the case of a Canadian subsidiary of IBM. They found that the implementation of a set of modifications including the implementation of multi-skilled teams was associated with a sharp increase in productivity, in terms of improving manufacturing capacity without additional labour or capital requirements. Moreover, multi-skilling capabilities can promote improvements in efficiency. Houghton and Portougal (2005) found the implementation of multi-skilling practices in manufacturing to be related to the reduction of labour costs. Yazici (2005) found that the allocation of fewer, but more multi-skilled workers in manufacturing was associated with higher utilisation and lower lead-time. These studies provide valuable contributions, but there is very little research that evaluates how the use of multi-skilling or job rotation is associated with productivity from a competitive advantage standpoint. Additionally, there is a dearth of studies that investigate how this association may be moderated by innovation.

The role of innovation

Innovation can lead to the development of key capabilities that improve performance (Zahra and Garvis, 2000). Looking at the example of job rotation and multi-skilling, there are arguments favouring the view that there is an interactive relationship between these capabilities and innovation. First, the introduction of new or improved products frequently leads to the implementation of several changes in firms' operational processes. These changes are most often handled better by multi-skilled workers (Kim and Park, 2003). Hence, by dealing more efficiently with changes due to the implementation of new or improved products, innovative firms that use multi-skilling may maintain or achieve a competitive advantage.

Second, innovative firms can use job rotation as a means to improve the organisation's knowledge about person-activity fit (Ortega, 2001). In addition to functioning as a method for employees to acquire new knowledge, job rotation may be used by innovative firms as a procedure that enables them to learn which employees perform which activities better. Because innovative firms tend to face more changes in internal processes as a consequence of the introduction of new or improved products, learning about person-activity fit is more frequently a critical activity. In this way, the use of job rotation may help innovative firms gain a competitive advantage.

Third, some of the major benefits of job rotation and multi-skilling, such as the improvement of internal communication and the dissemination of knowledge, may be particularly important for innovative enterprises. Given that innovation introduces changes in a firm's knowledge base and operational processes, innovative enterprises that are able to distribute new knowledge faster and/or more effectively may outperform most other firms in the same industry. Thus, innovative firms that use job rotation or multi-skilling may maintain or gain a competitive advantage.

Following the arguments above, the association between job rotation or multi-skilling and competitive advantage is likely to vary according to whether firms are innovators. Winning firms are those that are able to combine innovation with the development of organisational capabilities (Teece *et al.*, 1997). For innovators, the use of multi-skilling and rotation is particularly valuable for distributing new knowledge within the organisation. The combination of these capabilities and innovation is thus especially important for innovative firms to achieve a competitive advantage. Similarly, among firms that use multi-skilling capabilities, the chances of achieving a competitive advantage are likely to be greater for innovative firms.

H2: *Among firms that use multi-skilling capabilities, firms that are innovators will have a greater probability of being top-performing firms than non-innovators.*

In the next sections, this paper examines the extent to which there is empirical evidence consistent with these hypotheses.

Method

Data and sample

This paper uses a longitudinal dataset from the Centre for Business Research (CBR) panels of SMEs (Cosh and Hughes, 2003, 2007). Our dataset focuses on firms that answered the survey in both 2002 and 2004. The panels cover independent enterprises with fewer than 500 employees in England, Wales, and Scotland. For the 2002 survey, the sampling frame was the Dun and Bradstreet (D&B) UK Marketing Database (Bullock, 2003). For the 2004 survey, the sampling frame was based on both the D&B database and the Financial Analysis Made Easy (FAME) company database (Bullock and Milner, 2007). The sample design was based on stratified random sampling, with the stratification criteria being firm size and industry. A letter and questionnaire were sent to the managing director, partner, or proprietor of the firms. Two to three reminder letters were sent to non-respondents between two to six weeks after the mailing of the first letter. The survey questionnaire covers organisational and managerial characteristics, business performance, competitive position, and innovation.

The response rate is calculated as follows. The dataset for the 2002 time point comprises data from two panels: an existing panel and a new one (Bullock, 2003). For the previous panel, the sampling frame included 1,579 firms, of which 521 returned usable responses, resulting in a response rate of about 33%. For the new panel, 11,365 firms were surveyed, of which 1,609 returned usable questionnaires, resulting in a response rate of about 14%. The dataset for the 2004 time point contains data from an existing panel that surveyed 2,612 firms, of which 891 provided usable responses, resulting in a response rate of about 34% (Bullock and Milner, 2007).[1] Compared to the small business literature on how firm performance is associated with innovation (e.g. Freel and Robson, 2004; Wolff and Pett, 2006), the response rates for this study fall within the range of response rates reported in the literature.

[1]The dataset for the 2004 time point also contains data from a new panel (Bullock and Milner, 2007). The new panel surveyed 8,976 firms, of which 1,246 returned usable questionnaires, with a response rate of nearly 14%. Nonetheless, no data from the new 2004 panel are used, because this study focuses on firms that have also answered the 2002 survey.

A non-response bias analysis found very few differences in relation to industry and firm size between respondents and non-respondents (Bullock, 2003; Bullock and Milner, 2007). For the 2004 panel, the non-response bias analysis notes no significant difference concerning industry and firm size between respondents and non-respondents (Bullock and Milner, 2007). A spatial analysis reveals that the 2002 and the 2004 samples generally reflect the distribution by region and industry for SMEs in UK (Bullock, 2003; Bullock and Milner, 2007). Although this study does not seek to represent the population of SMEs in UK, to ensure legitimate results when using data from the stratified sample, the analytical models include the two stratification sampling criteria, namely, firm size and industry, as control variables.

This study focuses on 780 firms that answered the survey in both 2002 and 2004. Thus, a potential bias is associated with the requirements of participating in both surveys and surviving from 2002 to 2004. The analytical sample excludes all observations that have missing values for one or more variables from the analytical models. Firms with an average of fewer than two employees are also excluded, because the analyses focus on organisations with a personnel structure that allows employee rotation. The analytical sample includes only manufacturers, because there are insufficient cases in the dataset to examine innovation other than that in manufacturing. The final analytical sample comprises 300 manufacturing firms. Comparing the analytical sample to the original sample of manufacturers in the dataset, there is no noteworthy difference in the proportion of firms in the high-technology and traditional industries and in the means for business age and size. According to these criteria, there is no evidence of bias in the selection of the analytical sample.

Analytical strategy

The analytical strategy centres on examining competitive advantage. As an economic convention, normal performance is defined at the industry level (Schulze, 1994). A common operational definition of above normal performance or performance in excess of the competitive level is performance above one standard deviation of the average by industry. For this definition, the analytical sample contains insufficient cases for the group of top performers. Consequently, this study operationally defines top performers as those in the upper quartile of performance (measured in terms of productivity) by industry.

Alternative operational definitions and modelling approaches to evaluate competitive advantage are the following. First, prior research (Bhatt and Grover,

2005; Douglas and Judge, 2001; Spanos and Lioukas, 2001) has used subjective measures of competitive advantage, by asking respondents to classify their firms' performance relative to competitors. This method has value when objective measures are unavailable. Its critical limitation is that the measurement of competitive advantage relies on respondents' perception and subjectivity. Second, previous studies have used objective measures, by assessing the difference in performance between the firm and its competitor (Dehning and Stratopoulos, 2003) or between the firm and the industry average (Wiggins and Ruefli, 2002). This method has the advantage of measuring competitive advantage objectively. It represents an approach useful for research interested in modelling the determinants of performance differentials.

For this study, the second method above is not suitable because the purpose of this research is not estimating the magnitude of the difference in productivity between firms with diverse use of capabilities (e.g., firms that use job rotation/multi-skilling compared to those not using this routine). The selection of the most appropriate dependent variable for empirical resource-based analyses can be challenging (e.g. Ray *et al.*, 2004). This selection depends on the goals of each study and typically involves a tradeoff of advantages and disadvantages. This study uses a binary dependent variable. A major limitation of using this type of dependent variable is that it represents less information than a continuous indicator. Indeed, prior research (e.g., DeVoe and Pfeffer, 2007; Watson, 2007) using the approach of aggregating continuous values into binary categories of the dependent variable recognises disadvantages, such as not taking full advantage of the information available in the dataset and constraining the existing variation in performance available for the analysis.

On the other hand, there are important advantages in using this approach. First, a binary dependent variable enables the analysis to distinguish a group of top firms (i.e., those with a competitive advantage) from the others, allowing a modelling approach that provides a clear-cut, direct assessment of competitive advantage. Subsequently, this dependent variable allows the evaluation of the probability with which the use of capabilities is associated with being a top firm, namely, having a competitive advantage. Thus, this modelling approach has the benefit of producing results that permit a more intuitive interpretation of competitive advantage and is likely to be useful for different audiences including small-business practitioners.

Second, if there is an association between multi-skilling capabilities and firm performance, it is most likely to be evident at the extreme ends of the performance spectrum (Watson, 2007). As a result, examining firms at the upper quartile of the performance distributions relative to all other firms is more likely to capture such an association. Prior research (Watson, 2007) on the relationship between organisational practices and firm performance used the strategy of discarding firms in the second and third quartiles, and using only firms in the upper and lower quartile in a binary logistic regression. This strategy and its findings are to some extent similar to that of prior research (Havnes and Senneseth, 2001) on organisational practices and firm performance. In our study, we use all firms divided into two groups: a group of top firms composed of those in the upper quartile of the productivity distribution and a group of non-top firms composed of all other firms. Compared to prior research (Watson, 2007), an advantage of the approach we use in our study is that we do not discard firms in the second and third quartiles of the performance distribution. We use this approach because it is particularly suitable to assess the extent to which firms that use multi-skilling capabilities are associated with a greater probability of being top-performing firms relative to most other firms in their industry.

Consistent with the notion that the process/capability school views performance as a probabilistic event (Schulze, 1994), the analytical strategy evaluates firms' probability of being a top performer. For this strategy, this study uses binary logistic regression. By assessing the probability of being a top performer in relation to most other firms in the same industry, this study evaluates the extent to which the use of job rotation/multi-skilling is significantly linked with competitive advantage.

The modelling strategy builds on Belderbos *et al.*'s (2004) and Cainelli *et al.*'s (2006) empirical approaches, which investigate the effects of innovation and organisational factors in a previous time point on subsequent firm performance. In agreement with Belderbos *et al.* (2004), the modelling strategy evaluates the importance of a consistent investment in capabilities and innovation. SMEs that have invested in capabilities or innovation in only one of the two time points may have done so because of temporary or discrete circumstances, such as the purchase of a new manufacturing machine. Additionally, a firm needs to continually invest in its technological capabilities (Wernerfelt, 1984), because these capabilities are a result of an accumulative process. Resources are created through a path-dependent process that builds on the firm's history and experience (Dierickx and Cool, 1989). Organisational capabilities cannot be purchased in factor markets because they constitute an asset that is a result of prior investment (Dierickx and Cool, 1989). Thus, this study operationally defines its independent variables to reflect a continuous investment in capabilities and innovation.

Analysing interactions among organisational resources and capabilities is core to the resource-based theory (Schulze, 1994). Interaction terms are particularly relevant for the process/capability school, which "focuses on the connection between economic performance and qualitative differences among resources and/or the interaction of organisational processes and resources" (Schulze, 1994, p. 144). Hypothesis 2 in this study directly assesses the interaction between capabilities and innovation.

Consistent with a focus on firms as the units of analysis within the resource-based theory (Barney, 2001), the units of analysis in this study are firms. Two models are used. Model 1 specifies the probability of being a top performer as a function of capabilities and control variables. This model is nested in Model 2, which introduces the moderator variables, namely, innovation and the interaction terms between capabilities and innovation.

Measures

Dependent variable. The dependent variable represents competitive advantage in productivity. A central advantage of using productivity as the basis of the dependent variable is that, in comparison with measures of financial performance such as profitability, productivity, reflects more directly gains in performance. This is because financial performance such as profitability depends on how productivity gains are distributed among shareholders, employees, customers, and other stakeholders (Fabling and Grimes, 2007; Lazear, 2000). The use of productivity indicators, in the form of continuous or categorical variables, is common in empirical research on firm performance (Cainelli *et al.*, 2006; Fabling and Grimes, 2007) and on performance among smaller firms (Walter *et al.*, 2006).

Productivity is measured as annual sales per employee. *Top performers* are defined as firms belonging to the upper quartile of the productivity distribution by industry in the 2004 time point.[2] Top performers are coded to 1 or to 0 otherwise. This binary indicator is the dependent variable for the logistic models. The dependent variable reflects competitive advantage in productivity in 2004.

Independent variable. The CBR questionnaire includes a question that identifies whether the firm currently uses "job rotation/multi-skilling". This variable is

[2]An alternative operational definition of top performance is productivity above one standard deviation of the average by industry. Nonetheless, there are insufficient cases in the sample according to this definition for the group of top performers.

coded to 1 if the firm has consistently used *job rotation or multi-skilling* in both 2002 and 2004 or coded to 0 otherwise.

Control variables. The control variables comprise five measures. First, a measure for the *percent of R&D employees* is used to control for firms' efforts to develop new or improved products. This measure is calculated as the average of the 2002 and the 2004 number of R&D employees for each firm, divided by the firm's average number of employees.

Second, two measures represent inter-firm collaboration. Firms can use inter-firm relationships as a means to gain experience and to obtain external information. The CBR questionnaire asks whether the firm in the last three years has engaged in formal or informal collaborative or partnership arrangements with any external organisations. The questionnaire also asks whether these arrangements were established with "suppliers", "customers", "higher education institutes", "private research institutes/consultants", "firms in your line of business", and "others". The analytical sample comprises sufficient observations only for collaboration with suppliers and customers. *Collaboration with suppliers* and *customers* are coded to 1 if the firm has consistently collaborated in both 2002 and 2004 or coded to 0 otherwise.

Third, *past performance* indicates whether the firm was a top performer in productivity in 2002. This measure is coded to 1 for top performers or to 0 otherwise. Past performance is expected to be positively associated with subsequent performance, because firms that have presented a competitive advantage in the past are likely to have developed strong resources and capabilities. Fourth, *firm size* reflects the average number of employees that SMEs reported in their 2002 and 2004 surveys. Fifth, an industry measure is *high-technology manufacturing*. The survey uses the Butchart (1987) classification of industrial codes to categorise high-tech firms (Bullock, 2003; Bullock and Milner, 2007). The high-tech manufacturing sectors in the analytical sample include: plastics; synthetic rubber; medicaments; computers; electric motors; electronic components; telephone equipment; television and radio receivers; medical and surgical equipment; measuring instruments; electronic industrial process control equipment; optical precision instruments; photographic and cinematographic equipment; and aircraft and spacecraft. This measure is coded to 1 for firms operating in the high-tech industry or to 0 otherwise, with the reference category being traditional manufacturing. Because firm size and industry are items of the stratified sampling criteria, it is essential to include these items as control variables.

Moderator variable. The moderator variable reflects innovation. The CBR survey questionnaire asks whether firms have introduced the following types of innovation during the last three years. "Technologically new or significantly improved manufactured product", "technologically new or significantly improved methods of producing manufactured product", "technological improvements in supply, storage, or distribution systems for manufactured product", "new or significantly improved service product", and "new method to produce and deliver your service product". For each type of innovation, the questionnaire asks whether the innovation was new to the firm but not to the industry or new to the firm and to the industry. Of these types of innovation, only the first two items (i.e., manufacturing product and process innovation, respectively) from the category "new to firm but not the industry" present sufficient cases for the analysis of Model 2 (i.e., the model with the interaction term). Because there are differences in the characteristics of product innovation compared to process innovation in manufacturing (e.g., Preissl, 2000), it is useful to model each type of innovation separately.

For research focused on manufacturing industries, it is relevant to consider some differences between product and process innovation (e.g., Preissl, 2000). Whereas product innovation refers primarily to the creation of new marketable goods, process innovation typically represents the development of new methods such as new manufacturing techniques and operational procedures for a factory or assembly line. Principally in the manufacturing sector, product and process innovation normally differ with regard to drivers, implementation procedures, and performance effects (e.g., Antonnucci and Pianta, 2002). Generally, a major goal of process innovation is to improve the efficiency of operational processes, while for product innovation a key goal is to increase sales by generating new products. Thus, statistical models assessing the effects of these innovation categories may require different specifications. In the analytical sample, there is somewhat a low correlation ($r = 0.40$; $p < 0.01$; findings not shown in table format) between consistent process and product innovation. This finding implies that there is a considerable amount of unique information in each of these innovation categories. This result reiterates the relevance of studying process and product innovation separately. Indeed, using the same data source as this study (the CBR panels of UK SMEs), Hughes and Wood (2000) conduct empirical analyses separating process from product innovation.

For the purpose of this study, the scope is product innovation, in order to assess a major expected consequence of this innovation type, namely, the notion that the generation of new marketable goods will potentially benefit the firm's sales. (Accordingly, the dependent variable in this study represents competitive advantage in productivity measured in terms of sales per employee.) The item *manufactured product new to the firm but not to the industry* is coded to 1 if the firm has consistently introduced this type of innovation in both 2002 and 2004 or coded to 0 otherwise.

Results

Descriptive results

Table 1 describes the characteristics of the analytical sample. The proportion of firms that consistently use job rotation or multi-skilling in both 2002 and 2004 is about 39%. With regard to innovation, about 14% of the enterprises report in both the 2002 and 2004 surveys that they have continuously introduced manufactured products new to the firm.

Other characteristics of the sample are the following. About 9% of the firms consistently collaborate with suppliers and 9% with customers. The percentage of R&D employees ranges from 0 to 85 (findings not shown in table format), with an average of about 5%, whereas firm size varies between 2 and 431 employees (findings not shown in table format), with an average of 43 workers. High-tech manufacturers comprise roughly 37% of the firms.

As noted earlier, competitive advantage is viewed within the resource-based theory as an outcome associated with a value-creating strategy that is rare among competitors (Barney, 1991). Congruent with this view, in the sample only about 8% of the high-tech manufacturers and 9% of the traditional manufacturers (findings not shown in table format) report in both the 2002 and 2004 surveys that they have not only innovated but also used job rotation or multi-skilling. Thus, among the manufacturing SMEs in the sample, the strategy of consistently combining product innovation and these capabilities is rare.

Table 2 displays the characteristics of the firms that consistently use job rotation or multi-skilling and introduce product innovation. This study compares these characteristics to those of other firms by performing a t test for equality of means. The tests indicate that organisations continuously introducing product innovations and using rotation or multi-skilling present a larger average firm size (average of about 81 employees compared to 40 employees for other firms; $p < 0.001$). Moreover, a higher proportion of these firms tend to engage in collaboration with suppliers (about 27% compared to 7% of other firms; $p < 0.01$) and with customers (19% versus 8% of other firms; $p < 0.05$). On the other hand, there is no significant difference in means with regard to the percentage of R&D employees and to the proportion of firms that operate in the high-technology industry. In the next section, this study evaluates the extent to which the strategy

Table 1. Descriptive statistics and correlations for all variables.

Variable	Mean	SD	1	2	3	4	5	6	7	8	9
Dependent variable											
1 Top performers	0.25	0.43	—								
Independent variable											
Capability											
2 Job rotation/multi-skilling	0.39	0.49	−0.18	—							
Control variable											
3 Percentage of R&D employees	0.05	0.12	0.02	−0.05	—						
Collaboration											
4 With suppliers	0.09	0.29	0.04	0.20	0.12	—					
5 With customers	0.09	0.28	0.02	0.21	0.10	0.48	—				
6 Past performance	0.25	0.43	0.57	−0.10	−0.02	0.06	0.04	—			
7 Firm size	43.11	55.11	0.07	0.18	−0.08	0.16	0.24	0.06	—		
Industry											
8 High-technology manufacturing	0.17	0.37	−0.01	0.04	0.19	0.02	0.05	−0.01	−0.04	—	
Moderator variable											
Product innovation											
9 Manufactured product new to the firm	0.14	0.35	0.10	0.18	0.25	0.10	0.04	0.05	0.15	0.02	—

Note: $n = 300$. SD = Standard deviation. Correlations greater than $|0.10|$ are significant at $p < 0.05$ (two-tailed tests).

Table 2. Characteristics of firms that consistently use job rotation/multi-skilling and introduce product innovation compared to other firms.

Variable	Firms that Consistently Use Job Rotation/Multi-Skilling and Introduce Product Innovation			Other Firms			Difference in Means
	Minimum	Maximum	Mean	Minimum	Maximum	Mean	
Percentage of R&D employees	0.00	0.25	0.07	0.00	0.85	0.05	0.02
Collaboration							
With suppliers	0.00	1.00	0.27	0.00	1.00	0.07**	0.20**
With customers	0.00	1.00	0.19	0.00	1.00	0.08*	0.12*
Past performance	0.00	1.00	0.23	0.00	1.00	0.25	−0.02
Firm size	5.00	232.00	80.67	2.00	431.00	39.55***	
Industry							41.13***
High-technology manufacturing	0.00	1.00	0.15	0.00	1.00	0.17	−0.01

Note: $n = 26$ for firms that consistently use job rotation/multi-skilling and innovate. $n = 274$ for other firms. Total $n = 300$.
$^{*}p < 0.05$, $^{**}p < 0.01$, $^{***}p < 0.001$ (two-tailed tests).

of consistently combining job rotation/multi-skilling and product innovation is associated with competitive advantage.

Effects of product innovation and capabilities on competitive advantage

Table 1 shows the correlations for all variables. The maximum bivariate correlation between explanatory variables is about 0.25, with most correlations being lower than that. Therefore, there is insufficient evidence of a potential problem with multicollinearity. Table 3 displays the results from the logistic regression.

Table 3. Logistic regression of capability, product innovation, and control variables predicting the probability of top performance in productivity.

Variable	Model 1			Model 2		
	B	OR	SE	B	OR	SE
Independent variable						
Capability						
Job rotation/multi-skilling	−1.01	0.36**	0.38	−1.64	0.19**	0.47
Control variable						
Percentage of R&D employees	0.81	2.25	1.35	0.79	2.21	1.51
Collaboration						
With suppliers	0.23	1.26	0.66	0.02	1.02	0.67
With customers	−0.10	0.91	0.71	−0.02	0.98	0.71
Past performance	2.86	17.44***	0.34	2.99	19.94***	0.36
Firm size	0.00	1.00	0.00	0.00	1.00	0.00
Industry						
High-technology manufacturing	−0.05	0.95	0.46	−0.03	0.97	0.47
Traditional manufacturing (reference)						
Moderator variable						
Product innovation						
Manufactured product new to the firm				−0.40	0.67	0.72
Manufactured product new to the firm × Job rotation/multi-killing				2.35	10.48*	0.99
Constant	−2.00	0.14***	0.29	−1.98	0.14***	0.29
−2 Log likelihood		237.23			228.47	
Δ − 2 Log likelihood					8.75*	

Note: $n = 300$. B = unstandardised coefficient. OR = Odds ratio. SE = Standard error.
$^*p < 0.05$, $^{**}p < 0.01$, $^{***}p < 0.001$ (two-tailed tests).

The findings are shown in terms of odds ratios, calculated as the exponential of the logistic coefficient for the explanatory variables. Model 2 (the full model) tests Hypothesis 2 by assessing not only the significance of the coefficient for the interaction term, but also the significance of the change in the log likelihood compared to Model 1 (the nested model). Thus, compared to the nested model, the full model evaluates whether the inclusion of the moderator variables significantly increases the level of explanation for the probability of top performance relative to most other firms in the same industry.

Model 2 suggests that product innovation and the use of job rotation or multi-skilling predict top performance in productivity. For this model, the interaction term is significant ($p < 0.05$) and the change in the log likelihood for the inclusion of the moderator variables is also significant ($p < 0.05$). Firms that only use multi-skilling capabilities, but do not innovate, have a lower probability of being top-performing enterprises, compared to non-innovators that do not use these capabilities. That is, among non-innovators, the odds of top performance for firms using multi-skilling capabilities are about 81% lower than those for firms not using these capabilities. Stated differently, firms that have consistently used multi-skilling capabilities in both 2002 and 2004, but that have not consistently innovated in both years, have a lower probability of being top-performing enterprises in 2004.

In contrast, firms that not only use multi-skilling capabilities, but also innovate, have a greater probability of being top-performing enterprises, compared to innovators that do not use these capabilities. The results from Model 2 indicate that, among innovators, firms using multi-skilling capabilities are about two times more likely to be top performers than firms not using these capabilities. In other words, firms that have not only consistently used multi-skilling capabilities in both 2002 and 2004, but also consistently innovated in both years, have a greater probability of being top-performing enterprises in 2004. The results from Model 2 imply that product innovation moderates the effect of multi-skilling capabilities on competitive advantage.

The following sensitivity tests indicate that the results from Model 2 are stable. First, this study includes firm age in natural log units as an additional control variable. The results are similar (in direction, magnitude, and significance for all independent and moderator variables) to those from Model 2. Second, this study excludes two observations that present standardised residuals from the logistic regression above the benchmark of 3. Again, the results are similar to those from Model 2.

The results from the logistic models indicate the following conclusions in relation to the hypotheses. Hypothesis 1 is not supported, because the findings

Table 4. Predicted probabilities of top performance in productivity by capability and product innovation.

Variable	Product Innovation	
	Not Consistently Introducing Manufactured Product New to the Firm	Consistently Introducing Manufactured Product New to the Firm
Capability		
Consistently using job rotation/multi-skilling	7%	33%

from the logistic regression are consistent with Hypothesis 2, by indicating that competitive advantage in productivity depends on both capabilities and innovation. Hypothesis 2 is supported, because the findings indicate that firms consistently innovating and using job rotation or multi-skilling are more likely to be top performers.

Table 4 displays the results in terms of predicted probabilities. For job rotation, the probabilities of top performance are calculated based on the unstandardised coefficients from Model 2, by holding all other variables constant at their mean, and varying only the capability and innovation variables. For firms consistently using job rotation or multi-skilling but not consistently innovating, the probability is estimated by setting the value of the mean for job rotation/multi-skilling to 1 and the value for innovation (and for the interaction term) to 0. This probability is 7%. In contrast, for firms consistently using job rotation or multi-skilling and innovating, the probability is estimated by setting the values of the mean for job rotation/multi-skilling and for innovation (and for the interaction term) to 1. This probability is 33%. These findings suggest that manufacturing SMEs consistently combining job rotation/multi-skilling and product innovation are more likely to be top performers relative to most other firms in the same industry.

Effects of control variables

Most control variables have non-significant coefficients. Model 2 in Table 3 indicates that using a higher percentage of R&D employees or operating in the high-technology industry is not significantly associated with the probability of top performance. In contrast, past performance is positively associated with subsequent performance ($p < 0.001$). The results from Model 2 imply that, compared to non-top performers in 2002, top performers in 2002 are nearly 20 times more likely to be top performers in 2004.

With respect to inter-organisational collaboration and firm size, Model 2 shows no significant effects on the probability of top performance. According to the bivariate results from Table 2, the group of firms that continuously innovate and use rotation is significantly associated with firm size and collaboration. After controlling for all other variables, the results from the multiple regression indicate no apparent relationship between size and collaboration with competitive advantage.

Conclusion

This study focuses on examining the extent to which product innovation moderates the association between job rotation/multi-skilling capabilities and competitive advantage in productivity among manufacturing SMEs. This research draws on the resource-based theory to derive predictions, which are empirically tested with a dataset of 300 manufacturing SMEs from UK. To test the predictions, this paper uses logistic regression to examine the probability of top performance in productivity relative to most other firms in the same industry.

This paper uses a probabilistic approach in agreement with the process/capability school, which views firm performance from a probabilistic standpoint (Schulze, 1994). This approach identifies patterns of competitive advantage in the sample and has the potential of being useful for future empirical research with a process/capability orientation to evaluate the importance of firm capabilities. The results from the logistic analyses indicate that firms consistently introducing product innovations and using job rotation or multi-skilling capabilities are more likely to be top performers.

The logistic models identify a notable unexpected finding. In the absence of consistent innovation, the continuous use of job rotation or multi-skilling is negatively associated with the probability of top performance. The key to understanding this counter-intuitive finding may be the notion that consistently innovative firms more frequently create and introduce new knowledge in the organisation. Job rotation and multi-skilling constitute firm capabilities or routines that serve not only as a repository of organisational knowledge, but also as a mechanism to distribute more effectively this knowledge within the company. Firms that are not consistent innovators create new knowledge in a less frequent way, and thus have fewer opportunities to leverage the potential of rotation and multi-skilling to distribute knowledge. Hence, these routines may bring fewer contributions to competitive advantage for ventures that do not continuously innovate.

Moreover, the negative relationship between the use of multi-skilling capabilities and firm competitive advantage among non-innovators seems coherent with the following considerations. First, Rogerson (1993) emphasises that cross-trained workers may lose part of their new skills with time if they do not use them regularly. Second, Hoyt and Matuszek (2001) highlight that in many situations cross-trained workers do not have the opportunity to fully use their new skills. Given that innovation introduces several changes in firms' internal processes, and these changes are most often handled better by multi-skilled workers (Kim and Park, 2003), innovators may more frequently provide opportunities for multi-skilled workers to not only use but also develop their new skills. Another factor that should be considered in understanding this negative relationship is the potential adverse effect of a lack of specialisation. Hoyt and Gerloff (1999) highlight that cross-training may go against the career goals of many engineers who oppose the goal of becoming generalists. Hsieh and Chao (2004) found that, in some high-tech industries, increases in professional efficacy and reductions in employee job burnout were related not to job rotation, but to job specialisation. Future studies are needed to examine potential industry-specific effects of multi-skilling capabilities on competitive advantage, and how these effects may differ between innovators and non-innovators. Given the scarcity of research on the importance of capabilities in the context of innovation for SME competitive advantage, new studies are needed to examine capabilities other than job rotation and multi-skilling and to shed additional light on the interactive relationship between capabilities and innovation.

This paper provides a unique empirical assessment of SME competitive advantage, because it takes into account firm behaviour in a previous time point. Two key findings are evident. First, the continuity of organisational routines is critical, once the results imply that consistent innovators using job rotation or multi-skilling routines in both 2002 and 2004 tend to outperform most other firms in the same industry. Second, because the use of these capabilities in the absence of consistent innovation is not associated in the sample with a greater probability of top performance, the combination of capabilities and innovation is essential.

UK's Department for Business, Enterprise, and Regulatory Reform (the former Department of Trade and Industry) has treated SME innovation as a crucial area of action, through policies such as advice on intellectual property and greater access to R&D grants and to public procurement opportunities for small businesses (Department of Trade and Industry, 2003; Department for Business,

Enterprise, and Regulatory Reform, 2007). These policies are based on the view that "innovation is the successful exploitation of new ideas. It is the key to maintaining a competitive edge in an increasingly competitive global economy (...)" (Department for Business, Enterprise, and Regulatory Reform, 2007). Empirical research on SME competitive advantage examining the interaction between organisation capabilities and innovation is rare. Given the relevance of organisational capabilities for firms to gain competitive advantage, and the economic importance of SMEs in terms of representing the majority of firms and employing a large proportion of the workforce in many countries (e.g., Organization for Economic Cooperation and Development, 2005), future studies should investigate the connections among capabilities, innovation, and competitive advantage in SMEs from UK and from other countries. Additionally, given that research covering service industries has traditionally been sparser than research on manufacturing industries (e.g., Cainelli *et al.*, 2006; Callon *et al.*, 1999), future studies will benefit from examining how capabilities and innovation are connected with competitive advantage among service SMEs.

The empirical findings from this study suggest the following theoretical implications. First, the findings are coherent with the resource-based view that organisational capabilities follow an accumulative process and depend on firms' continuous effort. In the sample, enterprises that innovate and use job rotation or multi-skilling consistently over time, compared to those that do not perform these same activities continuously, tend to outperform most other firms in the same industry. Second, the findings are consistent with the process/capability school, by indicating the importance of an interaction between capabilities and innovation for competitive advantage. The findings support a theoretical model according to which the relationship between capabilities and competitive advantage is moderated by innovation, such that the combination between capabilities and innovation is positively associated with competitive advantage.

Acknowledgements

We thank Alan Hughes, Gerard George, Bronwyn Hall and the participants in the Seminar of Innovation at the University of California Berkeley, and two anonymous reviewers for their helpful comments on earlier versions of this paper.

References

Antonnucci, T and M Pianta (2002). Employment effects of product and process innovation in Europe. *International Review of Applied Economics*, 16(3), 295–307.

Barney, J (1991). Firm resources and sustained competitive advantage. *Journal of Management*, 17(1), 99–120.

Barney, JB (2001). Is the resource-based "view" a useful perspective for strategic management research? Yes. *Academy of Management Review*, 26(1), 41–56.

Becker, MC (2004). Organizational routines: A review of the literature. *Industrial and Corporate Change*, 13(4), 643–677.

Becker, MC (2005). A framework for applying organizational routines in empirical research: Linking antecedents, characteristics and performance outcomes of recurrent interaction patterns. *Industrial and Corporate Change*, 14(5), 817–846.

Belderbos, R, M Carree and B Lokshin (2004). Cooperative R&D and firm performance. *Research Policy*, 33, 1477–1492.

Bhatt, GD and V Grover (2005). Types of information technology capabilities and their role in competitive advantage: An empirical study. *Journal of Management Information Systems*, 22(2), 253–277.

Bullock, A (2003). Survey design, response bias, and sample characteristics in the 2002 CBR SME survey. In Enterprise Challenged: Policy and Performance in the British SME Sector 1999–2002, A Cosh and A Hughes (eds.), pp. 114–127. Cambridge, UK: Center for Business Research, University of Cambridge.

Bullock, A and I Milner (2007). Survey design, response bias, and sample characteristics in the 2004 CBR SME survey. In British Enterprise: SME Growth, Innovation, and Public Policy 2001–2004, A Cosh and A Hughes (eds.), pp. 107–117. Cambridge, UK: Center for Business Research, University of Cambridge.

Butchart, RL (1987). A new definition of the high technology industries. *Economic Trends*, 400, 82–88.

Cainelli, G, R Evangelista and M Savona (2004). The impact of innovation on economic performance in services. *Service Industries Journal*, 24(1), 116–130.

Cainelli, G, R Evangelista and M Savona (2006). Innovation and economic performance in services: A firm-level analysis. *Cambridge Journal of Economics*, 30(3), 435–458.

Callon, M, R Coenen, WM Cohen, C Freeman, F Kodama, F Meyer-Krahmer, KLR Pavitt and GP Pisano (1999). Retrospective evaluation (1971–1999). *Research Policy*, 28, 911–919.

Campion, MA, L Cheraskin and MJ Stevens (1994). Career-related antecedents and outcomes of job rotation. *Academy of Management Journal*, 37(6), 1518–1542.

Cefis, E and M Ciccarelli (2005). Profit differentials and innovation. *Economics of Innovation and New Technology*, 14(1–2), 43–61.

Cefis, E and O Marsili (2005). A matter of life and death: Innovation and firm survival. *Industrial and Corporate Change*, 14(6), 1167–1192.

Conner, KR (1991). A historical comparison of resource-based theory and five schools of thought within industrial organization economics: Do we have a new theory of the firm?*Journal of Management*, 17(1), 121–154.

Cosh, A, A Hughes and E Wood (1999). Innovation in UK SMEs: Causes and consequences for firm failure and acquisition. In Entrepreneurship, Small and Medium-Sized Enterprises, and the Macroeconomy, ZJ Acs, B Carlsson and C Karlsson (eds.), pp. 329–366. Cambridge, UK: Cambridge University Press.

Cosh, A and A Hughes (eds.) (2003). Enterprise Challenged: Policy and Performance in the British SME Sector 1999–2002. Cambridge, UK: Center for Business Research, University of Cambridge.

Cosh, A and A Hughes (eds.) (2007). British Enterprise: SME Growth, Innovation, and Public Policy 2001–2004. Cambridge, UK: Center for Business Research, University of Cambridge.

Dehning, B and T Stratopoulos (2003). Determinants of a sustainable competitive advantage due to an IT-enabled strategy. *Journal of Strategic Information Systems*, 12(1), 7–28.

Department for Business, Enterprise, and Regulatory Reform (2007). *Innovation Report — Actions*. Retrieved November 1, 2007 from http://www.berr.gov.uk/innovation/innovation-dti/Innovation%20Report/page10853.html.

Department of Trade and Industry (2003). *Competing in the Global Economy: The Innovation Challenge*. Retrieved November 1, 2007 from http://www.berr.gov.uk/files/file12093.pdf.

DeVoe, SE and J Pfeffer (2007). Hourly payment and volunteering: The effect of organizational practices on decisions about time use. *Academy of Management Journal*, 50(4), 783–798.

Dierickx, I and K Cool (1989). Asset stock accumulation and sustainability of competitive advantage. *Management Science*, 35(12), 1504–1511.

Douglas, TJ and WQ Judge (2001). Total quality management implementation and competitive advantage: The role of structural control and exploration. *Academy of Management Journal*, 44(1), 158–169.

Dyck, R and N Halpern (1999). Team-based organizations redesign at Celestica. *Journal for Quality and Participation*, 22(5), 36–40.

Eriksson, T and J Ortega (2006). The adoption of job rotation: Testing the theories. *Industrial and Labor Relations Review*, 59(4), 653–666.

Fabling, RB and A Grimes (2007). Practice makes profit: Business practices and firm success. *Small Business Economics*, 29(4), 383–399.

Freel, MS and PJA Robson (2004). Small firm innovation, growth, and performance: Evidence from Scotland and Northern England. *International Small Business Journal*, 22(6), 561–575.

Galende, J (2006). Analysis of technological innovation from business economics and management. *Technovation*, 26, 300–311.

Galende, J and I Suárez (1999). A resource-based analysis of the factors determining a firm's R&D activities. *Research Policy*, 28(8), 891–905.

Grant, RM (1991). The resource-based theory of competitive advantage: Implications for strategy formulation. *California Management Review*, 33(3), 114–135.

Havnes, PA and K Senneseth (2001). A panel study of firm growth among SMEs in networks. *Small Business Economics*, 16(4), 293–302.

Houghton, E and V Portougal (2005). A model for labour pooling in batch manufacturing. *International Journal of Production Research*, 43(4), 671–685.

Hoyt, J and EA Gerloff (1999). Organizational environment, changing economic conditions, and the effective supervision of technical personnel: A management challenge. *Journal of High Technology Management Research*, 10(2), 275–293.

Hoyt, J and T Matuszek (2001). Testing the contribution of multi-skilled employees to the financial performance of high-tech organizations. *Journal of High Technology Management Research*, 12(2), 167–181.

Hsieh, AT and HY Chao (2004). A reassessment of the relationship between job specialization, job rotation and job burnout: Example of Taiwan's high-technology industry. *International Journal of Human Resource Management*, 53(4), 529–540.

Hughes, A and E Wood (2000). Rethinking innovation comparisons between manufacturing and services: The experience of the CBR SME Surveys in the UK. In Innovation Systems in the Service Economy: Measurement and Case Study Analysis, JS Metcalfe and I Miles (eds.), pp. 105–124. Boston, MA: Kluwer Academic Publishers.

Kim, YM and KS Park (2003). Multiskilling and firm performance. *Seoul Journal of Economics*, 16(4), 387–422.

Lane, OJ, BR Koka and S Pathak (2006). The reification of absorptive capacity: A critical review and rejuvenation of the construct. *Academy of Management Review*, 31(4), 833–863.

Lazear, E (2000). Performance pay and productivity. *American Economic Review*, 90(5), 1346–1361.

Organization for Economic Cooperation and Development (2005). SME and Entrepreneurship Outlook. Paris, France: Organization for Economic Cooperation and Development.

Ortega, J (2001). Job rotation as a learning mechanism. *Management Science*, 47(10), 1361–1370.

Preissl, B (2000). Service innovation: What makes it different? In Innovation Systems in the Service Economy: Measurement and Case Study Analysis, JS Metcalfe and I Miles (eds.), pp. 125–148. Boston, MA: Kluwer Academic Publishers.

Ray, G, JB Barney and WA Muhanna (2004). Capabilities, business processes, and competitive advantage: Choosing the dependent variable in empirical tests of the resource-based view. *Strategic Management Journal*, 25, 23–37.

Rogerson, L (1993). Cover your bases. *Small Business Reports*, 18(7), 60–64.

Sanchez, R (2004). Understanding competence-based management: Identifying and managing five modes of competence. *Journal of Business Research*, 57(5), 518–532.

Schulze, WS (1994). The two schools of thought in resource-based theory: Definitions and implications for research. *Advances in Strategic Management*, 10A, 127–151.

Spanos, YE and S Lioukas (2001). An examination into the causal logic of rent generation: Contrasting Porter's competitive strategy framework and the resource-based perspective. *Strategic Management Journal*, 22(10), 907–934.

Teece, DJ, G Pisano and A Shuen (1997). Dynamic capabilities and strategic management. *Strategic Management Journal*, 18(7), 509–533.

Walter, A, M Auerb and T Ritter (2006). The impact of network capabilities and entrepreneurial orientation on university spin-off performance. *Journal of Business Venturing*, 21, 541–567.

Watson, J (2007). Modeling the relationship between networking and firm performance. *Journal of Business Venturing*, 22, 852–874.

Wernerfelt, B (1984). A resource-based view of the firm. *Strategic Management Journal*, 5, 171–180.

Wiggins, RR and TW Ruefli (2002). Sustained competitive advantage: Temporal dynamics and the incidence and persistence of superior economic performance. *Organization Science*, 13(1), 82–105.

Wolff, JA and TL Pett (2006). Small-firm performance: Modeling the role of product and process improvements. *Journal of Small Business Management*, 44(2), 268–284.

Yazici, HJ (2005). Influence of flexibilities on manufacturing cells for faster delivery using simulation. *Journal of Manufacturing Technology Management*, 16(8), 825–841.

Zahra, SA and DM Garvis (2000). International corporate entrepreneurship and firm performance: The moderating effect of international environmental hostility. *Journal of Business Venturing*, 15, 469–492.

Zahra, SA, HJ Sapienza and P Davidsson (2006). Entrepreneurship and dynamic capabilities: A review, model, and research agenda. *Journal of Management Studies*, 43(4), 917–955.

Chapter 4

INTELLECTUAL CAPITAL, INNOVATION, AND PERFORMANCE: EMPIRICAL EVIDENCE FROM SMEs

Karl-Heinz Leitner

Austrian Institute of Technology

This research paper examines the relationship between intellectual capital, product innovation and performance based on a study of Austrian firms covering a 10-year period. It is argued that intellectual capital enhances a firms ability to successfully realise innovations and thus contributes positively to its performance. Our study found that human capital and structural capital were both significantly associated with performance in product innovating firms, but that each had a different impact on this performance. While human capital had a positive impact on profitability and growth in the long run, contrary to expectations, structural capital had a negative effect on profitability and growth indicating that apparent strength can turn into a weakness over time. In addition, the study found that human capital and structural capital had no joint effect on the performance of product innovating firms.

Keywords: Intellectual capital; human capital; structural capital; product innovation; longitudinal study; small and medium enterprises.

Introduction

The product innovation process is frequently interpreted and analysed from the resource-based view (RBV) (e.g., Nelson, 1991; Nonaka, 1994; Verona, 1999; Teece, 2000; Terziovski, 2010) examing different forms of resources and competences in relation to corporate product innovation activities. In doing so, this strand of literature focuses on factors which go beyond those traditionally

This paper was first published in *International Journal of Innovation Management*, Vol. 19, No. 5 (October 2015), 1550060 (27 pages). © Imperial College Press. DOI: 10.1142/S1363919615500607 regarded as crucial for successful innovation activities, such as investments in R&D, market orientation or product strategy (e.g., Cooper and Kleinschmidt, 1987; Han *et al.*, 1998). Innovation is hence a knowledge creating process, and the ability to innovate is closely related to a firms intangible resources.

Some authors (e.g., Barney, 1991; Teece, 2000; Subramaniam and Youndt, 2005) differentiate between various forms of intangible resources, often referred to as knowledge assets or intellectual capital, to study their impact on innovation, competitiveness and performance. Accordingly, they contend that firms should invest in human resources, relationships and organisational procedures to raise their innovation capabilities and build up the important complementary assets that assure the success of improved or new products. In this study, we refer to the strand of literature which distinguishes different forms of intellectual capital and investigate its role on the innovativeness and performance of firms. We argue that intellectual capital facilitates the successful deployment of product innovations in several ways, including design and manufacturing quality, congruence with customer needs and timely product launches, a view which is, for instance, also adopted by Thornhill (2006). We divide intellectual capital into two components: Human capital and Structural capital and therefore adopt a taxonomy suggested by early proponents of intellectual capital literature (e.g., Roos *et al.*, 1997). Human capital is defined in this paper as the capabilities and attitude of a firms employees, while Structural capital is its processes, structures, brands and relationships with customers. This basic distinction is also similar to that proposed by scholars of the RBV; Barney (1991), for instance, distinguishes between human resources and organisational resources.

The empirical evidence about the relationship between intellectual capital and innovation is still fragmented, which is particularly true for SMEs. Most studies to date investigate the role of different forms of intellectual capital in isolation and, for instance, examine the impact of human capital, social capital or close customer relationships on product innovation (e.g., Hsieh and Tsai, 2007; Olavarrieta and Friedmann, 2008). In extension to many other studies we see intellectual capital not only as having a direct effect on product innovation activities but consider intellectual capital as factor which moderates the relationship between innovation and firm performance. This question is also of interest for practice as one can ask whether companies should invest rather in human capital (e.g., training of employees) or in structural capital (e.g., organisational processes), or in both, to improve the innovation performance of a company. Moreover, intellectual capital is a phenomenon based on interactions, combinations, and complementarities (Galunic and Rodan, 1998) and only a very few studies to date have examined the joint effect of different forms of intellectual capital (e.g., Subramaniam and Youndt, 2005; Huang and Liu, 2005). Our study therefore contributes to the debate by investigating the combined effect of human and structural capital on performance in product innovating firms.

In this paper, we first analyse the role of human and structural capital for product innovating firms, these are firms which at least once introduced a significantly improved product on the market covering a 10-year period. Secondly, we examine the group of firms which continuously develop new products and are considered to be "highly innovative" for the purposes of this paper. We assume that the more innovative a firm is, the more crucial intellectual capital will be in the successful development of new products, and that regular product development also contributes to the development of intellectual capital. Thus, both activities are path dependent and reinforce each other (Danneels, 2002) Third, we investigate whether human capital and structural capital jointly interact to contribute to the performance of product innovating firms.

There is less literature available that considers the impact of intellectual capital on performance over a longer period of time (e.g., Keller, 2004). In this respect, Wiggins and Ruefli (2002), for instance, emphasise the importance of a longitudinal dataset covering at least 10 years to reliably assess the impact of resources and capabilities on performance. This is particularly relevant, as proponents of the RBV have argued that intangible investments should have a sustaining effect on performance (e.g., Barney, 1991). This issue is taken into account in this paper, as we carry out a study among manufacturing SMEs covering the period 1992–2002. The paper uses data from a sample of 91 Austrian SMEs which provided data on their performance and innovation behaviour in this period in two surveys carried out in 1995 and 2003, this long term perspective is a further contribution of the paper.

In the next section, we develop hypotheses based on literature on the RBV of the firm, intellectual capital management and innovation economics. We then explain how the innovation, intellectual capital and firm performance variables are measured, present the results of our statistical analysis and conclude with a discussion of the main findings.

Background and Hypotheses

Different theories and perspectives from industrial economics, the RBV of the firm, evolutionary economics and organisational theory have all been used to explain the role of innovation in corporate development and performance. In the

RBV, firms are understood in terms of resources and routines and, accordingly, gain competitive advantage through their heterogeneous combination of resources, rather than through the product market conditions and positioning activities (e.g., Barney, 1991; Grant, 1996). In the RBV context, innovation, and product development are the result of unique resources and the accumulation of knowledge (e.g., Iansiti and Clark, 1994; Verona, 1999; Danneels, 2002; Branzei and Vertinsky, 2006).

Intellectual capital concepts and taxonomies help to explain the different roles played by intangible resources in the entrepreneurial innovation process. Many authors in this field argue that different forms of intellectual capital, such as human, structural or customer capital, contribute to a firms competitiveness and innovativeness (e.g., Stewart, 1997; Sveiby, 1997; Nahapiet and Ghoshal, 1998; Subramaniam and Youndt, 2005).

Human capital is suggested as one important element of intellectual capital and is part of all taxonomies proposed in the literature (e.g., Stewart, 1997; Roos *et al.*, 1997; Nahapiet and Ghoshal, 1998; Subramaniam and Youndt, 2005). Since innovation is viewed as the ongoing pursuit and harnessing of new and unique knowledge, the individual knowledge, skills and abilities of a firms employees, that is its human capital, are an important element in innovation. Agile and well trained individuals are seen as fundamental to innovation and form the basis for the creation, integration and transfer of knowledge within a firm (e.g., Michie and Sheehan, 1999; Shrader and Siegel, 2007).

Organisational capital or structural capital (e.g., Edvinsson and Malone, 1997; Roos *et al.*, 1997), both terms are often used interchangeably, is another frequently discussed and important form of intellectual capital which may be positively associated with the innovativeness and performance of firms. Organisational structures, processes and routines are important elements of organisational or structural capital (Bontis, 1998) which, for instance, enable an efficient product planning process, facilitate communication within new product development teams and provide the necessary information for product development. In general, quality management has become one of the key types of management system implemented within SMEs across Europe (van der Wiele and Brown, 1998), and it can be assumed that particular quality management systems are positively associated with innovativeness and performance (Chu and Pucik, 2005). Systems like ISO 9000 or TQM allow systematic process management and build up routines which facilitate product development processes.

The communication and sharing of knowledge both within the organisation and with external parties (in particular customers) is a further form of intellectual capital. Consequently, Edvinsson and Malone (1997) or Roos *et al.* (1997) divide structural capital into two subcategories, namely organisational capital and customer capital. In contrast, other authors define relational or social capital as separate forms of intellectual capital on a par with the other intellectual capital categories (e.g., Stewart, 1997; Nahapiet and Ghoshal, 1998), a view which is not adopted in this paper. Relationships and networks with partners (e.g., customers, suppliers and research institutions) are highly important for innovation activities as they allow firms to combine new forms of knowledge in a unique way and enhance information exchange not only within an innovation project team, but also with external actors. Good customer relationships are particularly important for product innovation activities, as they serve to raise customer trust, increase customer retention and produce superior performance (Baker and Sinkula, 2005).

For the purposes of the research presented in this paper, we classify intellectual capital into two forms, namely human capital and structural capital, and include both organisational and customer capital in the latter (Roos *et al.*, 1997). Human capital captures the capabilities and attitudes of employees. Structural capital is defined in this paper as covering processes, routines, structures, brands and relationships with customers.

In literature, there is no consensus as to the relationship between different forms of intellectual capital and innovation. Some authors (e.g., Evenson and Westphal, 1995; Roos *et al.*, 1997; McElroy, 2003) regard innovation as a specific aspect or subcategory of structural or organisational capital and occasionally use the term innovation capital. A few authors (e.g., Subramaniam and Youndt, 2005) consider innovation capabilities as the result of good intellectual capital while others maintain that innovation capabilities strengthen intellectual capital (e.g., Marqués *et al.*, 2006). Marqués *et al.* (2006) propose that innovation capabilities have an impact on the different forms of intellectual capital. In line with some authors (e.g., Youndt *et al.*, 2004), we conceptualise innovation as a separate construct, thus permitting us to study its interaction effects with intellectual capital.

Despite the appealing logic and convincing arguments put forward in the RBV, empirical evidence regarding the role of the different forms of intangibles, their interrelatedness and links to innovation is still limited, particularly for SMEs. In general, research in this field has so far examined the contribution of different intangible resources or intellectual capital mainly in isolation. A large number of publications examine how human resources or human capital is directly associated with the innovation activities of firms. In a study of Canadian SMEs, Baldwin and Johnson (1996), for instance, find evidence that more innovative firms are more likely to offer their employees access to formal and informal training or make use of innovative compensation packages than their less innovative counterparts. However, some studies also argue that human assets enable a firm to leverage innovation into performance thus emphasising the moderating effect of human capital (Thornhill, 2006). Thornhill (2006) reports that the education level of the workforce and training investments (defined as knowledge assets) and innovation jointly interact to have an positive effect on firm performance based on a Canadian survey.

Huang and Liu (2005) employ multiple regression models to examine relationships between innovation and/or structural capital and firm performance. They find that investment in structural capital has a positive effect on performance, but that this influence can become negative when the investment exceeds an optimal level. Further evidence on the role of different knowledge assets is provided by Smith *et al.* (2005), who study the rate of new product launches in high-tech firms. Their results indicate that existing employee knowledge and an organisational environment which supports knowledge exchange both contribute positively to the number of new products launched by a firm.

Drawing on the arguments provided, we assume that intellectual capital and innovation jointly interact to be positively associated with firm performance in the long run and propose the following hypotheses:

H1a: *The stronger the human capital of the firm, the higher the positive effect of product innovation on firm performance in the long run.*

H1b: *The stronger the structural capital of the firm, the higher the positive effect of product innovation on firm performance in the long run.*

There is no common understanding in the literature whether firms which continuously develop new products deploy specific forms of intellectual capital to be successful. We therefore distinguish in our study between less innovative and highly innovative firms, defining the latter as firms which regularly develop and launch new products.

Some scholars argue that human capital is more strongly linked to the ability to develop more (radical) innovative products and include in particular those capabilities which allow organisational barriers to innovation to be overcome (e.g., Rothaermel and Hess, 2007; Branzei and Vertinsky, 2006). Individuals and their associated human capital may encourage the questioning of existing norms and develop new ways of thinking (Subramaniam and Youndt, 2005).

In addition, some empirical data delivers evidence that strong structural capital contributes positively to the deployment of investments in the development of new products. Cho and Pucik (2005), for instance, report that effective quality management systems are positively associated with the performance in firms which continuously develop new products. Appiah-Adu and Singh (1998) emphasise the importance of good customer relationships for the commercial success and firm performance of new product innovating firms. Thus, structural capital is more important for more innovative firms than for incremental product innovators.

The development of intellectual capital can be regarded as an accumulative process which not only creates incentives to be exploited by product innovations, but also delivers the required complementary assets. The development of new products expands a firms competence base, which in turn enables further product innovations (Danneels, 2002). Thus, in contrast to incremental innovations or product enhancements, continuous new product developments permit the renewal of organisational resources and can contribute to the building up of dynamic capabilities (Teece, 2007; Eisenhardt and Martin, 2000). In line with these arguments, we argue that continuous product development enhances a firms ability to alter its resource configurations and explore new knowledge. We hence propose that both human capital and structural capital are particularly significant for the success of highly innovative firms, defined as firms which continuously develop new products and formulate:

H2a: *The stronger the human capital, the higher the performance of highly innovative firms compared to less innovative firms.*

H2b: *The stronger the structural capital, the higher the performance of highly innovative firms compared to less innovative firms.*

Figure 1 summarises hypotheses 1 and 2.

The concept of resource bundles conveys the notion that competitive advantage frequently depends on the interaction of resources (Barney, 1991; Galunic and Rodan, 1998). Scholars argue that the different forms of intellectual capital are not found in neat, separate packages but instead both make use of individual and organisational knowledge (e.g., Subramaniam and Youndt, 2005; Nonaka *et al.*, 2006; Namasivayam and Denizci, 2006). There is the belief that focusing solely on one element would lead to inferior performance, since the best human capital, would not, for instance, be able to unfold its creative and productive potential if organisational structures and processes were too bureaucratic and inefficient (Edvinsson and Malone, 1997; Hitt *et al.*, 2001). Indeed, literature on this subject stresses that intellectual capital is a phenomenon based on interactions, combinations and complementarities. Hence, individual knowledge can only be transformed into value if it is supported by appropriate organisational

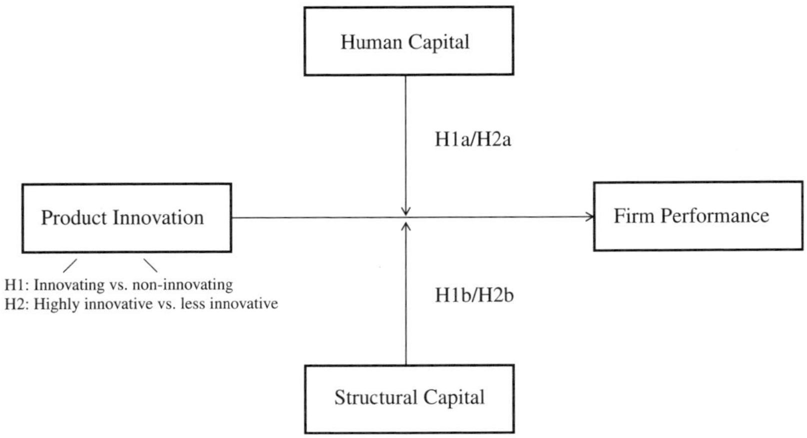

Fig. 1. Summary of hypothesis 1 and 2.

processes, such as codification, and if internal and external networks facilitate knowledge transfer. Stieglitz and Heine (2007) claim that firms which invest in human capital also tend to invest in complementary assets to bind employees to the firm. Thus, strong structural capital not only attracts talented employees, it may also facilitate knowledge creation and sharing.

The role of complementarity and interaction between different resources has been largely ignored in empirical studies. Most studies investigate the role and bundling of various human resource practices (e.g., Michie and Sheehan, 1999; Laursen and Foss, 2003), which does not account for structural or organisational capital as understood in our study. Branzei and Vertinsky (2006) assert that human capital has no direct effect on new product development success, but does catalyse the absorption of external ideas and enable the transformation of ideas into operations, thus indirectly contributing to innovation success. Hardly any empirical research has so far been carried out studying the relative impact and possible combined effects on innovation performance. In this context, we assume that human and structural capital to be complementary for product innovating firms and propose:

H3: *Human and structural capital jointly interact to have a positive effect on the relationship between product innovation and firm performance in the long run.*

Figure 2 summarises hypothesis 3.

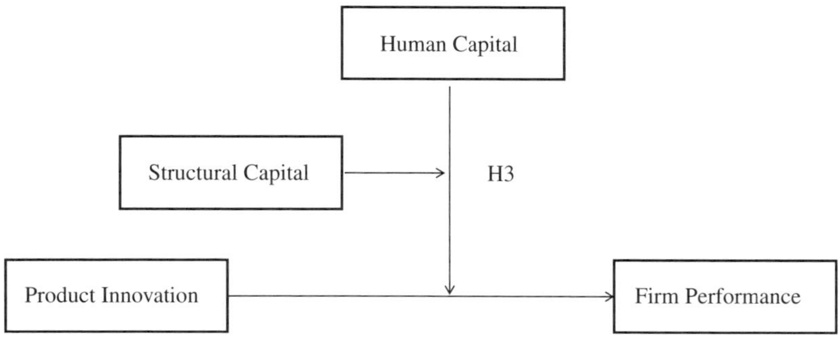

Fig. 2. Summary of hypothesis 3.

Sample

The data used to test the above hypotheses is taken from a longitudinal study of Austrian SMEs with 20 to 500 employees. The first survey was carried out in 1995 and was followed by a second survey of the same firm sample in 2003.

The firms were selected at random from the Dun & Bradstreet database, which covers all Austrian firms with more than 10 employees. The seven industries chosen represent about 30% of all SMEs in the country's manufacturing sector and cover the typical low- and medium-tech manufacturing industries which contribute to economic growth, technological change and productivity in Austria (ÖSTAT 1998). The total number of firms in these seven sectors was 2051 in 1995. The selected firms are distributed across the Austrian industry classification standards and cover "manufacture of wood and of products of wood" (NACE 20): 16%; "manufacture of furniture" (NACE 36): 10%; "manufacture of basic metals" (NACE 27): 11%; "manufacture of fabricated metal products" (NACE 28): 24%; "manufacture of machinery and equipment" (NACE 29): 19%; "manufacture of chemicals and chemical products" (NACE 24): 8%; "manufacture of rubber and plastic products" (NACE 25): 12%. This distribution accounts for the relative importance of these sectors across the Austrian manufacturing industry.

The interviews were carried out by phone as this method usually assures a higher response rate. It was also anticipated that the personal contact established by using this method would help to assure a high response rate in a subsequent interview. The interviews lasted about 90 min and were carried with the managing directors of the participating firms using a standard questionnaire. With the exception of those firms that had gone out of business in the meantime, we were able to convince all the firms to participate in our second survey in 2003, where interviews were again carried out by phone with the managing directors based on the same questionnaire. Nine of the firms initially interviewed had since gone out of business, leaving us with 91 firms for the second survey.

Measures and Methods

Innovation

We measured the innovation behaviour of the selected firms at both points in time by the type of innovation activities they had realised in the three years prior to the investigation, based on the information provided by the firms themselves. To obtain this information, we asked them if they had introduced incremental product innovations ("we have introduced a strongly improved product in the last three years") or developed new products ("we have introduced a new product which is new for the firm in the last three years"). Capturing two levels of innovativeness in this way is a common method of measuring a firms innovation output (e.g., Baldwin and Johnson, 1996; Vaona and Pianta, 2007). Following suggestions found in the literature on innovation (Noteboom, 1994), changes involving only minor design alterations were not considered innovative. Thus, we considered firms which had developed and launched new products between 1992 and 2002 to be more innovative than firms which had only incrementally improved their existing products.

Based on these measurements, we were then able to classify the firms into two large groups: Firms which had introduced a product innovation at least in one period were classified as "product innovating firms", firms which had not introduced any product innovation at all as "non-innovating firms". The group of the "product innovating firms" was further divided into two subgroups, firms which had developed a new product in both periods were categorised as "highly innovative firms", and firms which had developed new products irregularly or introduced incremental product innovations were classified as "less innovative firms". Thus, the group of innovative firms comprised those firms classified as "highly innovative firms" and "less innovative firms". All variables were dummy coded.

In addition, R&D expenditure as a percentage of the total turnover were measured. We calculated relative R&D expenditure (industry adjusted R&D) by subtracting the average industry R&D expenditure obtained from the Austrian Statistical Office) from the firms R&D expenditure to account for industry effects. The number of all patents granted to a firm in 1995 served as further variable indicative of product innovation activities.

Controls

Previous research suggests that firm size and firm age are important factors which influence performance (Birley and Westhead, 1990; Almus and Nerlinger, 1999). For instance, firm growth tends to decline with increasing firm size and age. Consequently, we used firm size (measured as the total number of employees) and firm age in 1995 (log transformed) as control variables.

In addition to firm-specific factors, we also accounted for industry environment, since this can also have an impact on firm performance. Since we had adjusted the growth rates by industry growth figures and asked the firms to compare their own financial performance (profitability) with that of their main competitors, we had already checked for a possible confounding effect of industry on performance and did not include industry as further control variable.

Intellectual capital

Empirical research on intangible resources and capabilities had not provided original scales for a comprehensive measurement of capabilities and intellectual capital when the first survey was designed in 1995. Although researchers have addressed the extent to which corporate resources and competences can indeed be measured at all, we followed an approach frequently encountered in the literature and measured the strength of the participating firms based on their self-assessment of their own strengths relative to those of their competitors (Miller and Roth, 1994). This approach is similar to the method used by Spanos and Lioukas (2001) to measure corporate assets, which captures a firms intellectual capital using the respondents' own assessment of their human and structural capital in different areas (see Appendix 1 for a definition of variables).

Human capital was a four-item measure assessed by a firms strength relative to its competitors in the areas "qualification of personnel", "flexibility of personnel", "well-informed personnel" and "ability to communicate" in line with typical features of human capital described in literature (e.g., Stewart, 1997; Roos *et al.*, 1997). This measure accounts in particular for those features of human capital that are said to affect knowledge creation and knowledge transfer and are crucial for innovation (e.g., Nonaka, 1991). To test validity and reliability, we checked factor loadings, item-total correlation and Cronbach's alpha. The factor loadings were all greater than 0.65 and significant, item-total correlations were very high, and Cronbach's alpha of the human capital factor was 0.71.

Structural capital was a five-item factor constructed of a firms strength in the areas "quality management system", "operational efficiency", "reputation", "customer relationships" and "distribution and marketing" which was derived from the discussion of components which may influence innovation performance in SMEs in Sec. 2 above. This is in line with suggestions regarding the basic characteristics of structural capital and its organisational and customer capital components (e.g., Edvinsson and Malone, 1997; Roos *et al.*, 1997). Quality management systems and operational efficiency were regarded particularly relevant for manufacturing firms. However, we did not account for databases or information systems as we assumed that databases still played a minor role in the mid-1990s in small manufacturing firms. The importance of patents is straightforward to measure and was captured as one aspect of innovation activities together with R&D investments. In the final structural capital construct the factor loadings were all higher than 0.59, while Cronbach's alpha was 0.69. In addition, item-total scale correlation analysis revealed significant high relationships in the anticipated direction, indicating that all items contributed meaningfully to the respective dimensions of structural capital in the study.

The alphas for both constructs exceeded the lowest alpha suggested by Naman and Slevin (1993) and were in the upper level proposed for broad constructs by Van de Ven and Ferry (1997). Both variables served as composite measures for hypothesis testing. Discriminant validity was assessed by checking the magnitude of the correlation coefficients between the two variables which at 0.51 were below the conventional cutoff point of 0.70 (Cohen and Cohen, 1983).

A potential weakness of a self-assessment of a firms strength compared to its competitors is that it may lead to a biased measurement since the assessment depends on the competitive environment and the firms benchmarking experience. We hence checked whether firm size, firm age or industry had an effect on the distribution of all nine human and structural capital items. This analysis showed that all correlation coefficients were lower than 0.20, indicating that there was no structural factor bias.

Performance

Various studies have addressed the relationship between different performance variables, suggesting that performance is a multi-dimensional construct. In general, product innovation strategies are said to be associated with both financial performance and growth. Moreover, some scholars (e.g., Baum *et al.*, 2001) maintain that growth is often necessary to survive in highly competitive environments, particularly for young and small firms, and is thus also a condition for financial performance.

Three performance indicators were used in our study in both time periods: average profitability, turnover growth and employment growth. It is often difficult to obtain data on the profit levels of small firms as they are, in many cases, not obliged to publish their results and are also often reluctant to provide financial information (Sapienza *et al.*, 1988). Profitability was measured in our study on the basis of a self-assessment by respondents, who were asked to compare themselves with their competitors using a 5-point scale (where 1 = much worse and 5 = much better). Respondents were asked to assess their annual performance for each year between 1992 and 2002. We then calculated the mean, obtaining an average value for profitability ranging from 1.00 (1 in each year) to 5.00 (5 in each year) for the period 1992–2002.

The figures for turnover and employment for the years 1992 and 2002 were reported by the interviewed firms themselves in both surveys. Firm growth rate can be operationalised in different ways (Delmar, 1997): We took the figures for 1992 and 2002 and calculated the average annual percentage growth. Given the differing growth rates in the industries used in the sample, the average annual percentage growth rate (turnover and employment) of a firms principal industry was subtracted from its real growth rate obtained by the *Austrian Statistical Office*.

Methods

Regression analyses were used to test the research question. The three performance variables — profitability, turnover growth, and employment growth — served as dependent variables. The different variables for innovation, intellectual capital along with the control factor served as independent variables. For the regression models, we calculated a base model with the control variable, innovation factors and intellectual capital, added the interaction term between innovation and intellectual capital in a second step and the interaction between human capital, structural capital and innovation in a third. We also calculated the variance inflation factors for the regression models to check for multicollinearity. See Appendix 1 for a definition of all variables and Table 1 for a summary of the statistics.

Results

We give first an overview of the general characteristics and innovation behaviour of the participating firms for the period in question. The size of the firms studied, which cover low-tech and medium-tech firms, ranged from 21 to 470 employees, with a mean of 128 (see also Table 1). The average age of the firms was 48 years, with actual values ranging from 2 to 183 years, indicating that the firms were fairly mature. The results of the study indicate that 58% of the participating SMEs were innovative and had introduced an incrementally improved or new product on the market in the three years prior to each of the two surveys. This ratio of innovative firms is comparable with other studies both in Austria and at an European level (Vaona and Pianta, 2007; Leo, 1999) and also indicates the external validity of our research. In addition, the correlation table (see Table 1) showed virtually no direct relationships between the indicators of innovation behaviour and human and structural capital.

In Hypotheses 1a and 1b, we proposed that intellectual capital positively moderates the relationship between product innovation and firm performance. To test this hypothesis, we initially performed a hierarchical moderated regression and constructed an interaction term between intellectual capital and innovation. The product innovation variable was a dummy variable indicating whether the firm had launched a new or incrementally improved product at least once in the period studied. This allowed us to divide the participating firms into two groups, namely innovative (= 58) and non-innovative (= 33) firms. We also incorporated a number of control factors and the amount of relative R&D expenditure to account for industry effects. As the significant beta of the interaction term suggests, models 2, 5 and 8 (see Table 2) revealed that human capital moderated the innovation-performance link (Model 2: $b = 0.50$, $p < 0.01$ for profitability; Model 5: $b = 0.54$, $p < 0.01$ for turnover growth). Contrary to our expectations, we found that structural capital had a negative impact on firm growth and performance (Model 5: $b = -0.55$, $p < 0.10$ for turnover growth; Model 8: $b = -0.52$, $p < 0.05$ for employment growth). However, as the rather low adjusted R^2 indicates (Table 2, Models 5–9) that the findings with respect to turnover and employment growth have only weak explanatory power. In addition, the analysis showed that R&D investment had no significant effect on performance. Moreover, the traditional measure patents had a significant effect only on turnover growth. We also included an additional interaction term between innovation and R&D to specifically test the effect of R&D for product innovation, but this delivered no substantive results (not shown). Thus, we found support for Hypothesis 1a, but had to reject Hypothesis 1b.

Hence, although there was a positive link between human and structural capital (i.e., firms with strong human capital are also more likely to have strong structural capital), each factor had a different effect on performance. As mentioned above, this relationship indicated a possible multicollinearity problem,

Table 1. Summary statistics and correlation matrix.

	Variable	Mean	S.D.	1	2	3	4	5	6	7	8	9	10	11
1	Firm size	133.9	97.50	1										
2	Firm age	3.46	1.10	0.178	1									
3	Patents	1.27	2.74	0.141	0.161	1								
4	Relative R&D intensity	−0.82	1.40	0.286**	−0.075	0.100	1							
5	Product innovation	0.65	0.48	0.119	0.111	0.138	0.342**	1						
6	Highly innovative	0.25	0.44	0.126	−0.029	0.133	0.036	0.470**	1					
7	Human capital	3.93	0.62	0.036	−0.080	0.106	0.164	0.124	0.211*	1				
8	Structural capital	3.72	0.67	0.107	−0.173	0.044	0.257*	0.013	0.036	0.512**	1			
9	Profitability	3.64	0.67	0.075	0.093	0.058	−0.076	−0.080	−0.027	0.232*	0.162	1		
10	Turnover growth	2.66	8.46	0.076	0.053	0.159	0.005	−0.008	−0.048	0.082	−0.018	0.444**	1	
11	Employment growth	0.56	6.81	0.102	0.084	0.092	0.073	−0.057	0.003	0.178	0.096	0.422**	0.778**	1

*$p < 0.05$; **$p < 0.01$.

$N = 91$, except for the variable "highly innovative" ($N = 58$).

Table 2. Regression analysis: Intellectual capital, product innovation and performance.

Dependent variable	All firms ($N = 91$)								
	Profitability			Turnover growth			Employment growth		
	Model 1	Model 2	Model 3	Model 4	Model 5	Model 6	Model 7	Model 8	Model 9
Control									
Firm size	0.00	−0.07	−0.08	0.04	−0.03	−0.04	0.04	0.01	0.01
Firm age	0.13	0.11	0.11	0.06	0.03	0.03	0.11	0.09	0.09
Innovation									
Relative R&D intensity	−0.07	−0.02	−0.02	0.09	0.14	0.14	−0.07	−0.07	−0.07
Patents	0.13	0.13	0.12	0.25*	0.26*	0.25*	0.16	0.17	0.17
Product innovation (= innovating firm)	0.08	0.10	0.04	0.12	0.12	0.07	0.04	0.05	0.07
Intellectual capital									
Human capital	0.18	−0.22	−0.22	0.10	-0.26^{+}	-0.27^{+}	0.22^{+}	0.07	0.07
Structural capital	0.13	0.29	0.30	−0.04	0.31	0.31	−0.03	0.35	0.35
Interaction[1]									
Innovation × human capital		0.50**	0.50**		0.54**	0.60**		0.25	0.25
Innovation × structural capital		-0.41^{+}	-0.41^{+}		−0.55*	−0.56*		−0.52*	−0.52*
Innovation × human capital × structural capital			0.11			−0.11			−0.01
Adj. R^2	0.04	0.10	0.10	0.04	0.07	0.07	0.06	0.09	0.08
Model F	1.53	2.22*	2.18*	1.05	1.95^{+}	1.83^{+}	1.90^{+}	2.08*	1.82^{+}

Note: Standardised coefficients are reported; $^{+}p < 0.10$; $^{*}p < 0.05$; $^{**}p < 0.01$.

[1]In cases of interaction, the information conveyed by the coefficients for the independent terms that make up these interactions is not meaningful, and possibly misleading (Cohen and Cohen, 1983).

although this proved to be more analytical than theoretical in nature. This type of multicollinearity is common in innovation research when studying the different determinants of performance and innovation (Crepon *et al.*, 1998). However, the correlation coefficients between all other independent variables used in the regression models were all below 0.40 (Kennedy, 1992) and none of the variance inflation factors for the models were greater than 4.6, which is the guideline figure (10) suggested by Chatterjee and Price (1991). Thus, it was unlikely that multicollinearity between the independent variables affected the findings.

We also carried out some additional *post-hoc* analysis to explore the reasons for the negative impact of structural capital on growth in the long term. This involved additional analysis using performance variables covering only the period from 1992 to 1994 (prior to the firms' assessment of their intellectual capital) to test whether the impact of structural capital might be positive if considered only for a short period of time. However, the findings of this analysis were no different to those obtained in the initial analysis and confirmed the results. As we also measured the importance of different innovation barriers (by asking respondents to assess this importance on a 5-point scale in the first survey), we tested for specific patterns in firms with strong structural capital. Interestingly, our research revealed that innovating firms which assessed their structural capital to be strong were more likely to complain about higher innovation barriers, such as perceived commercial risks ($r = -0.26$, $p < 0.5$).

Hypotheses 2a and 2b propose that a strength in structural and human capital positively contributes to the firm performance of firms which continuously develop new products. To test this proposition, we used a second innovation variable (dummy coded) to measure whether a firm was a continuous new product developer ("highly innovative") or an irregular or incremental innovator ("less innovative"). 23 of the participating SMEs had launched a new product in both periods, and an analysis of these firms showed that the majority had launched incremental innovations, too. We employed a hierarchical moderated regression analysis within the innovative firms subgroup (= 58 firms), constructing an interaction term between structural capital and the continuous new product development variable.

The base models (1, 3, 5) (see Table 3) confirmed the results from above, they showed the positive impact of human capital and negative impact of structural capital for the entire innovative firm group (irrespective of product innovativeness). The regression equation integrating the interaction term showed that the interaction coefficients between product innovativeness and intellectual capital were generally negative with respect to human capital and positive with respect to

Table 3. Regression analysis: Intellectual capital, continuous new product development and performance.

Dependent variable	Product innovating firms ($N = 58$)					
	Profitability		Turnover growth		Employment growth	
	Model 1	Model 2	Model 3	Model 4	Model 5	Model 6
Control						
Firm size	−0.07	0.06	−0.11	−0.04	−0.09	−0.04
Firm age	−0.08	−0.07	0.08	0.09	0.08	0.05
Innovation						
Relative R&D intensity	0.02	0.04	0.06	0.08	0.10	−0.11
Patents	0.08	0.08	0.37**	0.32*	0.26*	0.26+
Highly innovative (= continuous new product development)	−0.05	−0.04	−0.09	−0.08+	0.10	0.11
Intellectual capital						
Human capital	0.57**	0.73**	0.36**	0.43*	0.39**	0.39*
Structural capital	−0.15	−0.49+	−0.21*	−0.32	−0.31*	−0.31*
Interaction[1]						
Highly innovative × human capital		−0.17		−0.09		−0.32
Highly innovative × structural capital		0.42		0.18*		0.55
Adj. R^2	0.18	0.19	0.12	0.09	0.15	0.16
Model F	2.63*	2.56*	2.50*	1.56	2.21*	2.21*

Note: Standardized coefficients are reported; $^{+}p < 0.10$; $^{*}p < 0.05$; $^{**}p < 0.01$.

[1]In cases of interaction, the information conveyed by the coefficients for the independent terms that make up these interactions is not meaningful, and possibly misleading information (Cohen and Cohen, 1983).

structural capital (see Table 3, Models 2 and 4). The coefficients were significant for structural capital regarding the effect on turnover growth (Model 4: $b = 0.18$; $p < 0.05$). Given that the innovation variable is dummy coded, it is apparent that structural capital is negatively associated with firm growth, particularly for the less innovative firm group. With respect to human capital, the interaction coefficient between human capital and continuous product development (= highly innovative) was negative, albeit not significant. We also performed split sample analysis to check the results and hence we compared the impact of human capital and structural capital separately within the less and highly innovative firm group. These models revealed that the negative relationship of structural capital is significantly associated with growth for less innovative firms. In general, the positive influence of human capital and the negative influence of structural capital are stronger in the less innovative firm group than in the highly innovative firm group. Thus, we found no support for Hypothesis 2a and only weak support for Hypothesis 2b.

Overall, continuous product developers were neither more profitable nor had they grown more strongly than irregular product developers or incremental innovators. Again, traditional R&D expenditure had no impact in this model, while the number of patents had a positive impact only on turnover growth (Model 3 and 4: $b = 0.37$; $p < 0.01$).

Hypothesis 3 deals with the question of whether innovative firms with strengths in both human and structural capital can be associated with higher performance. To test this hypothesis, we constructed a three-way interaction to identify trade-offs or synergies between both elements and their effect among product innovating firms (see Table 2, Models 3, 6, and 9). We found no clear evidence of either a positive or a negative association between human and structural capital and the three performance variables. The regression analysis revealed that integrating this interaction term did not increase the predictive power of the model. Hence, no evidence was found to support Hypothesis 3.

Discussion

The RBV of the firm sees innovation and new product development as a knowledge generating and competence building process. Our study dealt with the long-term links between human capital, structural capital, innovation and performance. We did not explicitly address innovation capabilities or innovation capital, but were instead interested in the impact of intellectual capital on performance in product innovating firms, assuming that intellectual capital does actually moderate the innovation-performance link. Accordingly, we argue that innovation cannot be fully understood by focusing on input measures such as R&D expenditure, but must also take account of complementary resources. Furthermore, this study contributes to research by examining the role of both human and structural capital and seeking to explain the interconnectedness between these different resources.

Our study delivered evidence that the innovation persistence of firms is very high; we had a considerable number of firms which continuously introduced product innovations. We found that the classic innovation indicator R&D expenditure had no influence, while intellectual capital had both a direct and an indirect effect on performance. This suggests that intellectual capital may have a stronger effect on performance in innovative firms than the level of relative R&D

expenditure. The number of patents had a positive impact only on turnover growth. Both R&D investments and the number of patents are sometimes also interpreted as measures of a firms innovation capital, but were treated separately in our study as our interest lay in the complementary relationship between intellectual capital and innovation (e.g., Thornhill, 2006).

At the same time, neither human capital nor structural capital had an effect on performance for the whole group of firms studied, while intellectual capital moderated the innovation–performance link. Human capital was generally associated positively with both profitability and turnover growth. While human capital had a positive impact on performance — in line with the findings in the extensive literature on this topic (e.g., Hayton, 2003) — structural capital had a negative impact on employee and turnover growth over a 10-year period, which was contrary to our expectations. However, these results do not take into account the product innovativeness of a firm, i.e., whether a firm is less or highly innovative.

Structural capital was represented in our study by strengths in operational efficiency, quality management systems, customer relationships, marketing, brands and reputation. Hence, it incorporates important aspects of organisational and customer capital used by other scholars to define intellectual capital (e.g., Roos and Roos, 1997). Thus, product innovating firms with what might normally be described by accountants as a high level of "goodwill" found it difficult to grow. Although we did not investigate the reasons or environmental context for this growth barrier, we speculate that firms are constrained by their own strengths and in some way lose their ability to think creatively and develop new strategies beyond their traditional industry or market paradigms. This holds particularly true for firms which did not continuously develop new products, as this effect was particular strong for less innovative firms. Thus, highly innovative firms seem to be able to develop their structural capital more dynamically (Eisenhardt and Martin, 2000). Moreover, these results show that the product innovativeness has to be considered when assessing the role and contribution of structural capital for the successful creation and implementation of product innovations.

The negative association between structural capital and performance in the long term can also be explained by a phenomenon referred to by Leonard-Barton (1992) as "core rigidities". Even though literature stresses the importance of developing resources dynamically, structural capital may be interpreted as a rather static resource for firms and one that is hard to transform and associated with lock-in and inertia, particularly for less innovative firms. Thus, the existence of well-organised management systems, established routines, efficient processes or strong brands also make firms more reluctant to enter new product markets and leave their established trajectories for fear of placing their strengths, for example a good reputation, at risk. Accordingly, firms may opt to develop their strengths along their established trajectory, for instance, by optimising established systems or fostering existing customer relationships instead of re-engineering processes or developing new market segments. These routines limit their capacities to selectively search and acquire new knowledge (March, 1991; Brenner and Tushman, 2002). Thus, SMEs tend to focus their routines around those functions or competencies that have brought them success in the past ("competency trap"), but neglect new important technological areas or markets.

With particular regard to customer capital (treated as a component of structural capital in our study), authors like Christensen and Bower (1996) also claim that a strong focus on customer needs (measured in our study, for instance, by customer relationships) may negatively affect innovativeness and performance. By focusing on current customers, firms develop a tendency to self-confirm their mental models and limit their ability to absorb the knowledge from outside their customer base that would be important for more radical innovation. In addition, we found that firms with good structural capital also complained more frequently about barriers to innovation, thus making our results more plausible. However, as mentioned above, this is particularly valid for less innovative firms, these are incremental and irregular product developers.

We found no evidence that human capital and structural capital reinforce each other. Indeed, the converse would seem to be true, with human and structural capital showing mainly a negative interaction effect on the various performance variables, albeit not significant and not taking into account the product innovativeness. Hence, firms which were only strong in terms of human capital were more successful than firms with strengths in both dimensions. In this respect, Subramaniam and Youndt (2005) find no combined effect of organisational capital and human capital on incremental innovation capabilities. In their study of large firms, Youndt *et al.* (2004), for instance, show that a relatively small group of superior performing firms exhibited high levels of human and organisational capital, but that most firms tended to specialise and focus primarily on only one form of intellectual capital.

In interpreting the findings of this field study, also have to consider its limitations. The longitudinal survey was exploratory in nature with a relatively small sample size of low-tech and medium-tech firms. However, using this sample of firms allowed us to study the sustaining impact of intellectual capital and innovation SMEs over a 10-year period. Since nine of the participating firms went out of business between the two surveys, there was also a survival bias in the sample. We used a fairly simple measure for product innovativeness and did not deal with process and organisational innovations. Examining the links with other forms of innovation would be an interesting question for further research and might also deliver a better indication of the effects of intellectual capital on a firms product and organisational innovativeness. Furthermore, we did not deal with the reverse causality issue and assumed that innovation strategy and intellectual capital had an impact on performance. Similarly, intellectual capital may have a direct effect on the product innovation activities and innovativeness as studied in other papers (e.g., Subramaniam and Youndt, 2005), which is not specifically addressed in this paper, though. Moreover, we did not study the role of the industry setting, opting instead to focus on the manufacturing sector under the assumption that similar human and structural capital aspects would be relevant in this homogenous group. Structural and human capital may also have different effects in large firms, which would be another interesting topic for further research.

Our human and structural capital measure included some basic features of human and structural capital which addressed the specifics of manufacturing SMEs. As mentioned, we do not explicitly cover social capital (which is considered an integral element in human and structural capital), and focus instead on customer relationships. We also did not address the investment in and development of intellectual capital over the entire period, since we only examined intellectual capital strength in the first survey. The data for assessing the intellectual capital was collected in 1995. Since then, a number of publications has been reported which conceptualise different and more sophisticated forms of structural capital, for example dealing with organisational connectedness (e.g., Jansen *et al.*, 2006). This new development in the literature has not been accounted for in this paper, though.

The fact that we investigated the effects of intellectual capital and innovation in the long term may also explain the relatively low predictive power of the models (particularly for growth), although it should be noted that a myriad of factors beyond the scope of any single study influence the development of a firm (Thornhill, 2006).

The management conclusion for SMEs seeking to foster their innovation capabilities, yet which currently only have low or medium levels of innovation, is that they should invest primarily in human capital. When it comes to innovation, these investments are at least as important as any investments in R&D. Moreover, entrepreneurs and small business managers aiming to grow their firms should continuously and critically analyse whether their purported internal process and customer base strengths might inhibit their development and growth in the long run. It might also be of interest for further studies to examine what constitutes the optimal level of structural capital and how managers can determine the point in time at which they have to transform and reconfigure their structural assets for the long term.

Appendix 1. List of Variables

Variable	Scale	Definition, construction, coding, and date of survey
Control variables		
Firm size	Ordinal	Number of employees in 1995
Firm age	Metric	Log of the number of years since inception in 1995
Innovation		
Relative R&D expenditure	Metric	R&D intensity of the firm as % of turnover in 1995 relative to the industry level R&D on the 3-digit level
Patents	Ordinal	Number of patents in 1995
Product innovation (= product innovating firm)	Dichotomous	1: Firm has introduced a new product or a strongly improved product in the three years prior to 1995 or 2002; 0: other (= not innovative)
Highly innovative (= continuous product development)	Dichotomous	1: Firm has introduced a new product in the three years prior to both 1995 and 2002; 0: other
Intellectual capital		
Human capital	Metric (alpha = 0.71)	Firms strength (1995) relative to competition in: — Qualification of personnel (factor loading = 0.77) — Flexibility of personnel (factor loading = 0.65) — Well-informed personnel (factor loading = 0.76) — Individual ability to communicate (factor loading = 0.73)
Structural capital	Metric (alpha = 0.69)	Firms strength (1995) relative to competition in: — Operational efficiency (factor loading = 0.60) — Quality management system (factor loading = 0.77) — Distribution and marketing (factor loading = 0.76)

(*Continued*)

Variable	Scale	Definition, construction, coding, and date of survey
		— Customer relationships (factor loading = 0.59) — Reputation (factor loading = 0.76)
Performance		
Profitability	Metric	Average score of the sequence of values in every year from 1992 to 2002: self-assessment in comparison to main competitors (1: much worse, 5: much better)
Turnover growth	Metric	Average annual growth in % between 1992 and 2002 (industry adjusted)
Employment growth	Metric	Average annual growth in % between 1992 and 2002 (industry adjusted)

Acknowledgements

I thank Per Bukh, Leif Edvinsson, Jan Mouritsen, Paloma Sanchez, David O'Dennell and Claus Varnes for their helpful comments on earlier versions of the paper.

References

Almus, M and EA Nerlinger (1999). Growth of new technology-based firms: Which factors matter. *Small Business Economics*, 13(2), 141–154.

Appiah-Adu, K and S Singh (1998). Customer orientation and performance: A study of SMEs. *Management Decision*, 36(6), 385–402.

Baker, WE and JM Sinkula (2004). Market orientation and the new product paradox. *Journal of Product Innovation Management*, 22, 483–502.

Baldwin, JR and J Johnson (1996). Business strategies in more- and less-innovative firms in Canada. *Research Policy*, 25(5), 785–804.

Barney, J (1991). Firm resources and sustained competitive advantage. *Journal of Management*, 17(1), 99–120.

Baum, JR, EA Locke and KG Smith (2001). A Multi-dimensional model of venture growth. *Academy of Management Journal*, 44(2), 292–303.

Benner, MJ and M Tushman (2002). Process management and technological innovation: A study of the photography and paint industries. *Administrative Science Quarterly*, 47(4), 676–706.

Birley, S and P Westhead (1990). Growth and performance contrasts between "types" of small firms. *Strategic Management Journal*, 11(7), 535–557.

Bontis, N (1998). Intellectual capital: An exploratory study that develops measures and models. *Management Decision*, 36(2), 63–76.

Branzei, O and I Vertinsky (2006). Strategic pathways to product innovation capabilities in SMEs. *Journal of Business Venturing*, 21(1), 75–105.

Cabrita, MR and JL Vaz (2005). Intellectual capital and value creation: Evidence from the Portuguese banking industry. *The Electronic Journal of Knowledge Management*, 4(1), 11–20.

Chatterjee, S and B Price (1991). *Regression Analysis by Example*. New York: John Wiley.

Cohen, J and P Cohen (1981). *Applied Multiple Regression/Correlation Analysis for the Behavioral Scientist*, Erlbaum, Hillsdale, New York.

Cooper, RG and EJ Kleinschmidt (1987). New products: What separates winners from losers? *Journal of Product Innovation Management*, 4(3), 169–184.

Cho, H-J and V Pucik (2005). Relationship between innovativeness, quality, growth, profitability, and market value. *Strategic Management Journal*, 26(6), 555–575.

Christensen, CM and JL Bower (1996). Customer power, strategic investment, and the failure of leading firms. *Strategic Management Journal*, 17(4), 197–219.

Crepon, B, E Duguet and J Mairesse (1998). Research, innovation and productivity: An econometric analysis at the firm level. *Economics of Innovation and New Technology*, 7(2), 115–158.

Danneels, E (2002). The dynamics of product innovation and firm competences. *Strategic Management Journal*, 23(12), 1095–1121.

Delmar, F (1997). Measuring growth: Methodological considerations and empirical results. In *Entrepreneurship and SME Research: On its Way to the Next Millennium*, R Donckels and A Miettinen (eds.), pp. 190–216, Bookfield, VA: Aldershot.

Edvinsson, L (1997). Developing intellectual capital at Skandia. *Long Range Planning*, 30(3), 366–373.

Edvinsson, L and MS Malone (1997). *Intellectual Capital: Realizing Your Company's True Value by Finding its Hidden Brainpower*. New York: HarperBusiness.

Eisenhardt, KM and JA Martin (2000). Dynamic capabilities: What are they? *Strategic Management Journal*, 21(10/11), 1105–1121.

Evenson, RE and LE Westphal (1995). Technological change and technological strategy. In *Handbook of Development Economics*, J Behrman and TN Srinivasan (eds.), Amsterdam: North-Holland.

Galunic, DC and S Rodan (1998). Resource combinations in the firm: Knowledge structures and the potential for Schumpeterian innovation. *Strategic Management Journal*, 19(12), 1193–1201.

Han, JK, N Kim and RK Srivastava (1998). Market orientation and organizational performance: Is innovation a missing link? *Journal of Marketing*, 62(4), 30–45.

Hayton, JC (2003). Strategic human capital management in SMEs: An empirical study of entrepreneurial performance. *Human Resource Management*, 42(4), 375–391.

Heirman, A and B Clarysee (2007). Which tangible and intangible assets matter for innovation speed in start-ups? *Journal of Product Innovation Management*, 24(4), 303–315.

Helfat, CE and RS Raubitschek (2000). Product sequencing: Co-evolution of knowledge, capabilities and products. *Strategic Management Journal*, 21(10/11), 961–979.

Hitt, M, L Bierman, K Shimizu and R Kochhar (2001). Direct and moderating effects of human capital on strategy and performance in professional service firms: A resource-based perspective. *Academy of Management Journal*, 44(1), 13–28.

Huang, CJ and CJ Liu (2005). Exploration for the relationship between innovation, IT and performance. *Journal of Intellectual Capital*, 6(2), 237–252.

Iansiti, M and KB Clark (1994). Integration and dynamic capability: Evidence from product development in automobiles and mainframe computers. *Industrial and Corporate Change*, 3(3), 557–605.

Jansen, JJP, F Bosch and HW Volberda (2006). Exploratory innovation, exploitative innovation, and performance: Effects of organizational antecedents and environmental moderators. *Management Science*, 52(11), 1661–1674.

Keller, RT (2004). A resource-based study of new product development: Predicting five-year later commercial success and speed to market. *International Journal of Innovation Management*, 8(3), 243–260.

Kennedy, P (1992). *A Guide to Econometric Methods*. Cambridge: MIT Press.

Laursen, K and NJ Foss (2003). New human resource management practices, complementarities, and the impact on innovation performance. *Cambridge Journal of Economics*, 27(2), 253–263.

Leo, H (1999). *Die Innovationstätigkeit des produzierenden Sektors in Österreich. Community Innovation Survey 1996*, Working Paper, WIFO, Vienna.

Leonard-Barton, D (1992). Core capabilities and core rigidities: A paradox in managing new product development. *Strategic Management Journal*, 13(51), 111–125.

Marqués, DP, FJG Simón and CD Caranana (2006). The effect of innovation on intellectual capital: An empirical evaluation in the biotechnology and telecommunications industries. *International Journal of Innovation Management*, 10(1), 89–112.

McElroy, MW (2003). *The New Knowledge Management: Complexity, Learning, and Sustainable Innovation*. Burlington: Butterworth-Heinemann.

Michie, J and M Sheehan (199). HRM practices, R&D expenditures and innovative investment: Evidence from the UK's 1990 workplace industrial relations survey (WIRS). *Industrial and Corporate Change*, 8(2), 211–234.

Miller, JG and A Roth (1994). A taxonomy of manufacturing strategies. *Management Science*, 40(3), 285–304.

Murphy, GB, JW Trailer and RC Hill (1996). Measuring performance in entrepreneurship research. *Journal of Business Research*, 36(1), 15–23.

Nahapiet, J and S Ghoshal (1998). Social capital, intellectual capital, and the organizational advantage. *Academy of Management Review*, 23(2), 242–256.

Nelson, RR (1991). Why do firms differ, and how does it matter. *Strategic Management Journal*, 12(52), 61–74.

Naman, JL and DP Slevin (1993). Entrepreneurship and the concept of fit: A model and empirical test. *Strategic Management Journal*, 14(2), 137–154.

Namasivayam, K and B Denizci (2006). Human capital in service organizations: Identifying value drivers. *Journal of Intellectual Capital*, 7(3), 381–393.

Nonaka, I (1994). A dynamic theory of organizational knowledge creation. *Organization Science*, 5(1), 4–37.

Nonaka, I, G von Krogh and S Voelpel (2006). Organizational knowledge creation theory: Evolutionary paths and future advances. *Organization Studies*, 27(8), 1179–1208.

Olavarrieta, S and R Friedmann (2008). Market orientation, knowledge-related resources and firm performance. *Journal of Business Research*, 61(6), 623–630.

Roos, G and J Roos (1997). Measuring your company's intellectual performance. *Long Range Planning*, 30(3), 413–426.

Roos, J, G Roos, NC Dragonetti and L Edvinsson (1997). *Intellectual Capital: Navigating the New Business Landscape*. Basingstoke: MacMillan Press.

Roper, S (1997). Product innovation and small business growth: A comparison of the strategies of German, UK and Irish companies. *Small Business Economics*, 9(6), 523–537.

Rothaermel, FT and AM Hess (2007). Building dynamic capabilities: Innovation driven by individual, firm, and network level effects. *Organization Science*, 18(6), 898–921.

Shrader, R and DS Siegel (2007). Assessing the relationship between human capital and firm performance: Evidence from technology-based new ventures. *Entrepreneurship Theory and Practice*, November, 893–908.

Smith, KG, CJ Collins and KD Clark (2005). Existing knowledge, knowledge creation capabilities and the rate of new product introduction in high technology firms. *Academy of Management Journal*, 48(2), 346–357.

Spanos, YE and S Lioukas (2001). An examination into the causal logic of rent generation: Contrasting porter's competitive strategy framework and the resource-based perspective. *Strategic Management Journal*, 22(10), 907–934.

Stewart, TA (1996). *Intellectual Capital: The New Wealth of Organizations*. New York: Currency Doubleday.

Stieglitz, N and K Heine (2007). Innovations and the role of complementarities in a strategic theory of the firm. *Strategic Management Journal*, 28(1), 1–15.

Subramaniam, M and MA Youndt (2005). The influence of intellectual capital on the types of innovative capabilities. *Academy of Management Journal*, 48(3), 450–463.

Sveiby, KE (1997). The new organizational wealth: Managing and measuring knowledge-based assets. San Francisco: Berrett-Koehler.

Teece, DJ (2000). *Managing Intellectual Capital*. Oxford: Oxford University Press.

Teece, DJ (2007). Explicating dynamic capabilities: The nature and microfoundations of (sustainable) enterprise performance. *Strategic Management Journal*, 28(13), 1319–1350.

Terziovski, M (2010). Innovation practice and its performance implications in small and medium-sized enterprises (SMEs) in the manufacturing sector: A resource-based view. *Strategic Management Journal*, 31(8), 892–902.

Thornhill, S (2006). Knowledge, innovation and firm performance in high- and low-technology regimes. *Journal of Business Venturing*, 21(5), 687–703.

van der Wiele, T and A Brown (1998). Venturing down the TQM path for SME's. *International Small Business Journal*, 16(2), 50–58.

Van de Ven, A and D Ferry (1997). *Measuring and Assessing Organizations*. New York: The Free Press.

Vaona, A and M Pianta (2007). Firm size and innovation in european manufacturing. *Small Business Economics*, 30(3), 283–299.

Verona, G (1999). A resource-based view of product development. *Academy of Management Review*, 24(1), 132–142.

Wiggins, RR and TM Ruefli (2002). Sustained competitive advantage: Temporal dynamics and the incidence and persistence of superior economic performance. *Organization Science*, 13(1), 82–105.

Youndt, MA, M Subramaniam and SA Snell (2004). Intellectual capital profiles: An examination of investments and returns. *Journal of Management Studies*, 41(2), 335–361.

Chapter 5

FAST-GROWING SMEs AND THE ROLE OF INNOVATION

Christina Grundström, Roland Sjöström,
Anders Uddenberg and Anna Öhrwall Rönnbäck

Industrial Marketing
Department of Management and Engineering
Linköping University, Sweden

In a study of 409 Swedish SMEs, the difference between the highest growing, which can be characterized as gazelles, and the lowest growing companies were examined regarding performance change over a four year period and what contributed to the growth from an innovation perspective. It was concluded that, besides growing, the highest growing companies also showed high profitability, increased number of employees, and significantly higher markets shares locally, nationally and internationally than the lowest growing companies. Several traits were found to contribute to this. The highest growing companies had a significantly higher portion of new products as part of the turnover during the four years studied and they perceived themselves as differentiating from their competitors concerning: (1) offering better products, (2) understanding customer needs better, (3) having a higher pace or being more agile, and (4) keeping costs down. They also found it more important to take risks, reinvest any profit, and to focus on growth than the lowest growing companies, and this regardless of industry.

Keywords: Gazelles; innovation; SMEs; market share growth; perceived competitive advantages; risk-taking.

Why Gazelles and Innovation

While SMEs play a very important role in most economies (Tether and Storey, 1998), "gazelles," which are fast-growing SMEs (Birch, 1981) typically constituting 2–5% of all firms in a country (Sims and O'Regan, 2006) and existing

This paper was first published in *International Journal of Innovation Management*, Vol. 16, No. 3 (June 2012), 1240003 (19 pages). © Imperial College Press. DOI: 10.1142/S1363919612400038 within all industries (Parker *et al.*, 2010), contribute disproportionately much to the creation of wealth and jobs (Birch, 1981; Acs *et al.*, 2008; Parker *et al.*, 2010). This since they are considered to handle weakening of the market better than other companies by, to a greater extent, working with new business missions, new markets and new customers (di.se, 2010). The general idea is that gazelles grow regardless of industry growth, not just grow in line with other companies in a growing industry (Storey and Greene, 2010). Furthermore, a Swedish study (Deloitte and SEB, 2010) showed that 80% of the responding companies believed in growth the following year and 90% believed in growth over the next three years. The same results were concluded in a more recent report by the Swedish Agency for Economic and Regional Growth (2011). In other words, growth is seen as very important from the perspective of the businesses themselves.

How can then growth be achieved? Several sources link growth to innovation, and that innovation might be of crucial importance for fast-growing firms (Coad, 2009). Laforet (2010) writes that today's market leaders, besides having visionary growth strategies, also need to focus heavily on innovation. On the other hand, Coad (2009), in his large review of theories and empirical evidence on the growth of firms, states that although theoretical economists more or less "take it for granted that firms with higher performance will reinvest their profits into growth" (p. 49), the empirical evidence, he argues, tends to show contradictory results, i.e., that profit is not reinvested in R&D. This indicates that much more work needs to be done investigating profit, productivity, and firm growth. Storey and Greene (2010) state that the innovation processes in SMEs can largely be depicted as two choices: either high-growth innovation strategies or low-growth ones, i.e., innovation is linked to high growth while low-growth means being satisfied with status quo. Moreover, there is a need to distinguish between product and process innovation; Coad (2009) concludes that product innovation tends to be positively associated with employment growth, while process innovation might have a contradictory effect. When taking all of this together, the aim of this paper becomes: to highlight in what ways innovation contributes to the prosperity of fast-growing companies in comparison with slow-growing ones.

Theoretical Foundation

Characterizations of gazelles do not suffice

Research on gazelles specifically is often of the type "this is what characterizes a gazelle." Sims and O'Regan (2006) concluded that gazelles have a good financial performance (seen as gross profit per employee), are likely to be less than 15 years old, apt to have a managing director younger than 50 years old, and are most probably privately owned and managed by their owners. St-Pierre *et al.* (2010) noted, when reviewing previous research on SMEs, that company age *in relation to size* often is omitted from studies of SMEs. In fact, many studies of SMEs (see e.g., Dobbs and Hamilton, 2007) include company age, and quite a few study the relationship between age and growth rate, in most cases resulting in negative relationship between age and growth, i.e., that older firms being more inert and perhaps less dynamic and capable of changing, but not without exceptions (Coad, 2009). This is supported by St-Pierre *et al.* (2010) who argue that the size and the age of the firm do not seem to be substitutable variables. In a review of other studies, they noted that firms often are grouped by size without taking the age of the firm into account. With such grouping, it is assumed, implicitly or not, that small companies are new in terms of number of years in business, while the larger ones have been around for years. This would misidentify the large group of small businesses without a desire to grow, i.e., these companies are small but not new (cf. Grundström, Öberg and Öhrwall-Rönnbäck, 2011), and the small group of rapidly growing gazelles, i.e., (often new) companies quickly becoming large(r).

In her study of American gazelles, Nicholls-Nixon (2005) found that they shared traits such as full knowledge, shared objectives, empowerment and agility. Parker *et al.* (2010) noted in their review that which variables and how they are used to identify gazelles vary just as much as the number of authors, and in these studies measures of growth in turnover and number of employees are more frequent than those related to profitability. Other studies (e.g., Verhees and Meulenberg, 2004) point at the managing director or entrepreneur in him or herself, i.e. that some individuals are more creative and open to new ideas than other individuals. Yet other managing directors or entrepreneurs are afraid of losing control of their companies if they allow the company to grow (see e.g., Wiklund *et al.*, 2003). Still *et al.* (2010) clearly showed that growth in company size has a strong positive impact on company survival. These studies, however, provide little guidance as to what contributes to growth and how it is achieved at the firm level.

Ways to achieve growth

Early studies of successful companies, with regards to revenue growth and profitability (e.g. Peters and Waterman, 1982; Buzzell and Gale, 1987) focused on big companies, and again the studies were more of characterization than how to get there (Mintzberg *et al.*, 1998). Success in these studies was contributed to selling the right product to the right customer, or expressed otherwise, customer closeness and innovation. This is often discussed in terms of positioning or branding (Porter, 1979; Laforet, 2010) and reaching out with products (Rogers, 2003), but also in terms of high-quality products, market share and market growth (Buzzell and Gale, 1987). A lack of exports can act as an obstacle to finding new markets and thus growth (NUTEK, 2005). Moore, (1999), Niosi (2002) and Terziovski (2003) point to the need for alliances during certain phases of growth. Business co-location is also seen as an advantage as it increases the likelihood for access to labour, knowledge sharing and cost advantages (Storey and Greene, 2010). Another aspect affecting a company's ability to control its development and growth is the rigidity of the supply chain (depicting all actors involved in the process to produce a certain product). The higher the rigidity, the more a company acting as a supplier has to adjust to the needs of its customer, while in a low rigidity supply chain, a company can plan its own future to a greater extent. (Teichert and Bouncken, 2011)

All of these factors have an external perspective. There is of course an internal perspective, which includes sufficient competence (Geroski *et al.*, 2010; Storey and Greene, 2010) and processes to match customer needs, as well as a willingness to take and an ability to evaluate risks in relation to fulfilling these (Storey and Greene, 2010).

To be able to sell anything on a market a company must deliver sufficient customer benefits. When summarizing Porter (1979); Treacy and Wiersema (1993); Hax and Wilde (1999) and Moore (1999) it is about to what degree the company is focused on:

(1) Delivering the best product;
(2) Optimizing the processes; and
(3) Customizing the offer.

On the other hand Stalk *et al.* (1992); Day (1994) and Blois and Ramirez (2006) argue that unique processes are sufficient as strategy. Although the strategy dimensions can seem obvious in retrospect, they are not necessarily deliberate. There are many indications that some of the corporate strategy is planned, while others are more dependent on chance or coincidence (Mintzberg and Waters, 1985; Mintzberg, 1987; Parker *et al.*, 2010). One factor that often recurs, in terms of planning, is proactivity, i.e., if the company lacks a clear focus on growth, the probability of growth is reduced (Charan and Tichy, 1999). Other frequently discussed dimensions related to growth are time-to-market, first-mover advantage, barriers to entry, brand loyalty, technological leadership, access to assets, buyer switching costs, ROI expenses and distribution channels (Porter, 1980; Lieberman and Montgomery, 1990; Lilien and Yoon, 1990; Utterback, 1994; Tellis and Golder, 1996; Robinson *et al.*, 2001; Rogers, 2003). Most of these relate to technology and innovation management.

Access to funding is seen as crucial to enable growth, since growth usually involves developing all sides of the business, including new products. Research has so far found no clear link between finances and business growth, in particular in small businesses, but there are indications that internally-funded companies tend to grow less fast. (Becchetti and Trovato, 2002; Storey and Greene, 2010)

The role of innovation in today's business environment

Research on company growth still heavily leans on early works such as Penrose (1959), and many of the earlier identified success factors based on research from the last 30 years still seem to prevail, although a clear pattern as to explain company growth is still to be found (Coad, 2009). According to Laforet (2010), today's competition is characterized by a need for companies to have a proactive orientation. In addition, companies need to provide high quality and produce products at the lowest possible cost, while maintaining a high speed. They also need to continuously meet changing customer needs through innovation, and even understand customers better than they do themselves. In fact, innovation is seen as the core capability of a firm, and it involves, besides new products, business processes and building new markets to meet customer needs (Laforet, 2010); Bessant and Tidd, 2010). When building these new markets, it is important to create entry barriers to avoid high entry rates from other firms and that the overall macro factors for these markets are favorable. This to avoid a quick exit from the market. (Geroski *et al.*, 2010)

While flexibility towards customers' changing demands is often put forward as a major competitive advantage for SMEs (Sims and O'Regan, 2005; Dobbs and Hamilton, 2007; Bessant and Tidd, 2010; Storey and Greene, 2010), the relation between firm size and degree of innovation is unclear (Bessant and Tidd, 2010) and it has been argued that innovation per se might not be an apparent success factor for the small firm (Storey and Greene, 2010). Previous studies show that considerable investments in R&D and high innovation rates might not in the short run lead to profitable growth, although it seems that the exceptionally fast-growing companies tend to be more innovative (Coad, 2009; Storey and Greene,

2010). Instead of expensive and risky R&D projects, it is suggested that an SME should actively engage itself in developing products and markets to have a constant awareness of new technologies, competition, and to be able to broaden its customer base. It can do so either by launching new products into the market, or reaching new markets with existing products, to achieve long-term sustainable growth (Dobbs and Hamilton, 2007; Bessant and Tidd, 2010). This is done while focusing more on product than process innovation (Coad, 2009; Bessant and Tidd, 2010) and being specialized with regards to technological competencies and product range, i.e. serving niche markets (Bessant and Tidd, 2010). The actual innovation or new product development (NPD) process is likely to be more formalized in large companies (Cooper and Kleinschmidt, 1995), while usually involving the whole firm if SMEs (Strerlacchini, 1999). Still, regardless of size, companies able to manage the early (informal) phases of innovation and NPD, often referred to as the "fuzzy front end," usually come up with ideas that lead to products (radical or incremental, goods or services alike) in a shorter time (Jörgensen *et al.*, 2011). Other important dimensions of innovation and NPD are cooperation on the same level in the vertical market system, market orientation, i.e. companies having a closer relation to their customers and suppliers. Small companies usually get involved in some form of open innovation, which can vary with regards to degree of interaction (Chesbrough, 2003), since the internal resources do not suffice when innovating or conduction NPD. Depending on with whom the company is interacting, various types of openness is required. This openness is depicted as either supplier absorptive capacity (inside-out) or customer absorptive capacity (outside-in), which requires managing a wide array of roles. (Newey, 2010). For some companies outsourcing of development work may be the case (Quinn, 2000). Companies with a market orientation are likely to grow more since they are usually able to augment their products (Verhees and Meulenberg, 2004; Capelleras and Greene, 2008).

With market orientation comes market characteristics. A market recently created or shaken by a radical or architectural innovation is likely to see higher investments in product innovation than one where a dominant design has developed where investments in process innovation are higher (Henderson and Clark, 1990; Utterback, 1994; Schilling, 2010), yet some industries never develop a dominant design (Srinivasan *et al.*, 2004). Windows of opportunity can also develop due to changes in macro factors other than technological, such as political and environmental (Van de Ven and Garud, 1994, for applications see e.g., Huge-Brodin and Anderson, 2008; Sundin, 2009).

How the Study was Carried Out and Results

Company selection

For the initial company selection, a gazelle was defined as a company that had increased its turnover with at least 100% from 2006 to 2009. From Affärsdata (Business Data) and registers on the number of private limited companies it was estimated that about 10% of all companies in Sweden would meet the criteria to be a gazelle.

The criteria to identify a gazelle are in line with those most frequently used for identifying gazelles: a growth of 20% during three or more consecutive years (Sims and O'Regan, 2006). In Sweden, the company UC AB presents yearly a list of companies that meet the above mentioned criteria. From such lists for the years 2001–2010 a total of 4,655 companies were identified. The organization numbers for these, i.e., unique identity numbers, were used to identify gazelles for each year in the 2006–2009 period. Data, such as financials and ownership information from the service Affärsdata on all Swedish private limited companies with at least 10 employees, were collected to create a reference group. As of January 16, 2011, and after removing companies part of a big company group, and companies not having closing of accounts each year during the 2006–2009 period, 11,224 companies were listed in the reference group. After merging the two lists and removing duplicates, the net list held 14,587 companies. An email search resulted in 9,146 email addresses to an individual company (of which 4,287 related to companies that had been gazelles any time during 2001–2010), primarily the managing director or owner of each company, or, if this was lacking, to a general email address (such as info@companyname.se).

Design of the survey

With some 9,000 companies and an interest in highlighting the differences between the most and least growing companies, a tool enabling statistical evaluation was desired. Hence, a survey was designed to include some 80 different variables across 24 questions covering the most of the previously discussed aspects relating to growth and innovation. These questions were designed to provide information on what characterizes each company regarding its business, market, products (goods or service) and its governing priorities complemented with information about the respondent, such as position, gender and age. For the questions involving self-evaluation, a seven-point Likert scale was chosen to increase the chances for sufficiently high resolution, i.e. variances (similar to Wiklund *et al.*, 2003). The questions were formulated by the authors of this paper together with colleagues from the Department of Management and Engineering at Linköping University and an industry reference group. A pilot study with eight respondents was conducted, resulting in rewording of some questions.

Distribution of the survey

The survey was distributed on January 28th, 2011, using the web-based survey tool Webropol. Table 1 is a summary of the first distribution.

Three reminders were sent out during February 2011, and the survey was closed on March 7. As of March 7, a total of 417 companies had completed the survey. Out of these, 8 were removed (2 had more than 250 employees, 3 lacked a closing of accounts for the entire period, 1 had inaccessible financial information, and 2 had too small a turnover in 2006 resulting in an growth rate 500 times larger than other gazelles). For each reminder, approximately 1,600–1,700 emails bounced due to either failure of deliverance, i.e. wrong address, vacation replies, or with information that the managing director had left her or his position. The last type of bounce was due to the lack of updates in Affärsdata. Since there is no reliable way to determine the exact number of the emails that reached the intended recipient, an exact response rate for the survey is difficult to calculate. However, if only bounces are taken into account, the response rate is about 5% (417/9,146).

Description of sample

Out of the final 409 companies, 71 were companies previously identified as gazelles by UC AB (ref. "Company selection" above). Among the 409 respondents, 311 were managing directors, 44 were chairman of the board, 169 were owners and 40 had other positions (these positions were not mutually exclusive, sum therefore exceeds number of respondents). Basic data on the companies is provided in Table 2. The sample contained companies from all industries and the

Table 1. Recipients of first survey distribution.

Group	Managing director or owner address	General company address	Total
Gazelles	2,382	1,905	4,287
Reference	3,049	1,810	4,859
Total			9,146

Table 2. Basic data on 409 companies.

Variable	
Turnover in 2009 average/median	4.6/2.5 M€
No of employees in 2009 average/median	27.5/18
Age in 2009 average/median	20/18 years
Turnover increase 2006–2009 average/median	68.7%/28.0%
Employee increase 2006–2009 average/median	6.6/3.0
Managing director age in 2011, average/median	50/50
Managing director years with company in 2011, average/median	14/11
Managing director gender in 2011, female/male	22/290

Source: CAM Database, Linköping University. (1€ = 10 SEK.)

distribution is similar to the aggregated industry distribution (i.e., circa 20% were manufacturing companies, 15–20% were in trade, 15% in construction, another 15% in the technology industry and rest was spread across all other industries).

Selecting two groups: "High growers" and "Low growers"

The first step after having compiled the list of 409 companies was a test of homogeneity. The economic data for the 409 companies was compared with the same economic data for the entire population of the 9,146 companies, that is, turnover, number of employees, profitability, solvency, revenue growth, profitability growth, and growth in number of employees. None of these variables differed between the 409 companies and the underlying population, which was interpreted as the 409 companies being representative of the entire population.

Since the aim was to compare high-growth companies to low-growth ones, and with the insight that growth per se is more central for such a comparison than strictly following definitions (for gazelles) which present a large variation, the 409 companies were divided into quartiles based on turnover growth during 2006–2009 to obtain the largest possible contrast in the data collected. The quartile with the best performers is henceforth called "high growers" and contains 103 companies of which all but seven had increased their turnover at least 100% during the studied four-year period. The quartile with the lowest growth is henceforth called "low growers" and contains also 103 companies, of which 43 were former gazelles. This means that the respondents to a higher degree (compared to the estimated 10% average for Sweden) could be classified as gazelles. Despite having different turnover and growth during 2006–2009, "high growers and "low growers" share some features at the end of 2009 (calculated as means for each group):

- 30% market share in Sweden, 10% outside Sweden
- The biggest competitor has a 25% market share
- The two largest customers contribute to 30–35% of the turnover
- 20% of the turnover through distributor or collaboration partner
- Even geographical distribution over Sweden, i.e., no specific regional concentration
- Low percentage of female managing directors (6 female managing directors in each group)

Some basic data on differences between these two quartiles can be found in Table 3.

Table 4 shows the sectorial distribution of the two groups.

The next step was to test the questions in the survey (the hypotheses), i.e., any differences between "High growers" and "Low growers" were examined using SPSS.

How the "high growers" and "low growers" perceived themselves

Out of the self-perception variables concerning how the company management thinks and acts, eleven (11) showed statistical significance ($p < 0.05$ level) and three (3) tendency ($0.05 < p < 0.10$ level) when comparing the "high growers" and the "low growers" using the t-test. The results can be found in Table 5.

To test the rigidity of these findings, the companies showing the most extreme values were removed from the data set and the calculations were done once again with the same outcome.

Findings

From Table 3 it can be noted that profitability correlates with growth in turnover, and that the "high growers" significantly increased their employees during 2006–2009 more than the "low growers". This is noteworthy in comparison to other high-growth studies, which usually highlight only turnover and employee growth (Coad, 2009; Parker *et al.*, 2010).

The "high growers" in this study were found to be significantly younger than the "low growers" ($p < 0.001$) and could be found within all industries, which confirm the findings by Sims and O'Regan (2006) and Parker *et al.* (2010) respectively.

Table 3. Descriptive data on "high growers" and "low growers" with significant differences between groups marked.

Variable	"High growers"	"Low growers"
From business data:		
Average turnover change	From 2.2 to 6.0 M€	From 5.5 to 4.1 M€
Average profit change (profit before depreciation)	From 0.18 to 0.55 M€	From 0.52 to 0.18 M€
Company age in 2009 average/median (span)***	14/11 (4–50) years	24/22 (5–65) years
Employee change 2006–2009 average***	+37	−2
From survey:		
Managing director age in 2011, average/median***	47/46 years	53/52 years
Managing director years with company in 2011, average/median***	11/8 years	17/16 years
Managing director gender in 2011, female/male	6/71	6/68
% change in market share locally over the last 4 years***	39	12
% change in market share nationally over the last 4 years***	40	11
% change in market share internationally over the last 4 years*	22	8
Existing**/new*** customers' share of turnover growth	44/56	56/42
% of turnover invested in NPD	15	11
% of NPD completely new products	35	30
% of sales 2006–2009 from products launched 2006–2009*	27	20

Source: CAM Database, Linköping University. (1€ = 10 SEK.) *** = $p < 0.001$, ** = $p < 0.01$, * = $p < 0.05$.

With regards to managing director age, the "high growers" had significantly ($p < 0.001$) younger managing directors than the "low growers," but the average of 47 years for the"high growers" clearly indicates that several of their managing directors were over 50 years of age, contrasting the "recommendation" by Sims and O'Regan (2006). There was a significant difference between "high growers" and "low growers" regarding "time of managing director with company" ($p < 0.001$),

Table 4. Industry sectorial distribution for "high growers" and "low growers" (from business data) with significant differences with regards to growth marked.

Industry sector	"High growers"	"Low growers"
Manufacturing*	13%	36%
Trade***	24%	28%
Construction***	11%	10%
Technology*	38%	15%
Other	14%	11%

Source: CAM Database, Linköping University. $*** = p < 0.001$, $** = p < 0.01$, $*p < 0.05$.

where the managing director in "high-growers" had worked a significantly shorter time period in the company (median 8 years compared to 16 years for "low-growers"), but not regarding the gender of the managing director (about 8% women in both groups). That the relation between company size and degree of innovation is unclear (Bessant and Tidd, 2010) found support in this study, as no correlation between company size and innovativeness measured as a percentage of totally new products, and how much is put into NPD and going for new markets could be established. No geographical clustering could be detected. However, as no business network evaluation was performed, no evaluation of business co-location, an advantage highlighted by Storey and Greene (2010), could be made.

A number of the findings concerns innovation in relation to marketing. As seen in Table 3, the market share growth locally, nationally and internationally, share of new customers and the percentage of sales 2006–2009 from products launched 2006–2009 were significantly higher for "high growers" than "low growers," while the "low growers" had significantly higher percentage of turnover from existing customers. The "high growers" thus seem to better manage market share growth (Buzzell and Gale, 1987), reaching out with products (Rogers, 1993) and overcome the fear of going international as highlighted by NUTEK (2005).

It could be expected that differences in growth would depend on sectorial differences. During the studied time period there has been state-support for the construction industry (a program with influential subvention for tax-payers contracting construction companies). Also, there could have been better opportunities for the technology sector, which includes more high-tech oriented firms, than for manufacturing firms facing hard global competition and cost-pressure from low-wage countries. That is, companies in some sectors thus could have been expected

Table 5. Variables with statistical significance and tendency when comparing "High growers" and "Low growers" using the t-test.

No.	Variable	"High growers" (mean)	"Low growers" (mean)	Sign.
1	The competition in the industry considered to be (1 = neglectable, 7 = very strong)	5.41	5.73	†
2	Supplier influence considered to be (1 = neglectable, 7 = very strong)	4.23	4.63	†
3	Differentiation of goods/services compared to competitors (1 = competitors much better, 7 = our company much better)	5.74	5.26	**
4	Differentiation of understanding customer needs compared to competitors (1 = competitors much better, 7 = our company much better)	5.92	5.51	**
5	Differentiation in terms of pace and agility compared to competitors (1 = competitors much better, 7 = our company much better)	5.77	5.37	*
6	Cost compared to competitors considered to be (1 = much lower, 7 = much higher)	3.75	4.17	*
7	To continue to grow as regards turnover the last 4 years considered to be (1 = not at all important, 7 = very important)	5.85	5.03	*
8	To take risks the last 4 years considered to be (1 = not at all important, 7 = very important)	5.30	4.88	***
9	Prioritizes growth to profitability (1 = growth, 7 = profitability)	3.60	4.90	***
10	Prioritizes market share to profitability (1 = market share, 7 = profitabiltiy)	4.19	5.11	***
11	Prioritizes to reinvest compared to showing profit (1 = reinvest, 7 = profit)	2.91	3.92	***

Table 5. (*Continued*)

No.	Variable	"High growers" (mean)	"Low growers" (mean)	Sign.
12	Prioritizes own financing compared to venture capital (1 = own financing, 7 = venture capital)	1.98	2.46	*
13	Development work is conducted in collaboration with suppliers (1 = not at all, 7 = completely)	2.97	3.44	†
14	Development work is conducted in collaboration with other partners (1 = not at all, 7 = completely)	2.91	3.44	*

Source: CAM Database, Linköping University. *** $= p < 0.001$, ** $= p < 0.01$, *$p < 0.05$, † $= p < 0.10$.

to have faced changes in macro factors (Van de Ven and Garud, 1994) or technology changes opening windows of opportunity (Henderson and Clark, 1990; Schilling, 2010). However, when dividing the sample into subgroups, the groups become smaller, increasing the average error and reducing the likelihood for observing statistically significant differences. Still, from Table 4, which shows the companies divided into industry sectors, "high growers" and "low growers," there was a significant difference in growth for all sectors but "others." Nevertheless, due to the small sample, any sectorial differences between "high growers" and "low growers" could not be confirmed from this study.

As seen in Table 5, it seems like the "high growers" very much understand what is required to meet today's competition as described by Laforet (2010). At the $p < 0.01$ level the "high growers" perceived themselves as offering better products and services and better understanding customer needs than the "low growers." Also significant ($p < 0.05$) was the perception of "high growers" as having a higher pace or being more agile and having lower costs in relation to competitors. The proactive orientation with regards to growth pointed out by Charan and Tichy (1999), the focus on market share growth (Buzzell and Gale, 1987) and the need for and ability to take risks in relation to this (Storey and Greene, 2010), show in the significantly higher values obtained for "high growers" as regards desire to grow continuously ($p < 0.001$), importance to take risks continuously ($p < 0.05$), the priority put on growth rather than profitability ($p < 0.001$), the priority put on market share rather than profitability ($p < 0.001$) and the priority put on reinvesting rather than showing profit ($p < 0.001$). The "high growers" viewed the competition in the industry somewhat lower than the "low growers" ($p < 0.10$), indicating to some degree more of a "we are in control"-view of the market as a whole on part of the "high growers."

With regards to development work, the "high growers" preferred to a lesser degree than the "low growers" to collaborate with suppliers ($p < 0.10$) and other partners ($p < 0.05$). In general, the "high growers" viewed the influence of the suppliers to be lower than the "low growers" ($p < 0.10$), which could contribute to the lower importance "high growers" put to suppliers as regards development work. There was no significant difference between the groups with regards to collaboration with customers and the formality of the NPD. Despite this, the "high growers" perceived themselves as being better at understanding customer needs and having a higher pace or being more agile as discussed earlier, thereby tending to differ from competitors on customer value ($p < 0.107$). This latter finding (higher perceived customer satisfaction despite preferring to collaborate less) somewhat opposes Verhees and Meulenberg (2004) and Capelleras and Greene (2008), who state the importance of collaborating with suppliers and customers to develop products better satisfying customer needs and hence grow more.

The "high growers" prioritized their own funding to venture capital much more than the "low growers" ($p < 0.05$) with regards to access to capital. This prioritization somewhat contradicts Storey and Greene (2010) and Becchetti and Trovato (2002) but may be explained by the priority to reinvest rather than show profitablity, and the fact that the "high growers" may be sufficiently profitable to finance their growth through their own funding.

Conclusion and Contribution

Despite the fact that the study covers the years 2008 and 2009 when the world faced a severe recession, the "high growers" had as an average a larger turnover and profit in 2009 than the "low growers" had as an average in 2006. Indeed, during this time period of three years, the "high growers" managed to grow at least 78%, which is well above the frequently used criteria to identify gazelles. On the other hand, the advantage of a good profitability the "low growers" had in 2006, they were not able to sustain it over the period studied. What could be determined in this study is that there is a large and highly significant correlation between growth in turnover and growth in profitability, a development seen by the "high growers." Expressed otherwise: if the company focuses on growth, which was indeed more the self-perception of the "high growers" compared to the "low growers," then profitability will automatically be higher. Most likely this is due to the increased focus on cash-flow and costs. The "high growers" expressed this as having a cost advantage in relation to competitors.

Even though the "high growers" in this study to a large degree coincide with the characterization found in the literature (i.e., average 14 years old in line with results presented by Sims and O'Regan, 2006), among the "high growers" there were companies as old as 48 and 50 years old. This means that age is not always a determining factor for a company but rather the ability to reinvent itself as regards products, processes and markets. Thus, this study raises doubts concerning the relatively robust feature the negative dependence age has on growth rate, as concluded from mainstream growth studies (Coad, 2009).

The "high growers" had a significantly greatershare of new products launched during 2006–2009 as part of the sales during 2006–2009 compared to the "low growers." This could be attributed to traits such as a perception of offering better products or services, understanding customer needs better, having lower costs and having higher agility and pace, a higher focus on continuously taking risks and a higher focus on growth in itself (supporting the ideas of Penrose, 1959), but all of these in combination with preferring to rely on own capital and own capabilities with regards to innovation and NPD. This view on how to manage the company could be an effect of that the "high growers" are good at the fuzzy front end of NPD and innovation as discussed by Jörgensen *et al.* (2011). In other words, these companies are good at sensing what is "buzzing in the air" and turn that into new products quicker than their competitors. Another explanation could be that these companies may be good at taking on the various roles required for effective and efficient supplier and customer absorptive capacity during innovation or NPD as highlighted by Newey (2010).

When taking all of these traits together, the "high growers" can be characterized to a greater degree as perceiving themselves as "being in control" both with regards to what is going on on the market and what focus the innovation and NPD should have while wanting to stay independent. This could be a result of the "high growers" being in markets with less rigidity, as discussed by Teichert and Bouncken (2011), or with fewer competitors or new entrants, higlighted as important when entering a market (Geroski *et al.*, 2010). Markets with such characteristics could be markets affected by major changes in macro factors (e.g., Schilling, 2010) or be niche markets (Bessant and Tidd, 2010), where the latter often are recommended to SMEs.

The overall conclusion from this research is that fast-growing companies (or gazelles) grow and become profitable thanks to a will or desire to grow and to take the risks associated with that in combination with an awareness of the competitive situation including market development. All of this is achieved despite contradicting some of the "rules" pointed out by previous research with regards to how to achieve customer value and satisfaction. Even though more than 80 variables on various aspects of how to achieve growth with the aid of proper innovation and NPD management were included in this research, more research is needed in order to understand how gazelles truly achieve this since the plausible explanations to the observed fast growth are somewhat speculative at this stage. This outcome is in line with e.g., Coad (2009:151), concluding from previous studies on firm growth, that efficient allocation of growth opportunities remains a challenge for the economy.

Managerial Implications

A high rate of innovations and/or new products, elaborate innovation and NPD methods, and high expenditure on innovations per se are no guarantee for company growth and profitability. However, this paper shows that a manager in addition needs a keen understanding of the competitive environment and the trends thereof, an ability to take risks in relation to innovation and NPD and markets, and a strong belief in the company's own abilities and capabilities to lead it to success.

Acknowledgements

The authors of this paper would like to thank The Center for Applied Management for SMEs at Linköping University for financing, Dagens industri for support on gazelle selection, and Gunnar Almesåker, Staffan Brege, Ronny Mårtensson and Christoffer Rutgersson for their invaluable contribution during the work with the survey.

References

Acs, Z, W Parsons and S Tracy (2008). *High-Impact Firms: Gazelles Revisited.* Report published by Corporate Research Board, LLC Washington, DC 20037, accessed from Office of Advocacy, the United States Small Business Administration. Available at: http://www.sba.gov/adv.

Becchetti, L and G Trovato (2002). The determinants of growth for small and medium sized firms. The role of the availability of external finance. *Small Business Economics*, 19(4), 291–302.

Bessant, J and J Tidd (2011). *Innovation and Entrepreneurship*, 2nd ed. Chichester: John Wiley & Sons.

Birch, DL (1981). Who Creates Jobs? *The Public Interest*, 65, 3–14.

Blois, K and R Ramirez (2006). Capabilities as marketable assets. *Industrial Marketing Management*, 35, 1027–1031.

Buzzell, RD and BT Gale (1987). *The PIMS Principles: Linking Strategy to Performance.* New York: Free Press.

Capelleras, JL and FJ Greene (2008). The determinants and growth implications of venture creation speed. *Entrepreneurship and Regional Development*, 20(4), 317–343.

Charan, R and NM Tichy (1999). *Every Business is a Growth Business.* Chichester: John Wiley & Sons Ltd.

Chesbrough, H (2003). *Open Innovation: The New Imperative for Creating and Profiting from Technology.* Boston, MA: Harvard Business School Press.

Coad, A (2009). *The Growth of Firms: A Survey of Theories and Empirical Evidence.* Cheltenham: Edward Elgar.

Cooper, RG and EJ Kleinschmidt (1995). Benchmarking the firm's critical success factors in new product development. *Journal of Product Innovation Management*, 12, 374–391.

Day, GS (1994). The capabilities of market-driven organizations. *Journal of Marketing*, 58, 37–52.

Deloitte and SEB (2010). *Tillväxtbarometern 2010.* [online] Deloitte AB. In English (authors' translation): The Growth Barometer. Available at: http://www.deloitte.com/assets/Dcom-Sweden/Local%20Assets/Documents/se_Deloitte_Tillvaxtbarometern_2010_web.pdf.

di.se (2010). Hårda krav på årets Gaseller [online] updated 2010-08-18. In English (authors' translation): Tough requirements on this year's gazelles. Available at: http://www.di.se.

Dobbs, M and RT Hamilton (2007). Small business growth: recent evidence and new directions. *International Journal of Entrepreneurial Behaviour & Research*, 13(5), 296–322.

Geroski, PA, J Mata and P Portugal (2010). Founding conditions and the survival of new firms. *Strategic Management Journal*, 31, 510–529.

Grundström, C, C Öberg and A Öhrwall-Rönnbäck (2011). View and management of innovativeness upon succession in family-owned SMEs. *International Journal of Innovation Management*, 15(3), 1–24.

Hax, A and DL Wilde II (1999). The delta model: Adaptive management for a changing world. *Sloan Management Review*, 40(2), 11–28.

Henderson, RM and KB Clark (1990). Architectural innovation: The reconfiguration of existing product technologies and the failure of established firms. *Administrative Science Quarterly*, 35, 9–30.

Huge-Brodin, M and H Anderson (2008). Recycling calls for revaluation. *Supply Chain Management*, 13(1), 9–15.

Jörgensen, JH, C Bergenholtz, RC Goduscheit and ES Rasmussen (2011). Managing inter-firm collaboration in the fuzzy front-end: Structure as a two-edged Sword. *International Journal of Innovation Management*, 15(1) (February), 145–163.

Laforet, S (2010). *Managing Brands — A Contemporary Perspective.* Maidenhead: McGraw-Hill.

Lieberman, MB and DB Montgomery (1990). To pioneer or follow? Strategy of entry order. Working Papers (Faculty), *Stanford Graduate School of Business*, 1–39.

Lilien, GL and E Yoon (1990). The timing of competitive market entry: An exploratory study of new industrial products. *Management Science*, May, 36(5), 568–585.

Mintzberg, H, B Ahlstrand and J Lampel (1998). *Strategy Safari — A Guided Tour Through The Wilds of Strategic Management.* New York: The Free Press.

Mintzberg, H (1987). Crafting strategy. *Harvard Business Review*, 65(4), 66–75.

Mintzberg, H and JA Waters (1985). Of strategies, deliberate and emergent. *Strategic Management Journal*, 6(3), 257–272.

Moore, GA (1995). *Inside the Tornado*. New York: Harper-Business.

Newey, L (2010). Wearing different hats: How absorptive capacity differs in open innovation. *International Journal of Innovation Management*, 14(4), 703–731.

Nicholls-Nixon, CL (2005). Rapid growth and high performance: The entrepreneur's "impossible dream"? *Academy of Management Executive*, 19(1), 77–89.

Niosi, J (2002). National systems of innovations are "x-efficient" (and x-effective): Why some are slow learners. *Research Policy*, 31, 291–302.

NUTEK, (2005). *Företagens Villkor Och Verklighet*. Stockholm: NUTEK.

Parker, SC, DJ Storey and A van Witteloostuijn (2010). What happens to gazelles? The importance of dynamic management strategy. *Small Business Economics*, 35, 203–226.

Penrose, E (1959). *The Theory of the Growth of the Firm*, 3rd ed. Oxford: Oxford University Press.

Peters, T and R Waterman (1982). *In Search of Excellence: Lessons From America's Best Run Companies.* New York: Harper and Row.

Porter, ME (1979). How competitive forces shape strategy. *Harvard Business Review*, 57(2), 137–145.

Porter, ME (1980). *Competitive Strategy*. New York: Free Press.

Quinn, JB (2000). Outsourcing innovation: The new engine of growth. *Sloan Management Review*, Summer, 13–28.

Robinson, KC and P Phillips McDougall (2001). Entry barriers and new venture performance: A comparison of universal and contingency approaches. *Strategic Management Journal*, 22, 659–685.

Rogers, EM (2003). *Diffusion of Innovations*, 5th ed. New York: Free Press.

Schilling, MA (2010). *Strategic Management of Technological Innovation*, 3rd ed. New York City: McGrawHill.

Sims, MA and N O'Regan (2006). In search of gazelles using a research DNA model. *Technovation*, 26, 943–954.

Srinivasan, R, GL Lilien and A Rangaswamy (2004). *Emergence of Dominant Designs.* ISBM Report 9-2004.

Stalk, G, P Evans and LE Shulman (1992). Competing on capabilities: The new rules of corporate strategy. *Harvard Business Review*, 70(2), 57–69.

Storey, DJ and FJ Greene (2010). *Small Business and Entrepreneurship.* Harlow: Pearson Education Limited.

St-Pierre, J, P-A Julien and M Morin (2010). L'effet de l'âge et de la taille sur la performance financière et économique des PME (The effect of age and size on SME financial and economic performance). *Journal of Small Business & Entrepreneurship*, 23(2), 287–306.

Strerlacchini, A (1999). Do innovative activities matter to small firms in non-R&D intensive industries?. *Research Policy*, 28, 819–832.

Sundin, E (2009). Everyday innovation in elderly care and healthcare. In M Döös and L Wilhelmson (Eds.), *Organising Work for Innovation and Growth.* Vinnova Report VR 2009:22, 101–114.

Swedish Agency for Economic and Regional Growth (2011). *The Situation and Conditions of Enterprises 2011*, published October, ISBN 978-91-86987-11-4, Info 0359 (report in Swedish titled "Tillväxtmöjligheter och tillväxthinder för svenska små och medelstora företag" with summary in English).

Teichert, T and RB Bouncken (2011). Rigidities considered: Supplier strategies for integrated innovation. *International Journal of Innovation Management*, 15(1), 95–119.

Tellis, GJ and PN Golder (1996). First to market, first to fail? real causes of enduring market leadership. *Sloan Management Review*, 37(2), 65–75.

Terziovski, M (2003). The relationship between networking practices and business excellences: A study of small and medium enterprises (SMEs). *Measuring Business Excellence*, 7(2), 78–92.

Tether, B and D Storey (1998). Smaller firms and Europe's high technology sectors: A framework for analysis and some statistical evidence. *Research Policy*, 26, 947–971.

Treacy, M and F Wiersema (1993). Customer intimacy and other values disciplines. *Harvard Business Review*, 71(1), 84–93.

Utterback, JM (1994). *Mastering the Dynamics of Innovation — How Companies Can Seize Opportunities in the Face of Technological Change.* Boston, MA: Harvard Business School Press.

Van de Ven, AH and R Garud (1994). The coevolution of technical and institutional events in the development of an innovation. In Baum, JAC and JS Singh (Eds.), *Evolutionary Dynamics of Organizations.* New York: Oxford University Press.

Verhees, FJHM and MTG Meulenberg (2004). Market orientation, innovativeness, product innovation, and performance in small firms. *Journal of Small Business Management*, 42(2), 134–154.

Wiklund, J, P Davidsson and F Delmar (2003). What do they think and feel about growth? An expectancy-value approach to small business managers' attitudes toward growth. *Entrepreneurship Theory and Practice*, 27(3), 247–270.

Chapter 6

PERCEIVED COMPETITION AND INNOVATIVE INTENTIONS IN DUTCH SMALL AND MEDIUM-SIZED ENTERPRISES

Jeroen P. J. De Jong

RSM Erasmus University, EIM Business and Policy Research

This paper explores the complex relationship between competition and innovation by analyzing survey data of 2,281 small and medium-sized enterprises (SMEs) in the Netherlands. We develop and test hypotheses on the relationships between three dimensions of perceived competition (internal rivalry, supplier power and buyer power) and firms' intentions to engage in product and process innovation. Moreover, we analyze if firms' strategic attention for innovation is a moderating variable. We find that specific innovative intentions relate to different perceptions of competitive forces. Intentions to engage in process innovation correlate with the perceived bargaining power of suppliers, while intended product innovation correlates with perceived buyer power and internal rivalry between incumbent firms in the market — but the correlation with internal rivalry is significant only when firms report no strategic attention for innovation.

Keywords: Perceived competition; innovative intentions; product innovation; process innovation; SMEs.

Introduction

There have been many empirical studies on the relationship between competition and innovation, usually drawing on large datasets, but with mixed results (Cohen, 1995). In the industrial organization literature, the hypothesis that innovation takes place in the R&D laboratories of large firms benefiting of a situation of lack of competition is known as the "Schumpeterian hypothesis" — inspired by the

This paper was first published in *International Journal of Innovation Management*, Vol. 15, No. 4 (August 2011), 687–707. © Imperial College Press. DOI: 10.1142/S1363919611003283 late work of Schumpeter (1942). Some empirical studies indeed found that more innovation is found in industries with low competition, which may be regarded as evidence that for individual firms, positions of market power provide incentives to innovate because of better appropriation opportunities. In contrast, other studies have found opposite evidence, showing greater innovative activities in industries with high competition. Here the interpretation goes back to Schumpeter's early view that innovations are initiated by heroic entrepreneurs and more likely to be introduced in markets by start-up firms, in situations of nearly perfect competition (Schumpeter, 1934). In analogy with Schumpeter's views, these patterns are labeled as "Schumpeter Mark II" and "Schumpeter Mark I" regimes. Yet another set of studies found evidence of an inverted-U shaped relationship between innovation and market concentration, indicating that there is an "optimal" level of competition for innovation (Scherer, 1967; Aghion *et al.*, 2005).

This paper contributes to the research on competition and innovation in three respects. Firstly, there is a debate on how to measure competition and innovation (Tang, 2006; Kemp *et al.*, 2004). Common competition indicators like market concentration or price cost margins are criticized for not capturing the underlying dynamic processes that influence decision-making. Besides, these indicators are at the level of industries, implying that they do not capture the actual competitive domain of the firm (Kemp *et al.*, 2004). To overcome some of these limitations, we draw on perception-based competition measures collected from individual managers and small business owners. As for innovation, this is usually measured with R&D expenditures or patents, but these indicators are less suitable for SMEs (which are central in our analysis here). Here, we focus on firms' intentions to engage in innovation, rather than using traditional indicators which reflect inputs or outcomes of past innovation processes. This helps to mitigate the problem of reversed causality in the relationship between competition and innovation.

Secondly, inconclusive evidence in previous work suggests that the relationship may vary with the nature of competition and/or specific types of innovation involved (Tang, 2006). We here explore if there a more subtle relation between both variables, by arguing that competition becomes manifest in different forms which may affect various aspects of firms' innovation behavior. In the current paper, our data allowed us to distinguish between three types of competition that firms perceive in their value chain — including internal rivalry (among incumbent firms within industries) and external sources of competition due to the bargaining power of suppliers and buyers. As for innovation, we distinguish between products and process innovation — which have been demonstrated to differ in their developmental requirements and antecedents (e.g., Cohen and Klepper, 1996).

Thirdly, the relationship between perceived competition and innovative intentions may be contingent on other variables. More specifically, we explore the moderating influence of firms' strategic attention for innovation. Some firms may have a clear "entrepreneurial orientation" (Lee *et al.*, 2001), i.e., they are more or less continuously dedicated to innovation, which is to a certain degree independent of their competitive environment and may diminish the influence of external competition. Other firms may lack such systematic orientation, and introduce innovations while adapting to changes in their external environments. In what follows we will argue why a moderating influence of strategic attention for innovation may be anticipated, and empirically test if such moderation is present.

In sum, this paper draws on perception-based measures collected directly from managers and small business owners. It distinguishes between various types of competition and innovation, and analyzes the potential moderating influence of firms' strategic attention for innovation. Drawing on survey data of 2,281 SME, we find a subtle view on the relation between perceived competition and their intentions to engage in innovation. Intentions for process innovation appear to correlate with the perceived bargaining power of suppliers. Intentions to engage in product innovation are associated with perceived buyer power, and also with perceived internal rivalry — but only if firms report no strategic attention for innovation. In the next sections we provide a theoretical background and develop hypotheses, discuss our data, present our regression models and findings, which we will be discussed.

Background and Hypotheses

This section elaborates on the three proposed contributions of this paper, i.e., applying alternative measures, distinguishing between various forms of competition and innovation, and exploring the potential moderating influence of strategic attention for innovation.

Measures

One fundamental problem in the analysis of competition and innovation is that the former concept is rather abstract and cannot be measured directly. Conventional approaches in industrial organization literature are dominated by models in which competition is measured using structural variables indicative of the outcomes of competitive processes, including indicators such as market concentration (e.g., Herfindahl-Hirschmann index, C4-ratios), price-cost margins and numbers of entrants in industries (Tang, 2006). These have been criticized because they reflect the outcome of competitive processes, and not the underlying dynamic processes which influence firms' decision to innovate (Kemp *et al.*, 2004). Another fact of life is that the current indicators are at the industry level. They rely on the standard industrial classifications and accordingly fail to capture the actual competitive domain of the firm. This is most evident when firms specialize in small niches, below the digit level available in the data or, to the other extreme, when firms are highly diversified and active in multiple industries with different classification codes. Moreover, market structure explains only a small proportion of variation in innovative activities (Cohen, 1995; Scott, 1984).

In this paper we alternatively use data on firms' perceptions of their competitive environment (cf. Tang, 2006; Kemp *et al.*, 2004). What might be the value added of adding new measures of competition to the long list already available? Our central argument is the trivial observation that, at the end of the day, competitive decisions are made by human beings. The connection between competition and innovation can be better understood by incorporating insights from psychology i.e., entrepreneurs' perceptions of competition and the contingencies that influence the connection between competition and innovation behavior. Such a "humanized" view of decision-making is far from new. It has in fact a long tradition in business literature, a classical example being the behavioral theory of the firm as developed in the 1950s and 1960s (e.g., Cyert and March, 1963) — a theory that inspired modern evolutionary economics to introduce a routine-driven conception of the firm in models of competition, where satisfying — rather than maximizing — behavior rules firm-level decision-making (e.g., Nelson and Winter, 1982).

Yet, in the analysis of competition and innovation such an approach is relatively new. Tang (2006) proposed to use perception-based indicators of competition which are directly collected from firms' decision-makers. He stresses several advantages of such measures. Being firm specific, data may vary across firms active within the same industry, because of differences in cognitive processes across small business managers or because of specialization in various niche markets. Another advantage is that the definition of the market, when it is given by the firm instead from the national statistics, includes both national and international competitors of the firm. Perception-based measures allow responding managers and small business owners to implicitly define where their markets end in terms of geographical distance and similarities with other firms in their market.

With regard to the measurement of innovation, there are well-known limitations in the use of R&D and patent statistics, especially when SMEs are the subject of investigation (as in the current paper). Few SMEs practice R&D, and besides, R&D expenditures do not adequately capture the range of sources of innovation from within or outside the firm (e.g., Laestadius, 1998; Muscio, 2007). With regard to patents as a measure of innovative output, limitations include that firms may patent for various reasons, not necessarily related to gaining economic rents from innovation. In addition, the propensity to patent differs across firms of different sizes and industries (Cohen *et al.*, 2000). Also, the value of patents is highly skewed, with few patents generating most of the revenues, and several patents never being commercialized as new products or processes (Scherer *et al.*, 2000).

We here focus on the innovative outputs of firms, but rather than patents or R&D we draw on indicators for product and process innovation. These types differ in terms of their specific developmental patterns and the returns they can generate (Cohen and Klepper, 1996). Likewise, Tang (2006) demonstrated that the likeliness of realizing product and process innovations varies with different types of competition. We continue in this line of research, but rather than realized innovations we focus on firms' intentions to engage in product and process innovation. In doing so we obtain a more direct connection with our independent variable of perceived competition, i.e., decision-makers in firms first intend and then act. In this vein, past entrepreneurship studies have demonstrated that business owners' intentions are the most important predictor of their future behavior and results (Shane, 2003). Likewise, social psychological models have shown that intentions to act are a very important determinant of decision-making (Ajzen and Fishbein, 2005). In contrast, implemented innovations are related to past behaviors, making it difficult to claim that innovation is a consequence of current, perceived competition.

Competition and innovative intentions

Before developing hypotheses, a first issue to address is whether we should expect a positive or negative connection between perceived competition and innovation indicators in our sample of SMEs. We recall that in his initial work Schumpeter (1934) implicitly suggested a positive relationship (innovation originates from competitive small firms challenging the status quo), while his later work predicted that competition diminishes firms' innovative efforts (as monopoly power provides better opportunities to achieve and appropriate results from innovative efforts). In our current research of SMEs however, we hypothesize a positive connection between perceived competition and innovative intentions. Our data include only firms with fewer than 100 employees (see the next section). Such firms are usually located in environments which can be characterized as fragmented markets with many incumbent firms. In other words, the relevant market form is monopolistic competition rather than an oligopoly or monopoly. In such environments, incentives to terminate innovative efforts as a consequence of losing monopoly power (from increased competition) are not very likely.[1]

As discussed in the introduction section previous work has provided inconclusive evidence, suggesting that the relationship may vary with the nature of competition and/or specific types of innovation involved. In this vein, Tang (2006), in a study of Canadian firms, distinguished between four types of perceived competition (product substitution, entry of competing products, rapid product obsolescence and changes in production technologies) and two types of innovation. He demonstrated that the likeliness of realizing product and process innovations varies with different types of competition, and we here continue with this line of research.

Perception-based measures of competition can take different perspectives. Rather than taking a holistic view of competition (e.g., Porac and Thomas, 1994; Hodgkinson, 1997) we managed to access and analyze data on three competitive forces that firms encounter in their value chain. We distinguish between internal rivalry (among incumbent firms within the industry), the bargaining power of suppliers and that of buyers. These forces are all part of Porter's (1979) framework of competition. Initially, he listed a number of structural characteristics of industries that indicate intense competition, such as a high number of incumbent firms, low entry barriers and a high market growth rate. Arguing that such indicators mainly reflect the outcomes of competition, he then suggested his well-known five forces as a framework to analyze the nature of competition. We now develop hypotheses on how the three above-mentioned forces are related to the innovative intentions of SMEs. Porter's other forces, i.e., the threat of new entrants and substitute products remain unexplored due to data restrictions (see Section 3).

[1]Of course, for large firms or specific industries such as manufacturers of medicines, the situation could be different. In Section 5, we elaborate on the generalizability of our results.

Internal rivalry is the extent to which incumbent firms in an industry engage in outwardly manifested competitive actions in their search for competitive advantage (Pecotich *et al.*, 1999). This can be done by either cost reductions or differentiating products (Porter, 1980). It can be argued that when incumbent rivalry is strong, their collusive power declines, and they will be more prone to differentiate their products or services to avoid costly price wars (Cabral, 2000). Such responses to internal rivalry can be especially appropriate or necessary for SMEs. In a study of SMEs in the Netherlands, it was found that a large majority of firms prefer, in general, to differentiate their products from incumbent competitors rather than engaging in price competition. More specifically, 84% of the Dutch SMEs preferred to distinguish themselves by developing new products or by offering extra services, while only 8% favored low-cost strategies (Bruins, 2006). As a consequence, intentions to engage in product innovation will be more likely.

A connection between perceived internal rivalry and intended process innovation is less evident. As indicated, process innovations aimed to realize cost savings will not be favored by small business owners, and due to their size, their have fewer opportunities to engage in process innovation anyway. Indeed, cost-spreading advantages are more favorable for firms performing process-related R&D, but these returns are more likely to be realized by large firms (Cohen and Klepper, 1996). These arguments suggest that in response to internal rivalry, SMEs are more likely to strive for product differentiation instead of cost-reductions by means of process-related innovations. Thus,

H1: *Perceived internal rivalry is positively related with SMEs' intentions to engage in product innovation.*

The bargaining power of suppliers is defined as the extent to which suppliers are able to exert influence and affect the firm's profitability and general well-being (Pecotich *et al.*, 1999). They may increase their input prices or lower the quality of their inputs (Porter, 1979). When firms operate in markets with price-sensitive customers, they will be unable to compensate for supplier power by increasing their own prices. Under such conditions, firms are anticipated to seek advances in production techniques that would save relatively costly inputs (Cabral, 2000), inducing process innovation. Besides, we reason that a connection between supplier power and product innovation is less likely, simply because such initiatives would target buyers rather than suppliers. Incumbent firms probably need to accept the restrictions that suppliers impose on them by adjusting their processes i.e., by attempting alternative cost savings or adjustments in their production processes to become less dependent on their suppliers in the long run. Furthermore, we reason that powerful suppliers are less willing to invest to meet the needs of their customers. Such failure has been demonstrated to be a major determinant of user-developed innovations, i.e., process-related innovations which firms develop by themselves because due to a lack of market supply (von Hippel, 2005). We hypothesize

H2: *Perceived supplier power is positively related with SMEs' intentions to engage in process innovation.*

Buyer power is the extent to which buyers can exert influence and affect the firm's profitability and general well-being (Pecotich *et al.*, 1999). It is present when buyers can impose pressures to lower prices or to improve the quality of products (Porter, 1979). As in the case of incumbent rivalry, when faced with powerful buyers firms probably try to avoid price-cutting moves that may lead to expensive price wars (Cabral, 2000). In contrast, they may be anticipated to differentiate their products from other incumbent firms. This approach can be especially convenient or even necessary for SMEs who are less able to sustain prolonged price-cutting. In line with this argument, a Dutch study found that SMEs strongly prefer to meet their customers' demands by developing new products or offering extra services, rather than pursuing a low-cost strategy (Bruins, 2006). Another argument to support a positive relationship between product innovation and buyer power is based on the "demand-pull" interpretation of technological change (Schmookler, 1966). Likewise, in the literature on new product development "tough customers" who demand improvements in product quality and specifications are regarded to be essential and active sources of ideas for new product design (Cooper, 2003). Thus,

H3: *Perceived buyer power is positively related with SMEs' intentions to engage in product innovation.*

Strategic attention for innovation

Mixed results in previous studies suggest that the relationship between competition and innovation is conditional on other factors (Cohen, 1995). In the industrial organization literature, there is for example emerging consensus that a straightforward link between innovation and competition is not to be found; the reason being that such a relationship is contingent on other variables such as technological regimes (Cohen, 1995) or technological distances between competitors (Aghion *et al.*, 2005). In this vein, Porter (1979) already discussed that the nature of competitive responses should be contingent on firms' internal strengths and weaknesses. Any plan of action should position the firm so that its capabilities provide the best response against relevant competitive forces. This creates opportunities for many hypotheses on specific factors that can interact with perceived competition to explain what kind of responses are most likely. An obvious example is the availability of innovation resources such as time, money and relevant knowledge. A lack of such resources would diminish the connection between any kind of competition and firms' innovative intentions (Gatignon and Reibstein, 1997).

This paper empirically explores the impact of one specific internal capability, namely if small business managers report to have strategic attention for innovation. The more they consider innovation of strategic importance, i.e., to be essential for the long-term performance of their business, the more likely that this guides their future innovation behavior. In this vein, it has been demonstrated that SMEs with innovation strategies are more likely to realize innovative outputs (Hadjimanolis, 2000; de Jong and Marsili, 2006). We propose that strategic attention for innovation moderates the connection between competition and innovation. Our reasoning is that if innovation is considered an important strategic objective, managers are more likely to be inclined towards the initiation and implementation of innovations. Such SMEs will innovate anyway, no matter what competitive forces are applicable. A common activity in any strategic exercise is an analysis of competitive forces (Campbell *et al.*, 1999). Accordingly, innovative responses will already be manifest in current strategic planning and objectives for the forthcoming period, leaving fewer possibilities for a direct association between competitive pressures and innovative intentions. A variation in the innovative response to perceived competition, with respect to an established strategy, may then reflect only unexpected external changes or earlier errors by the managers (Kirzner, 1997). We hypothesize

In SMEs reporting strategic attention for innovation the relationship between…

H4: … *perceived internal rivalry and intended product innovation…*

H5: … *perceived supplier power and intended process innovation…*

H6: … *perceived buyer power and intended product innovation…*
… *is weaker, and vice versa.*

Data

The data are derived from a survey conducted by EIM, a Dutch research institute specialized in entrepreneurship and SMEs. The survey was commissioned by the Dutch ministry of Economic Affairs, with the purpose to measure how SMEs organize their business processes, information that is not systematically recorded in publicly funded sources in the Netherlands. Although the survey was not specifically organized for this paper, some of its questions were suitable for our current purposes.

The survey was implemented in June 2006 over a period of four weeks, by means of computer-assisted telephone interviewing. All respondents were small business owners or general managers. An initial sample of 5,500 firms was drawn from the entire population of SMEs in the Netherlands, as available from the Chambers of Commerce trade register database. Following the Dutch definition of SMEs, only firms with less than 100 employees were sampled (Bangma, 2005). Responses were obtained from 2,281 persons, a response rate of 41%. Table 1 presents the distribution of respondents by industries and size classes.

A comparison of the distribution of respondents and non-respondents indicated no problems with response bias. We performed a χ^2-test to compare the joint distribution for industrial sector and size class, as shown in Table 1, with a similar table of non-respondents ($n = 3{,}219$), and found no significant difference ($p = 0.07$). We do remark however that the respondents are not representative for

Table 1. Number of respondents by industry and size classes.

Type of industry	NACE-codes	Size class (number of employees)			Total
		1–9	10–49	50–99	
Agriculture	01, 02, 05	109	50	6	165
Low-tech manufacturing	15–22, 25–28, other 34, 35–37	236	128	29	393
High-tech manufacturing	23–24, 29–33, 34.1	64	44	11	119
Construction	45	99	49	9	157
Wholesale trade	51	169	93	16	278
Retail trade	50, 52	156	74	8	238
Hotels, restaurants and catering	55	52	25	6	83
Transport	60–64	120	55	20	195
Knowledge-intensive services	72, 73, 74.2	68	44	7	119
Other business services	70–71, other 74	198	115	29	342
Personal services	92–93	71	38	6	115
Financial services	65–67	44	28	5	77
Total		1386	743	152	2281

the larger business population. The commissioners of the survey had requested a disproportional sample to enable comparisons between size classes and industries. Firms with 1–9 were for example over-sampled (i.e. they represent 83 percent of the full business population (Bangma, 2005)). Although the given stratification inevitably distorts the validity of the descriptive statistics presented hereafter, we do not expect that the legitimacy of the correlation analyses presented in the next section has been compromised.

Variables

Table 2 gives a detailed description of the variables that we used to test our hypotheses. We used indicators of firms' intentions to engage in product and process innovation, respectively. Both types of innovation were defined similarly as in the Community Innovation Surveys (see OECD, 2005). Respondents were asked for their intentions to invest in the forthcoming year.

The survey contained measures for the three forms of competition, i.e., Porter's forces of entry threat and substitute products were not recorded. Each force

Table 2. Variables.

Variable	Description and values$^{\Psi}$
Intended product innovation	Firm intends to invest in product innovation in the forthcoming year, i.e., develop or implement new products/services (yes/no)
Intended process innovation	Firm intends to invest in process innovation in the forthcoming year, i.e., develop or implement new techniques, equipment or software (yes/no)
Strategic attention for innovation	Firm reports that innovation is considered to be of strategic importance, evidenced by a business plan or strategy in which innovation is explicitly mentioned (yes/no)
Internal rivalry	Mean score of three items ($\alpha = 0.73$, mean r $= 0.47$, lowest IRC $= 0.51$): • "Firms in our market compete intensively, as a result of which our market share is threatened" (totally disagree/disagree/neither agree nor disagree/ agree/totally agree) • "Our margins are squeezed because of aggressive marketing by our competitors" (totally disagree/disagree/neither agree nor disagree/ agree/totally agree)

Table 2. (*Continued*)

Variable	Description and values[Ψ]
	• "Internal rivalry among incumbent firms in our market limits our profitability…" (not at all/marginally/some extent/large extent/very large extent)
Supplier power	Mean score of three items ($\alpha = 0.70$, mean r $= 0.44$, lowest IRC $= 0.47$): • "Our suppliers are powerful and do not hesitate to use it" (totally disagree/ disagree/neither agree nor disagree/agree/totally agree) • "We need to be loyal to our suppliers even if they raise their prices or lower the quality of their products" (totally disagree/ disagree/ neither agree nor disagree/ agree/ totally agree) • "Bargaining power of suppliers limits our profitability…" (not at all/marginally/some extent/large extent/very large extent)
Buyer power	Mean score of three items ($\alpha = 0.77$, mean r $= 0.53$, lowest IRC $= 0.57$): • "Our buyers are in a position to negotiate our terms of delivery and prices" (totally disagree/disagree/neither agree nor disagree/agree/ totally agree). • "Our buyers are tough negotiators" (totally disagree/disagree/neither agree nor disagree/agree/totally agree). • "Bargaining power of buyers limits our profitability…" (not at all/ marginally/some extent/large extent/very large extent).
Type of industry	Dummies for 11 industries (low-tech manufacturing; high-tech manufacturing; construction; wholesale trade; retail trade; hotels, restaurants and catering; transport; knowledge-intensive services; other business services; personal services; financial services) with agriculture as the reference group
Firm size	Number of employees in full-time equivalents, log transformed

[Ψ]Dichotomous responses and dummies are coded as "yes" = 1 and "no" = 0. Responses on 5-point Likert scales are coded from 1 (in case of "totally disagree" and "not at all") to 5 ("totally agree" and "very large extent").

was measured with a multiple-item Likert scale of three items. These measures were developed by Kemp *et al.* (2004) as part of a multi-dimensional measure of perceived competition for application by statistical offices. We calculated various reliability statistics to assess internal consistency: Cronbach's α, mean correlations between the items (mean r) and item-rest correlations (IRCs). Recommended critical values are 0.70 for α, 0.40 for mean r, and 0.30 for any item-rest correlation (IRC) (Hair *et al.*, 1998). As Table 2 shows, all these measures had satisfactory values. The three measures also reflected truly different aspects of perceived competition. To check this we applied exploratory factor analysis with oblique rotation to find three distinct factors with eigenvalues greater than one, while each item corresponded to its proposed competitive force only.

Respondents had also indicated if they considered innovation to be of strategic importance for their business, evidenced by a business plan or strategy in which innovation was explicitly mentioned. Previous work indicates that this indicator correlates significantly with many other innovation capability indicators (e.g. de Jong and Marsili, 2006). Finally, industry dummies and firm size were added as control variables. Industry effects are found to be important in explaining variance in innovative activity (Cohen, 1995; Scott, 1984). Firm size is another well-known determinant of the innovativeness of firms, for example by influencing the ability to finance innovative investments, to spread risks associated with innovation, to organize for innovation and to recruit specialized staff (Cohen, 1995; Nooteboom, 1994).

Descriptive statistics

Table 3 provides the means, standard deviations (SD) and correlations for the variables in our analyses (leaving out industry dummies due to space restrictions). Firm size has been log transformed because the raw data violated the assumed normal distribution (skewness = 2.5, kurtosis = 6.8).

As anticipated, we found significant, rather strong correlations between the innovation indicators and firms' strategic attention for innovation. The table also suggests that our independent variables do not suffer from multicollinearity, as the Pearson correlations do not exceed 0.42. None of the independent variables shares more than 20% of their variance with any other. Multicollinearity may be become problematic in case of correlations exceeding 0.80 (Hair *et al.*, 1998).[2]

[2] In the analyses presented hereafter, variance inflation factors and condition indices indeed indicated no problems with multicollinearity (results available on request).

Table 3. Descriptive statistics (n = 2,281).

Variable	Mean	SD	Correlations					
			1	2	3	4	5	6
Intended product innovation	0.56	0.50						
Intended process innovation	0.46	0.50	0.30**					
Strategic attention for innovation	0.54	0.50	0.42**	0.34**				
Internal rivalry	3.06	0.99	0.07**	0.02	0.06*			
Supplier power	2.72	0.96	0.03	0.10**	0.03	0.34**		
Buyer power	3.05	0.98	0.13**	0.11**	0.14**	0.38**	0.27**	
Log transformed firm size	1.91	1.17	0.23**	0.33**	0.26**	0.12**	0.01	0.20**

**$p < 0.001$; *$p < 0.01$.

Results

We set up a range of binary logistic regression models to test our hypotheses on intended product and process innovation.

Intended product innovation

Table 4 presents the results for all models in which intended product innovation is the dependent variable. Various specifications of the model were applied by entering the independent variables and interaction terms at successive steps. Before entering the variables into the model, the indicators of competitive forces were rescaled and centered on their means. The interaction terms were computed by multiplying these mean-centred values of competitive forces with the indicator for strategic attention (Jaccard, 2001). The results in Table 4 refer to our Hypotheses 1, 3, 4 and 6.

The goodness-of-fit of logistic regression models is assessed by comparing the transformed loglikelihood value −2LL with the previous model. The difference between both values follows a χ^2-distribution (Verbeek, 2004). Other fit measures include the hit rate (representing the share of correctly classified cases) and Nagelkerke's R^2 (indicating the strength of association in the overall model). The first model was an empty model (intercept only) to obtain baseline values for

Table 4. Binary logistic regression models of intended product innovation ($n = 2{,}281$).

	Models				
	I	II	III	IV	V
Parameter estimates:					
Constant	0.25**	−1.62**	−1.62**	−1.60**	−1.62**
Industry: low-tech manufacturing		0.46^	0.52^	0.52^	0.52^
Industry: high-tech manufacturing		0.75*	0.82*	0.81*	0.82*
Industry: construction		0.11	0.11	0.10	0.11
Industry: wholesale trade		0.68*	0.70*	0.68*	0.70*
Industry: retail trade		0.44	0.53^	0.51^	0.53^
Industry: hotels, restaurants and catering		0.86*	1.04*	1.04*	1.04*
Industry: transport		0.57^	0.59^	0.57^	0.59^
Industry: knowledge-intensive services		1.19**	1.25**	1.25**	1.25**
Industry: other business services		0.39	0.47^	0.46^	0.47^
Industry: personal services		0.91*	1.07**	1.06**	1.07**
Industry: financial services		0.15	0.22	0.21	0.23
Log firm size		0.29**	0.26**	0.26**	0.26**
Strategic attention for innovation (SA)		1.60**	1.57**	1.58**	1.57**
Internal rivalry (IR)			0.07	0.17^	0.07
Supplier power			0.00	0.00	0.00
Buyer power (BP)			0.15^	0.14^	0.15^
Interaction: SA * IR				−0.20^	
Interaction: SA * BP					−0.01
Model fit:					
−2LL	3128.10	2556.88	2544.31	2539.94	2544.29
Δ−2LL		571.22	12.57	4.37	0.02
Δdf		13	3	1	1
Significance		**	*	^	
Correctly classified cases	56.1%	71.0%	73.9%	74.4%	74.0%
Nagelkerke R^2		0.25	0.26	0.27	0.26

**$p < 0.001$; *$p < 0.01$; ^$p < 0.05$.

–2LL and the hit rate. It appears that 56.1% of the sampled firms is classified correctly, equal to the share of firms with intentions to invest in product innovation.

The second model added all control variables to the equation. This significantly diminished the transformed loglikelihood value (Δ–2LL = 571.22 with Δdf = 13, $p < 0.001$), implying that when taken together, the independents are linearly related to the odds of SMEs' intentions to invest in product innovation. The hit rate increased to 71.0% while Nagelkerke's R^2 was 0.25, suggesting a reasonable association with our control variables. Wald tests on the individual effect parameters indicated various industry differences, including for example that high-tech manufacturing and knowledge-intensive services firms are much more likely to have innovative intentions (compared to the reference group of agriculture). Besides, as we had anticipated firm size and strategic attention appeared to be strong predictors of the probability of intended product innovation.

The third model provided a test of our hypotheses 1 and 3 by entering measures of the three competitive forces. Goodness-of-fit again improved significantly (Δ–2LL = 12.57 with Δdf = 3, $p < 0.01$) while the hit rate and Nagelkerke's R^2 were slightly better. From the Wald tests we conclude that bargaining power of buyers makes a significant contribution to the odds of intentions for product innovation. More specifically, a one-unit increase in perceived buyer power makes it $\exp(0.15) = 1.16$ more likely that SMEs intend to invest in product innovation. Thus, Hypothesis 3 is supported. Besides, as we expected the parameter for supplier power is insignificant, suggesting that firms do not invest in new products to cope with bargaining power of their suppliers. Finally, we found no positive influence of internal rivalry. The logistic regression coefficient is positive ($b = 0.07$) but insignificant, so our first hypothesis is not supported.[3]

In the fourth model entered the interaction term between strategic attention and internal rivalry. The transformed log likelihood value decreased significantly (Δ–2LL = 4.37 with Δdf = 1, $p < 0.05$). Wald tests on the individual parameters revealed a negative sign for the interaction term ($b = -0.20$, $p < 0.05$). To further analyze this significant interaction effect, the regression equation was rearranged in simple regressions given conditional values of strategic attention. By evaluating the parameter estimates in case strategic attention is reported (versus lacking), different estimates are obtained for the connection between perceived internal rivalry and intended product innovation (Jaccard, 2001). When

[3]When we entered the competitive forces into the equation one-by-one, results were almost identical.

SMEs reported no strategic attention, the parameter estimate of internal rivalry was positive and significant ($b = 0.17$, $p < 0.05$). In contrast, if firms reported that innovation was of strategic importance, the estimated coefficient was insignificant ($b = -0.03$, $p > 0.05$). These results support Hypothesis 4.

The fifth model tested the interaction between strategic attention and buyer power. Here, adding the interaction term did not improve the goodness-of-fit. Buyers' bargaining power is apparently directly connected to firms' intentions to invest in product innovation, but no interaction effect is found. Hypothesis 6 is not supported.

Intended process innovation

Table 5 gives the results for all models with intended process innovation as the dependent variable. The table corresponds with Hypotheses 2 and 5.

Table 5. Binary logistic regression models of intended process innovation ($n = 2{,}281$).

	Models			
	VI	VII	VIII	IX
Parameter estimates:				
Constant	−0.17**	−1.71**	−1.74**	−1.74**
Industry: low-tech manufacturing		−0.12	−0.10	−0.10
Industry: high-tech manufacturing		−0.07	−0.07	−0.06
Industry: construction		−0.77*	−0.76*	−0.76*
Industry: wholesale trade		−0.03	−0.02	−0.02
Industry: retail trade		−0.51^	−0.47^	−0.47^
Industry: hotels, restaurants and catering		0.01	0.05	0.04
Industry: transport		−0.09	−0.07	−0.07
Industry: knowledge-intensive services		0.20	0.26	0.26
Industry: other business services		−0.01	0.04	0.04
Industry: personal services		−0.23	−0.16	−0.16
Industry: financial services		0.45	0.50	0.50
Log firm size		0.52**	0.52**	0.52**
Strategic attention for innovation (SA)		1.18**	1.17**	1.17**
Internal rivalry			−0.08	−0.08
Supplier power (SP)			0.11^	0.14^
Buyer power			0.05	0.05
Interaction: SA*SP				−0.05

Table 5. (*Continued*)

	Models			
	VI	VII	VIII	IX
Model fit:				
−2LL	3146.78	2637.10	2628.59	2628.36
Δ−2LL		509.68	8.51	0.23
Δdf		13	3	1
significance		**	^	
Correctly classified cases	54.1%	67.8%	70.5%	70.6%
Nagelkerke R^2		0.24	0.25	0.24

$^{**}p < 0.001$; $^{*}p < 0.01$; $^{\wedge}p < 0.05$.

Model VI was again an empty model telling us that 54.1% of the cases can be correctly classified (equal to the share of firms with no intentions to engage in process innovation). Model VII added the control variables which again significantly improved model fit (in terms of Δ-2LL, hit rate and Nagelkerke's R^2). The significances of individual effect parameters again demonstrated that firm size and strategic attention are strongly associated with SMEs' innovative intentions.

In model VII we added the three competition measures. Goodness-of-fit improved significantly, while the significance of the individual parameters showed that only the bargaining power of suppliers is associated with intended process innovation. Our second hypothesis is supported. Besides, the coefficients of internal rivalry and buyer power were not significant.

In the final model the interaction term between strategic attention and supplier power was entered. We found no moderating effect on the connection between supplier power and intended process innovation, so Hypothesis 5 is not supported.

Discussion

This paper explored how various types of perceived competition are connected with SMEs' intentions to engage in product and process innovation, and if SMEs' strategic attention for innovation moderates the relationship between perceived competition and innovative intentions. Rather than using structural competition indicators such as concentration indices, we here based our exploration on perception-based, multiple-item measures. Compared to previous work we found a more subtle relationship between competition and innovation. Perceived buyer power appeared to be directly related to firms' intentions to engage in product innovation, while no such a connection was found for process innovation. When confronted with powerful customers SMEs apparently prefer to avoid price wars; but rather differentiate their offerings by investing in product innovation. For perceived supplier power we found a direct association with firms' intentions to invest in process innovation, while there was no relationship with product innovation. This competitive force means that suppliers can enforce restrictions on the incumbent firm (such as price increases or quality reductions) without being punished. SMEs apparently need to adapt to such behavior by attempting to achieve alternative cost savings or adjustments in their production processes to become less dependent on the supplier in the long run. For internal rivalry, we found a positive connection with intended product innovation, but only when SMEs reported no strategic attention for innovation (evidenced by a business plan or strategy in which innovation was explicitly mentioned). This finding suggests that when firms perceive other incumbents as threatening, they prefer to develop new, differentiated products. However, if they consider innovation to be strategically important, the direct impact of internal rivalry is insignificant. Innovative intentions as a consequence of internal rivalry are then probably already taken accounted for in their strategic planning.

Our analysis comes with a number of caveats. Firstly, in the introduction section we discussed the Schumpeterian hypotheses, implying that competition and innovation can, in general, be either positively or negatively related. We here found some evidence for a positive connection, but remark that this may very well be an artifact of the empirical context of SMEs. In general such firms operate in monopolistically competitive markets with many other (entrepreneurial) firms. Following Schumpeter Mark I, in such a context a positive connection between competition and innovation is likely. This result, however, cannot be generalized to the broader business population. In oligopolistic or monopolistic markets the connection may in fact be negative.

Secondly, in terms of size, the estimated parameters of our competition measures are very modest. In this respect our findings do not add much to the incumbent industrial organization literature in which similarly modest correlations are found (Cohen, 1995). As an example, we found perceived buyer power increased the odds of intended product innovation by $\exp(0.15) = 1.16$. In comparison, when firms have strategic attention for innovation, the same odds increase by $\exp(1.57) = 4.81$. In this context we note that previous studies identified many alternative responses to perceived competition, including price reductions, accommodation (e.g., silent cooperation or mergers), ignorance (for instance when resources to respond are lacking) or even withdrawal from current markets (Gatignon and Reibstein, 1997). Likewise, Chen *et al.* (1992) argue that responses may be "strategic" versus "tactical". Strategic responses include significant investments in fixed assets and/or people and results in major changes to structure and products. In contrast, tactical responses do not involve such commitments and require relatively minor, routine changes. Given that product and process innovation are only two of out many potential responses and that many competitive moves will be routine changes rather than innovations, our results may be plausible. Future research should explore this further, for instance by capturing other competitive forces, objects of innovation and new interaction variables that may enhance the connection between the two.

Thirdly, some methodological issues certainly remain. There is a dark side to perception-based measures, including that small business owners may misinterpret their external environment or engage in oversimplified, heuristic interpretations of competition. Next, we recognize that some of our indicators were quite simple, including intended innovation (magnitude of intentions remains uncertain) and strategic attention (i.e. strategies can be very diverse). Besides, our data were cross-sectional. Despite the fact that our dependent variables were formulated in terms of intentions, this paper is unable to deal with the two-way causality problem that challenges most empirical studies. To be truly convincing on causality, time series data and an improved research design, with a convincing source of exogenous variation in competitive forces, would be required. For this reason we formulated our hypotheses in terms of associations rather than causal relationships.

Implications of our findings are related to research and policymaking. For research, perception-based measures of competition will be helpful to account for the fact that individual firms can assess competitive forces differently, and respond in various ways and with varied intensity, even when operating in identical markets. It also helps to distinguish between various types of competition and innovation, enabling more fine-grained analyses on how both variables are related. Moreover, echoing Cohen (1995), our findings confirm that competition and innovation are not monotonously related, but contingent on other variables. Our findings should be extended, for example by analyzing the threat of substitute products and new entrants – two other commonly accepted dimensions of competition for which we lacked data. Some may also analyze alternative objects of innovation — examples include organizational and marketing innovation, two forms of innovation which have been included in the Community Innovation Surveys relatively recently (OECD, 2005).

For policymaking, it is important to recognize that changes in regulatory policies may affect various types of competition which, in turn, have a differentiated impact on firms' innovation activities. Regulatory policy changes intended to promote innovation should be evaluated against different types of competition. Stimulating competition could probably be more effective if specific types of competition were taken into account or aimed for. Of course, this first demands more elaborated work to identify relevant competitive forces, their connection with different types of innovation and contingency factors that moderate these relationships.

Acknowledgements

The author appreciates Orietta Marsili (RSM Erasmus University) and Ron Kemp (Dutch Competition Authority) for their comments on earlier versions of this paper.

References

Aghion, P, N Bloom, R Blundell, R Griffith and P Howitt (2005). Competition and innovation: An inverted-u relationship. *Quarterly Journal of Economics*, 120(2), 701–728.

Ajzen, I and M Fishbein (2005). The influence of attitudes on behavior. In Albarracín, D, BT Johnson and MP Zanna Mahwah (Eds.). *The Handbook of Attitudes*. NJ: Erlbaum, pp. 173–221.

Bangma, K (2005). *Kleinschalig Ondernemen 2005: Structuur en Ontwikkeling van het MKB* (Entrepreneurship in small business 2005: Structure and development of Dutch SMEs), EIM: Zoetermeer.

Bruins, A (2006). *Ondernemerschap en Strategie in het MKB* (Entrepreneurship and strategy of small firms). EIM: Zoetermeer.

Cabral, LMB (2000). *Introduction to Industrial Organization*. Cambridge, MA: MIT Press.

Campbell, D, G Stonehouse and B Houston (1999). *Business Strategy*. Oxford: Butterworth-Heinemann.

Chen, M-J, KG Smith and CM Grimm (1992). Action characteristics as predictors of competitive responses. *Management Science*, 38(3), 439–455.

Cohen, WM (1995). Empirical studies of innovative activities. In Stoneman, P (Ed.). *Handbook of the Economics of Innovation and Technological Change*. Oxford, Basil Blackwell, pp. 182–264.

Cohen, WM and S Klepper (1996). Firm size and the nature of innovation within industries: The case of process and product R&D. *The Review of Economics and Statistics*, 78(2), 232–243.

Cohen, WM, RR Nelson and JP Walsh (2000). *Protecting their Intellectual Assets, Appropriability Conditions and Why US Manufacturing Firms Patent (or not)*. Working paper, National Bureau of Economic Research.

Cooper, RG (2003). Profitable product innovation: The critical success factors. In Shavinina, LV (Ed.). *The International Handbook on Innovation*, Elsevier Science, pp. 139–157.

Cyert, RM and JG March (1963). *A Behavioral Theory of the Firm*. Englewood Press, NJ: Prentice-Hall.

de Jong, JPJ and O Marsili (2006). The fruit flies of innovation: A taxonomy of innovative small firms. *Research Policy*, 35(2), 213–229.

Gatignon, H and D Reibstein (1997). Creative strategies for responding to competitive actions. In Day, GS and DJ Reibstein (Eds.). *Warton on Dynamic Competitive Strategy*, New York: John Wiley.

Hadjimanolis, A (2000). An investigation of innovation antecedents in small firms in the context of a small developing country. *R&D Management*, 30(3), 235–245.

Hair, JF, RE Anderson, RL Tatham and WC Black (1998). *Multivariate Data Analysis: Fifth Edition*. Englewood Cliffs: Prentice Hall.

Hodgkinson, GP (1997). The cognitive analysis of competitive structures: A review and critique. *Human Relations*, 50(6), 625–654.

Jaccard, J (2001). *Interaction Effects in Logistic Regression*. Sage: Thousand Oaks, CA.

Kemp, RGM, M Mosselman and A van Witteloostuijn (2004). *The Perception of Competition Index*. EIM: Zoetermeer.

Kirzner, IM (1997). Entrepreneurial discovery and the competitive market process: An Austrian approach. *Journal of Economic Literature*, 35(1), 60–85.

Laestadius, S (1998). The relevance of science and technology indicators: The case of pulp and paper. *Research Policy*, 27, 385–395.

Lee, C, K Lee and JM Pennings (2001). Internal capabilities, external networks, and performance: A study on technology-based ventures. *Strategic Management Journal*, 22, 615–640.

Muscio, A (2007). The impact of absorptive capacity on SME's collaboration. *Economics of Innovation and New Technology*, 16, 653–668.

Nelson, RR and SG Winter (1982). *An Evolutionary Theory of Economic Change*, Cambridge, MA: Bellkamp.

Nooteboom, B (1994). Innovation and diffusion in small firms: Theory and evidence. *Small Business Economics*, 6(5), 327–347.

OECD (2005). *Oslo Manual: Guidelines for Collecting and Interpreting Innovation Data*. 3rd Edition, OECD: Paris.

Pecotich, A, J Hattie and LP Low (1999). Development of industruct: A scale for the measurement of perceptions of industry structure. *Marketing Letters*, 11(4), 409–422.

Porac, JF and H Thomas (1994). Cognitive categorization and subjective rivalry among retailers in a small city. *Journal of Applied Psychology*, 79, 54–66.

Porter, ME (1979). How competitive forces shape strategy. *Harvard Business Review*, March-April, 137–145.

Porter, ME (1980). *Competitive Strategy: Techniques for Analysing Industries and Competitors*. New York: Free Press.

Scherer, FM (1967). Market structure and the employment of scientists and engineers. *American Economic Review*, 57, 524–531.
Scherer, FM, D Harhoff and J Kukies (2000). Uncertainty and the size distribution of rewards from innovation. *Journal of Evolutionary Economics*, 10, 175–200.
Schmookler, J (1966). *Invention and Economic Growth*. Cambridge, MA: Harvard University Press.
Schumpeter, JA (1934). *The Theory of Economic Development*. Cambridge, MA: Harvard University Press.
Schumpeter, JA (1942). *Capitalism, Socialism and Democracy*. New York: Harper & Row.
Scott, JT (1984). Firm versus industry variability in R&D intensity. In Griliches, Z (Ed.). *R&D, Patents and Productivity*, Chicago and London: Chicago University Press.
Shane, SA (2003). *A General Theory of Entrepreneurship: The Individual-Opportunity Nexus*. Aldershot, UK: Edward Elgar.
Tang, J-M (2006). Competition and innovation behaviour. *Research Policy*, 35(1), 68–82.
Verbeek, M (2004). *A Guide to Modern Econometrics*. John Wiley: Chichester.
von Hippel, E (2005). *Democratizing Innovation*. Cambridge, MA: MIT Press.

Chapter 7

THE IMPACT OF ENVIRONMENTAL UNCERTAINTY DIMENSIONS ON ORGANISATIONAL INNOVATIVENESS: AN EMPIRICAL STUDY ON SMEs

Cevahir Uzkurt

Eskisehir Osmangazi University
Faculty of Economics and Administrative Sciences

Rachna Kumar

Alliant School of Management
Alliant International University

Halil Semih Kimzan

Eskisehir Osmangazi University
Faculty of Economics and Administrative Sciences

Hanife Sert

Eskisehir Osmangazi University
Social Sciences Institute

The purpose of this study is to examine the effects of factors of environmental uncertainty on the innovativeness of small and medium enterprises (SMEs). Innovativeness is widely accepted as an important characteristic for firm competitiveness and it has been studied by both researchers as well as business managers. Environmental uncertainty is a measure of the complexity of changing external forces faced by an organisation and it crucially impacts the responses of organisations in order to stay competitive. Based on approaches in existing literature, this study conceptualises environmental uncertainty comprised three separate dimensions — competitive intensity, market/demand turbulence, technological turbulence.

This paper was first published in *International Journal of Innovation Management*, Vol. 16, No. 2 (April 2012), 1250015 (23 pages). DOI: 10.1142/S1363919611003647

Data for the study were collected from 156 SMEs in Turkey. SMEs are regarded as an important ingredient in the economic growth of nations and especially so in developing nations such as Turkey. The findings of the study reveal that market/demand turbulence and technological turbulence have a positive effect on the innovativeness of SMEs. Interestingly and contrary to popular belief, competitive intensity was not found to have significant effect on an SME's innovativeness. The implication of the results from this research is that the degree of organisational innovativeness for SMEs tends to increase and therefore should be supported in environments with greater technological and market/demand turbulence. This research makes an important contribution to the developing body of innovation literature and provides directions for managers and researchers in influencing innovativeness of firms.

Keywords: Innovativeness; innovative; innovation; environmental uncertainty; demand turbulence; technological turbulence; competitive intensity; market turbulence; SMEs; Turkey.

Introduction

Innovation has been recognised as one of the most important drivers of competition and economic growth in recent years (Tidd *et al.*, 2005; Salavou, 2004; Tushman *et al.*, 1997). Therefore, many researchers have focused on studying innovation and innovativeness with the objective of providing prescriptive recommendations to organisations for increasing the innovativeness of their workforce (Porter, 1985; Tushman *et al.*, 1997). Innovation and innovativeness have become a key tool for corporations that try to cope with the complexity of today's highly competitive environment by flexibility and openness to new ways to create knowledge-based solutions (Liu and Phillips, 2011; Ellonen *et al.*, 2008). Tidd *et al.* (2005:3) emphasises that innovation is driven by the ability to see connections, to spot opportunities and to take advantage of them in organisations. In this context, innovation has become prominent as a competitive instrument for a firm's long-term survival and growth especially in today's competitive and turbulent market conditions (Santos-Rodrigues *et al.*, 2010; Musteen *et al.*, 2010). The role of innovation has been investigated in several studies within the disciplines of psychology, economics, business, sociology, engineering and education. Previous research on innovation has focused on defining innovation, describing the organisational determinants of innovation, or understanding the processes, types and measures of organisational innovation (Damanpour, 1991; Damanpour and Schneider, 2006; Rogers, 1995; Subramanian and Nilakanta, 1996; Ellonen *et al.*, 2008; Wang and Ahmed, 2004).

This study researches the interplay between dimensions of environmental uncertainty and organisational innovation. Global events have caused several environmental stresses, changes and uncertainties. Volatility of the financial markets, increased cost and productivity pressures, and weakened demand and consumer confidence have contributed to the need for appropriate managerial interventions for staying afloat (Normann, 2001; Galbraith, 2002). While organisations have responded with a myriad of actions such as cost management, price reductions, downsizing, and quality improvements, there is also the need to think, work and produce in new and innovative ways. In this global environmental scenario, the importance of influencing the innovativeness of the organisation can have important benefits. In an attempt to better understand and influence the innovativeness of organisations, researchers have studied several determinants of organisational innovation. Some examples of determinants which have been found to influence innovation in organisations include centralisation, formalisation, size, slack resource, specialisation, functional differentiation, professionalism, managerial attitude toward change, technical knowledge resource, external communication, and vertical differentiation (Damanpour and Schneider, 2006). Studies in innovation literature have found that organisational innovation plays a key role on organisational performance and firm's competitiveness (Baker and Sinkula, 2002; Damanpour, 1991; Farley *et al.*, 2008; Jimenez-Jimenez and Sanz-Valle, 2011; Luk *et al.*, 2008).

An important perspective in the literature conceptualises innovation as a response and adaptation to the needs of the environment (Subramanian and Nilakanta, 1996; Damanpour, 1992; Ravichandran, 2000; Nystrom *et al.*, 2002; Mudrak *et al.*, 2005). Organisations innovate or adopt existing innovations in order to stay competitive and successful in changing environments. In this view of innovativeness, organisations respond to uncertainty, changes and turbulence in their environments by staying nimble on their feet and innovating their products, processes or structures. Recent studies have emphasised the critical importance of organisations having a market orientation, flexibility and adaptation of environmental changes. For example, Lin *et al.* (2008) study the relationship between market orientation and innovativeness and Martinez-Sanchez *et al.* (2009) study the relationship between organisational flexibility (internal, workplace, and strategic flexibility) and innovation. Montes *et al.* (2004) find that an organisation's ability to respond to changing environments will impact the organisation's innovativeness and performance. Environmental changes make organisations pay close attention to environmental conditions and encourage employees to get involved in helping organisations to adapt to the changing environmental conditions.

Yet, we found no research in the context of innovation that studies specific dimensions of environmental changes and uncertainties for organisations to respond to. Freel (2005) finds that there is limited research examining the effects of environmental uncertainty on innovativeness. Therefore, the aim of this study is to fill this gap in the literature. In this research, we study the important dimensions of environmental uncertainty and changes and have investigated their impact on the innovativeness of the firm. Researchers have studied several factors which capture the complexity of the external environment. We review literature as a basis to conceptualise environmental uncertainty and summarise that the three most important factors which describe the complexity and uncertainty of the external environment were related to the competition faced by the firm, the change in the market and demand faced by the firm and the changing landscape in the technology solutions available to the firm (Droge *et al.*, 2008; Ansoff and Antoniou, 2005; Phromket and Ussahawannitchakit, 2009). We propose these three dimensions as *competitive intensity, market/demand turbulence and technological turbulence*. This paper reports on research which examines the impact of these three environmental uncertainties on an organisation's innovation.

The context of the data for the study is small and medium enterprises (SME) in a developing region, Turkey. Previous studies have been conducted to understand the mechanisms which strengthen the innovativeness and entrepreneurial orientation of SME's in developing economies (Kumar and Uzkurt, 2010; Kumar and Kelly, 2006; Kelly and Kumar, 2009). In addition, researchers have also studied overcoming barriers to innovation in SMEs (Xie *et al.*, 2010), the relationship between types of innovation strategy and the export performance of SMEs (Man, 2010), the impacts of internal and contextual variables on innovation performance of SMEs (Kang and Lee, 2008) and the relationships between environmental factors (market turbulence, competitive intensity) and innovation outcomes, and performance (Phromket and Ussahawannitchakit, 2009). The studies emphasise that innovation is one of the critical characteristics that can contribute to a developing economy's competitiveness. In addition, they emphasise SMEs as another of the driving engines for growth in several economies, especially in developing nations. Understanding and improving the chances of progress and prosperity in developing economies has important advantages in the stability and engagement of the global community in the post-industrial era. Accordingly, the referent dataset for this study was obtained from SMEs in Turkey and we have as an additional objective to understand and help raise innovativeness of Turkish SMEs. The Turkish context provided us with a culture that has characteristics of both Europe and Middle East due to Turkey's regional location.

Results of our study indicate that high levels of technological and market/demand turbulence have a significant positive impact on the innovativeness of SME firms irrespective of the competitive intensity faced by the firms. Our results give important contributions to the innovation literature by empirically demonstrating the effects of external environmental dimensions on innovativeness. Our results also contribute insights for the SME research by confirming prior research that shows that competitive intensity is not necessarily related to the positive gains in innovation.

The next section will review the background literature relevant to these concepts. Our research model describing the relationship between the three dimensions of environmental uncertainties (*competitive intensity, market/demand turbulence and technological turbulence*) and innovation will be introduced in the next section. The last section will present data and results for testing this model.

Conceptual Background and Model Development

There are three streams of research literature which have informed the development of our model and hypotheses and we review them here: (1) organisational innovation; (2) environmental uncertainty and; (3) environmental uncertainty and innovation.

Innovation and innovativeness

The terms of innovation and innovativeness are used either interchangeably or distinctly depending on the authors' view (Damanpour, 1991). Damanpour and Schneider (2006) define innovation as the creation, development, adoption and use of new ideas or behaviours in organisations. Innovation includes an internally generated or purchased idea, device, system, policy, program, process, product, or service that is new to the creating or adopting organisation (Daft, 1982; Damanpour, 1991; Damanpour and Evan, 1984). Salavou (2004) emphasises that '*Innovation seems to incorporate the adoption or/and implementation of "new" defined rather in subjective ways, whereas, innovativeness appears to embody some kind of measurement contingent on an organisation's proclivity towards innovation*'. Hurley and Hult (1998) identified innovativeness as willingness of the employees to adopt innovation. Besides, Menguc and Auh (2006) emphasise that innovativeness '*implies a firm being proactive by exploring new opportunities rather than merely exploiting current strengths*'. Therefore, innovativeness could be considered as an organisation's propensity to engage in developing, implementing, and supporting new ideas, processes, products (Luk *et al.*, 2008; Salavou, 2004). In this sense, innovativeness refers to capability and skills of the organisation to be able to produce or adopt and use or commercialise a new idea, product, or service (Thompson, 1965).

Existing typologies have classified innovations in several ways: administrative or technical, product or process, and radical or incremental (Damanpour, 1991; Downs and Mohr, 1976; Gopalakrishnan and Damanpour, 1997). Administrative innovations have direct relationship with management oriented processes, basic work activities, and organisational structure, while technical innovations are associated with the process technology of production for products or services (Damanpour, 1991; Gopalakrishnan and Damanpour, 1997; Damanpour and Evan, 1984; Kimberly and Evanisko, 1981; Read, 2000). Radical versus incremental innovation refers to the degree of change involved in the innovation. While incremental innovations need little changes or minor adjustments, fundamental changes are required for radical innovations (Cooper, 1973; Ettlie *et al.*, 1984; Gopalakrishnan and Damanpour, 1997; Henderson and Clark, 1990; Nystrom *et al.*, 2002). Another useful distinction is whether the organisations innovate by adopting or by generating innovation. In the former, innovation is created by the organisation, while in the second route, the organisation imitates, adopts or purchases innovation from other organisations. Both generation as well as adoption of innovations are coveted and regarded as successful forms of innovation in the literature (Amabile, 1988; Daft, 1982; Rogers, 1995; Zaltman *et al.*, 1973). Ellonen *et al.* (2008) investigated six different types of innovative activity including developing new products, new services, new methods of production, opening new markets, finding new sources of supply, and new ways of organising. The lack of clarity and consistency in the dimensions of organisational innovativeness is clear.

In addition to the definitions and typologies for innovations, literature has studied the phases which are important for understanding the processes involved in innovating. The phases in the processes have been categorised by Rogers (1995) as initiation, adoption and implementation; by Hage and Aiken (1970) as evaluation, initiation implementation and routinisation; by Zaltman *et al.* (1973) as knowledge, awareness, attitudes formation, decision, and initial and sustained implementation; and by Klein and Sorra (1996) as awareness, selection, adoption, implementation, and routinisation.

Finally, identifying the determinants of organisational innovation has been another important stream of innovation literature in the past. Examples of determinants studied include characteristics such as centralisation, formalisation, size, slack resource, specialisation, functional differentiation, professionalism, managerial attitude toward change, technical knowledge resource, external communication, and vertical differentiation (Damanpour, 1991; Damanpour and Schneider, 2006; Garcia-Morales *et al.*, 2006; Germain, 1996; Kimberly and Evanisko, 1981; Nystrom *et al.*, 2002; Subramanian and Nilakanta, 1996). Sajeva and Jucevicius (2008) differentiated four major characteristics of innovative organisations including organisation's ability to create and adopt an idea or behaviour new to the organisation, organisational learning, environment that supports innovation, and the ability to build and sustain competitive advantage. Some studies have also highlighted the importance of the relationships between organisational learning and innovation (Baker and Sinkula, 1999; Campo and Skerlavaj, 2009; Hage, 1999; Jimenez-Jimenez and Sanz-Valle, 2011; Nonaka and Takeuchi, 1995; Sajeva and Jucevicius, 2008) as well as organisational culture and innovation (Yang, 2007).

The general interest in examining these determinants has led to an increased understanding of innovativeness in organisations and has at the same time raised some questions and some gaps in understanding innovativeness, pointing to the need for further research. The influence of environmental uncertainties on innovativeness is an area which would benefit from further research in order to adequately understand the impacts. Freel (2005) reported that there is limited research examining the effects of environmental uncertainty on innovativeness. The question to be investigated relates to the specific dimensions of environmental changes and uncertainties, which organisations could respond to in order to stay innovative. The current research investigates how organisations can influence the innovativeness of their responses by studying the relationship between innovativeness and the environmental uncertainties in which the organisations operate.

Environmental uncertainty and complexity

There is a general recognition by researchers and managers that effective assessment of the external environment faced by the organisation and an appropriate response to the external environment is crucially important for continued organisational success (Galbraith, 2002). Internal functioning, including operational, tactical and strategic effectiveness, is understood as only part of the equation in an organisation's continued success while responsiveness to the external environment in which the organisation is functioning is considered to be another crucial part of the equation (Normann, 2001; Eisenhardt and Brown, 1998).

There have been several researchers who have studied and conceptualised the organisation's external environment and its impact on organisational performance (Hart and Banbury, 1994; Ansoff and Antoniou, 2005; Camillus, 1982). The stream of research which has most extensively conceptualised environmental turbulence is the Ansoffian strategic success paradigm (Ansoff, 1987). Ansoff was one of the first researchers to identify and define environmental turbulence and how it affects business performance. The general idea in this research stream is that businesses must assess the turbulence of the environment in which they operate and match their aggressiveness, capabilities, and responsiveness to the environmental turbulence. Ansoff provides five levels of environmental turbulence — repetitive, expanding, changing, discontinuous, and surprising — which need to be synchronised with an organisation's internal responses in order to be most effective. At one extreme, the 'repetitive environment' is one where nothing changes, is placid and there is absolute predictability. At the other extreme, the 'surprising environment' is characterised by major technological change and social political upheavals. Ansoff's idea of environmental turbulence essentially highlights that a firm's potential is optimum when the firm's behaviour, responses and capabilities match the turbulence of its environment.

Other research streams have also recognised the importance of environmental uncertainty as a driver of organisational behaviour and responses. Mintzberg (1990) critiqued that environmental uncertainty and complexity could not be predicted and thus organisational planning based on an analysis of environmental uncertainty would be difficult. However, the importance of environmental uncertainty *per se* in impacting the organisation's internal responses and behaviour is accepted in these research streams as well (Brews and Hunt, 1999; Mintzberg, 1990; Grant, 1996; Camillus, 1987; Emery and Trist, 1965).

The impact of environmental uncertainty and complexity has been studied in several different industries and contexts. For example, researchers have studied impacts of external environmental uncertainties and changes on product development speed and quality, leadership characteristics, top management team capability, new product development, ethical behaviours, and export plans (Buganza *et al.*, 2009; Olli *et al.*, 2004), Roger *et al.*, 2003, (Morris *et al.*, 1996). Human resource flexibility has been studied as an organisational orientation in response to external environmental uncertainty and dynamism (Ketkar and Sett, 2010). Innovativeness and the closely linked construct of entrepreneurial orientation have been studied as proactive responses and organisational orientations to an increasingly uncertain and complex external environment (Davis *et al.*, 1991).

In this research, we propose to operationalise the external environmental uncertainty of an organisation with the dimensions of competitive intensity, technological turbulence and market/demand turbulence.

Environmental uncertainty and innovativeness

Researchers believe that innovation is one of the important organisational responses to environmental complexity (Christensen, 1997). Prior studies have found that a firm's orientation towards innovation is embedded in the firm's internal as well as external environmental context (Levinthal, 2009; Levinthal and March, 1993).

As discussed in the previous sections, a firm's environment requires organisations to adapt themselves to environmental uncertainties and the threat of changes. In order to stay competitive and grow, organisations need to respond as well as counter these environmental changes, leading organisations to either adopt or generate a change involving their product, structure or process. As a result, environmental uncertainty and complexity often leads to organisational innovation in some form or another. Previous literature has found that the changing and dynamic nature of the environment triggers organisational innovation (Clarke, 1994; Starkey and McKinlay, 1988; Verdu-Jover, 2006). Some empirical evidence also exists linking organisational innovation to environmental uncertainty and changes in general (Daft, 1992; Duncan, 1972; Damanpour, 1996). Researchers have argued that environmental uncertainty would affect all types of organisational innovation ranging from externally focused to internally focused as well as those that change organisational form (Huber *et al.*, 1993; Damanpour, 1996). When the complexity changes and uncertainty in the external environment increases, the level of innovation orientation also tends to increase (Damanpour, 1996; Aiken and Hage, 1971; Nystrom *et al.*, 2002).

Based on discussions of previous research in the domain of external environmental uncertainty and an organisation's innovativeness, we propose the following main research hypothesis:

H1: *The greater the environmental uncertainty faced by an organisation, the higher will be the firm's innovativeness.*

Several factors relevant to assessing the external environmental uncertainty and complexity have been studied (Droge *et al.*, 2008; Song *et al.*, 2005; Grewal and Tansuhaj, 2001; Jaworski and Kohli, 1993). The influence of external environment on outbound innovations has been studied by Lichtenthaler with results indicating that the environmental context and specifically its competitive intensity positively influence the firm's open innovations (Lichtenthaler, 2009; Robertson and Gatignon, 1986). *Competitive intensity* is the degree of competition that a firm faces from organisations within its industry and competitive reference set (Grewal and Tansuhaj, 2001; Jaworski and Kohli, 1993). A firm's innovativeness is associated with the competitive intensity it faces thus we expect a positive relationship between competitive intensity and innovativeness.

H1a: *The greater the competitive intensity in an organisation's external environment, the higher will be the firm's innovativeness.*

A third dimension which has been studied for assessing the external environment of organisations is the nature of the market and demand faced by the organisation. *Market/demand turbulence* relates to the degree and pace of instability and uncertainty within a firm's markets and customers (Helfat *et al.*, 2007; Jaworski and Kohli, 1993). An environment characterised by market turbulence shows changes in the combination of customers and their preferences, and increasing market turbulence is characterised by rapidly changing styles, models, and short product lifetimes (Phromket and Ussahawannitchakit, 2009). Firms will have to respond to this rapidly changing market. Thus, we expect that as market turbulence increases, so does a firm's innovativeness.

H1b: *The greater the market/demand turbulence in an organisation's external environment, the higher will be the firm's innovativeness.*

Technological turbulence is another dimension studied in assessing the external environment of organisations. *Technological turbulence* refers to the degree and rate of technological change relevant to the products and processes of the organisation (Jaworski and Kohli, 1993). Firms in high-technology markets allocate greater resources to technology to cope with the uncertainty created by technological changes (Slater and Narver, 1994). So, often firms tend to be technology oriented to be innovative and to build competitive advantage in the environment characterised technological turbulence. Therefore, we expect that technological turbulence has a positive role on the firm's innovativeness.

H1c: *The greater the technological turbulence in an organisation's external environment, the higher will be the firm's innovativeness.*

Figure 1 shows that research model including dependent and independent variables and relationships between them. The construct of environment uncertainty has been assessed with three independent variables, competitive intensity,

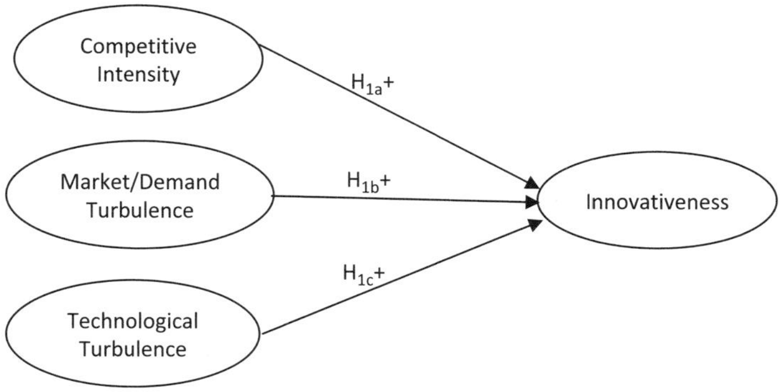

Fig. 1. The research model.

market/demand turbulence and technological turbulence while innovativeness of firms is the dependent variable. The model also expresses the hypothesised relationship that the dimensions of environmental uncertainty have positive effects on innovativeness, which is the dependant variable in this research.

Method

Sample and data collection

Data for this study were collected from the SMEs which are members of Eskisehir Chamber of Industry through a survey instrument. The study chose Eskisehir as a representative city in Turkey representing a diverse set of demographic, socio-economic and cultural characteristics. Moreover, Eskisehir is one of the largest and most cosmopolitan cities in Turkey. It is generally expected to represent a cross-section of Turkish professionals in respect to its demographic, socio-economic and cultural characteristics. Among the membership of the Chamber, SMEs having 10 to 250 employees were chosen for the research. The survey was mailed to 397 SMEs and 156 of the survey forms were returned. The respondents were top level managers in the sample SMEs. Data was analysed using the 156 usable responses and utilising SPSS 17.0 software. Table 1 shows the industry distribution profile of the SMEs in the sample. Majority of the sample consists of food products, textile, and wood products, paper products and furniture industry (34.6%), representing a wide representation of different industries and mirroring the national distribution of industries in Turkey.

Table 1. Distribution of industry affiliation of SMEs in sample.

Industries	Frequency	Percent
Food products industry	23	14.7
Textile	16	10.3
Machinery industry	15	9.6
Foundry	7	4.5
Wood products, paper products and furniture industry	5	3.2
Electrical machines	5	3.2
Agricultural machinery and equipments	4	2.6
Chemicals, rubber and plastic products industry	3	1.9
Minery and quarry	3	1.9
Metal goods industry	1	0.6
Other	74	47.4
Total	156	100.0

Measures

To measure innovativeness, we used a scale developed by Darroch and Jardine (2002). This scale contains 11 items and is scaled on a five-point Likert scale (1 = strongly disagree to 5 = strongly agree). To measure the three dimensions of the environment i.e., competitive intensity, market/demand turbulence, and technological turbulence, we adopted 10 items from Jaworski and Kohli (1993) for competitive intensity and technological turbulence and four items from Grewal and Tansuhaj (2001) for market/demand turbulence. The items of competitive intensity scale assessed the behaviour, competitor resources and ability of competitors to differentiate themselves. The market/demand turbulence scale assessed the stability of customer preferences which tend to change over time. Technological turbulence scale measured the extent of change and pace in technology relevant to the organisation. A five-point scoring format was used for all items (1 = strongly disagree; 5 = strongly agree).

Results

Arithmetic means of the items in innovativeness scale were between 3.15 and 3.93 except of two items (Table 2). That means that the degree of innovativeness of firms in the sample can be accepted as high. Arithmetic means of the items in the environmental uncertainties scale were between 3.34 and 4.17 for competitive intensity, 3.47 and 4.14 for demand turbulence, and 3.77 and 4.02 for

Table 2. Means and standard deviations.

Items	Mean	Std. deviation
Innovativeness		
INV1. We often improve or revise existing products or services	3.93	0.95
INV2. We often add new product or services to our existing ranges	3.90	0.92
INV3. We often introduce new ranges of products or services not previously offered by this company	2.78	1.12
INV4. We often reposition existing products or services.	3.28	1.08
INV5. We often change the way we make or deliver products or services.	2.98	1.13
INV6. We develop product or services that offer greater advantages to consumers than any other product or service currently available	3.41	1.11
INV7. We develop products or services that better meet the needs of consumers than any other product or service currently available	3.50	1.13
INV8. We have launched products or services that are the first of their kind in the world	3.15	1.16
INV9. We develop products or services that require consumers to substantially alter their behaviour	2.55	1.16
INV10. We have launched products or services for which this organisation lacks the technological knowledge	3.34	1.15
INV11. We have launched products or services for which this organisation lacks the business experience or knowledge	2.99	2.70
Competitive Intensity		
CI1. Competition in our industry is cutthroat	4.16	0.83
CI2. There are many 'promotional wars' in our industry	3.75	1.05
CI3. Anything that one competitor can offer, others can match readily	4.17	0.87
CI4. Price competition is the hallmark of our industry	3.99	0.84
CI5. One hears of new competitive moves almost everyday	3.54	1.01
CI6. Our competitors are relatively weak	3.34	1.02
Market/Demand Turbulence		
DT1. The demand of our customers varies a lot	4.14	0.82
DT2. In our industry the product and brand features vary a lot	4.13	0.90
DT3. In our industry the price/quality demanded by customers vary a lot	4.07	0.84
DT4. In our industry, customers often take unpredictable actions	3.47	1.18

Table 2. (*Continued*)

Items	Mean	Std. deviation
Technological Turbulence		
TT1. The technology in our industry is changing rapidly	3.98	0.79
TT2. Technological changes provide big opportunities in our industry	4.02	0.79
TT3. It is very difficult to forecast where the technology in our industry will be in the next two to three years	3.77	0.93
TT4. A large number of new product ideas have been made possible through technological breakthrough in our industry	3.78	0.94

INV3, INV9, INV11, DT1, DT4, CI1, CI2, CI3 were eliminated because of having the item total correlation lower than 0.40 which is the critical value based on reliability analyses
1. Strongly Disagree 5. Strongly Agree

technological turbulence. This means that the firms in the sample were functioning in environmental conditions with high turbulence, uncertainties, and changing rapidly over time.

Validity and reliability

Cronbach Alpha scores were calculated for each scale to test the reliability of the scales used in the study. Table 3 shows that the Cronbach Alpha scores for each scale were good with overall Alpha reported between 0.66–0.83. The original scale of innovativeness had Alpha scores between 0.78 and 0.86 (Darroch and Jardine, 2002). Alpha scores of the original environmental uncertainty scales were between 0.68 and 0.88 (Jaworski and Kohli, 1993). From the results of the reliability analysis, all items with a corrected-item-total correlation of less than 0.40 were eliminated. Three items in the innovativeness scale, two items in the demand uncertainty scale, and three items in the competitive intensity scale were dropped based on the reliability analysis.

Exploratory factor analysis using principle components of factor extraction and varimax rotation techniques was performed to examine the unidimensionality/convergence of each predefined multi-item construct. Factor analysis resulted in the eight independent items of the innovativeness scale being reduced to one factor/construct. The factor analysis of the environmental uncertainties with nine remaining items reduced to three factors which we analysed. The authors were in

Table 3. The results of factor and reliability analyses.

Items	Factor loadings	Cronbach's alpha	% Variance
Innovativeness (*KMO = 0.830)**			
INV7. We develop products or services that better meet the needs of consumers than any other product or service currently available	0.815	0.83	47.01
INV6. We develop product or services that offer greater advantages to consumers than any other product or service currently	0.788		
INV10. We have launched products or services for which this organisation lacks the technological knowledge	0.687		
INV2. We often add new product or services to our existing ranges	0.680		
INV1. We often improve or revise existing products or services	0.644		
INV5. We often change the way we make or deliver products or services	0.642		
INV8. We have launched products or services that are the first of their kind in the world	0.610		
INV4. We often reposition existing products or services	0.581		
Environmental Uncertainties (*KMO = 0.777)**			
Factor 1: Technological Turbulence			
TT1. The technology in our industry is changing rapidly	0.821	0.66	36.34
TT3. It is very difficult to forecast where the technology in our industry will be in the next two to three years	0.717		
TT2. Technological changes provide big opportunities in our industry	0.677		
TT4. A large number of new product ideas have been made possible through technological breakthroughs in our industry	0.668		
Factor 2: Competitive Intensity			
CI4. Price competition is the hallmark of our industry	0.481	0.70	14.56
CI5. One hears of new competitive moves almost everyday	0.806		

Table 3. (*Continued*)

Items	Factor loadings	Cronbach's alpha	% Variance
CI6. Our competitors are relatively weak	0.748		
Factor 3: Demand Uncertainty			
DT2. In our industry the product and brand features vary a lot	0.866	0.72	11.55
DT3. In our industry the price and quality demanded by customers vary a lot	0.734		

INV3, INV9, INV11, DT1, DT4, CI1, CI2, CI3 were eliminated because of having the item total correlation low than 0.40 based on reliability analyses.

KMO: Kaiser–Meyer–Olkin test to assess the appropriateness of using factor analysis on data.

agreement that the three factors can be usefully titled competitive intensity, demand turbulence, and technological turbulence. A cut-off loading of 0.40 was used and all factor loadings were above 0.50, exhibiting a high level of significance. The results from our factor analysis of the measurement items for each of the categories shows that measures used in this study have construct validity (Nunnally, 1978). Percent variances of innovativeness, competitive intensity, demand turbulence, and technological turbulence are 47.01%, 36.34%, 14.56%, and 11.55% respectively. This indicates that the sub-dimensions of the construct have good variances on the whole construct. To assess the degree of multi-colinearity for the factors in all the model is used the VIF (variation inflation factor) (Stapleton, 1995). It is seen that no problem of the multi-colinearity appears to be present in the study. Because the VIF values for the factors were among 1.32 and 1.73.

Hypotheses testing

Regression was used to test our research model and Table 4 shows the results of the regression analysis. This model has three independent variables (*competitive intensity, demand turbulence and technological turbulence*) and one dependent variable (innovativeness). The results of the regression analysis reveal that there is a meaningful effect of the environmental uncertainty on innovativeness of firms ($R = 0.573; F = 22.513; p < 0.001$). Demand turbulence ($\beta = 0.431; p < 0.001$) and technological turbulence ($\beta = 0.232$; $p < 0.001$) have a significant positive

Table 4. The results of regression analysis.

Independent variables	*B*	Std. error	β	*t*	*p*
(Constant)	0.563	0.383		1.470	0.144
Competitive intensity	0.025	0.083	0.023	0.299	0.766
Demand turbulence	0.425	0.077	0.431	5.511	0.000
Technological turbulence	0.268	0.091	0.232	2.953	0.004
$R = 0.573$ $R^2 = 0.329$					
F Change $= 22.513$ $p < 0.001$					
Dependent variable: Innovativeness.					

effect on innovativeness but the effects of the competitive intensity on innovativeness is not statistically meaningful.

These findings supported the research hypotheses (H_{1b} and H_{1c}). Therefore the main research hypothesis, the greater the environmental uncertainty, the higher will be firm's innovativeness is substantially accepted. The firms in our research sample experience higher levels of innovativeness when faced with high levels of technological turbulence and demand turbulence but do not necessarily see high levels of innovativeness when experiencing high levels of competitor intensity.

Discussion

Innovativeness has become one of the most important drivers to stay ahead of competition in today's marketplace. Therefore, researchers are increasingly studying innovation in an attempt to better understand its antecedents, moderators and impacts. Our research has viewed innovation as a critical means which businesses use for adapting themselves to changing, uncertain and complex external environmental conditions. The pace of change in today's business environment is very fast and businesses recognise that survival requires innovation as a means to survive and grow. So, it is important to examine the relationship between environmental changes and innovation. This study investigated the impacts of environmental uncertainties on firm innovativeness.

Results show that technological and market/demand turbulence have a significant impact on the innovativeness of the SME firms in Turkey. When firms face high technological and market/demand turbulence or uncertainties, they experience higher levels of firm innovativeness. High levels of technological turbulence in the environment tend to force firms to constantly assess and adopt new solutions for their businesses. This leads to opportunities for innovative breakthroughs in product or processes. High levels of market/demand turbulence tends to push the firm towards heightened customer responsiveness and most likely benefits the firm in the innovating means for improving the quality, features or price of their product offerings. In other words, SMEs which want to stay innovative and competitive must focus on being responsive to the environmental uncertainties.

Important managerial insights arise from these results. When an SME organisation is facing uncertain and complex external environments, it is likely that the internal mechanisms are poised for greater innovativeness. This would then be an ideal time for managers to further encourage innovation and reinforce messages which emphasise the importance and openness to innovation. In times of external environmental uncertainty, managers will benefit from the realisation that more aggressive championing and support for innovativeness is necessary. If not specifically promoted, then innovative responses may not naturally emanate from their organisations. Internal systems, policies and procedures which share innovative work, make innovation easy and coveted, and make innovation a priority would be essential during these times. Other research has also found that SME's which invest in high levels of internal R&D resources have high levels of innovation performance (Kang and Lee, 2008). Some may argue that encouraging innovativeness is equally important in calm and placid external environmental conditions. However, our results emphasise that in uncertain and turbulent conditions, the time is best for capitalising and leveraging the positive imperative and increased tendency for organisational innovativeness.

There are also important managerial implications from our findings on the lack of a relationship between competitive intensity and innovativeness. When SMEs are faced with increased competitive intensity, they do not necessarily respond by increased innovativeness. It may be that other responses such as price wars, marketing expenditure, and increased customer relationship tactics become the more natural priorities. This is an important result which managers can heed when faced with increased competitive intensity. A more careful study and analysis is needed to assess the most appropriate responses in situations when competitive intensity increases. It is interesting to note that other research has found similar results for the intensity of competition in that firm performance is positively impacted by higher levels of innovation, irrespective of the completion in the external environment (Gibb and Haar, 2010). It seems that entrepreneurs in the SME context recognise technology turbulence and market/demand turbulence as opportunities to be harnessed through internal innovation while competition is not viewed as a phenomenon which needs an innovative response.

Limitations and Future Research

While this research has made important contributions to the literature for understanding the relationship between external environmental uncertainties and organisational innovativeness, the study has some limitations. The data sample is from SMEs in Turkey and so the results strictly hold for the SME context and for the Turkish business environment in general. Future research should study SMEs from other cultural contexts in both developing as well as developed countries. A comparison of results would increase understanding on the cultural context of the results. In addition, the research studies three components of the external environment — competitive intensity, market/demand turbulence, and technological turbulence. There are other components of the external environment, such as governmental regulations, citizen demographics and political stability, which could also be included in future studies. In this research, control variables are not used. Future research could examine the effect of some control variables like firm demographics (firm size, age and, industry characteristic) on relationship between environmental factors and innovativeness. Finally, it would be a useful end goal to study a global model for innovation as the dependent variable which includes several other independent variables from previous literature in addition to the external environmental uncertainty.

In future research with our model, insights from our current study would provide us with interesting comparisons and managerial implications when designing innovation related interventions or training. In addition, Turkey and other Middle East nations have given special attention to innovativeness in recent years. Turkish businesses have realised that they need to be globally competitive in order to play in the same league as other developing but innovative countries such as India, China, and Korea.

References

Aiken, M and J Hage (1971). The organic organization and innovation. *Sociology*, 5, 63–82.

Amabile, TM (1988). A model of creativity and innovation in organizations. In *Research in Organizational Behavior*, BM Shaw and LL Cummings (Eds.), pp. 123–167. Greenwich, CT: JAI Press.

Ansoff, HI (1987). The emerging paradigm of strategic behavior. *Strategic Management Journal*, 8, 501–515.

Ansoff, HI and P Antoniou (2005). *The Secrets of Strategic Management: The Ansoffian Approach*. UK: Booksurge, LLC.

Baker, W and JM Sinkula (1999). Learning orientation, market orientation, and innovation: Integrating and extending models of organizational performance. *Journal of Market Focused Management*, 4(4), 295–308.

Baker, WE and JM Sinkula (2002). Market orientation, learning orientation and product innovation: Delving inside the organization's black box. *Journal of Market Focused Management*, 1.

Brews, PJ and MR Hunt (1999). Learning to plan and planning to learn: Resolving the planning school/learning school debate. *Strategic Management Journal*, 20(10).

Buganza, T, C Dell'Era and R Verganti (2009). Exploring the relationships between product development and environmental turbulence: The case of mobile TLC services. *Journal of Product Innovation Management*, 26(3), 308–321.

Camillus, J (1982). Reconciling logical incrementalism and synoptic formalism: An integrated approach to designing strategic planning processes. *Strategic Management Journal*, 3, 277–283.

Camillus, LC (1987). Contrasting planning systems in high, medium and low performance companies. *Journal of Management Studies*, 24(4), 363–385.

Campo, JDSPGC and M Skerlavaj (2009). The impact of organizational learning on innovativeness in Spanish companies. In *Proc. WSKS*, MD Lytras *et al.* (Eds.), pp. 67–76. New York: Springer.

Christensen, CM (1997). *The Innovator's Dilemma*. Boston, MA: Harvard Business School Press.

Clarke, L (1994). *The Essence of Change*. Englewood Cliffs, NJ: Prentice-Hall.

Cooper, AC (1973). Technical entrepreneurship: What do we know? *Research and Development Management*, 3(2), 59–64.

Daft, RL (1982). Bureaucratic versus nonbureaucratic structure and the process of innovation and change. In *Research in the Sociology of Organizations*, SB Bacharach (Ed.), pp. 129–166. Greenwich. GT: JAI Press.

Daft, RL (1992). *Organizational Theory and Design*. St Paul, MN: West Publishing Company.

Damanpour, F (1991). Organizational innovation: A meta-analysis of effects of determinants and moderators. *The Academy of Management Journal*, 34(3), 555–590.

Damanpour, F (1992). Organizational size and innovation. *Organization Studies*, 13(3), 375–402.

Damanpour, F (1996). Organizational complexity and innovation: Developing and testing multiple contingency models. *Management Science*, 42(5), 693–716.

Damanpour, F and WM Evan (1984). Organizational innovation and performance: The problem of organizational lag. *Administrative Science Quarterly*, 29(3), 392–409.

Damanpour, F and JD Wischnevsky (2006). Research on organizational innovation: Distinguishing innovation-generating from innovation-adopting organizations. *Journal of Engineering and Technology Management*, 23, 269–291.

Damanpour, F and M Schneider (2006). Phases of the adoption of innovation in organizations: Effects of environment, organization and top managers. *British Journal of Management*, 17, 215–236.

Darroch, J and A Jardine (2002). Combining firm-based and consumer-based perspectives to develop a new measure for innovation. In *3rd Int. Conf. Management of Innovation and Technology*, Hangzhou City, China.

Davis, D, M Morris and J Allen (1991). Perceived environmental turbulence and its effect on selected entrepreneurship, marketing, and organizational characteristics in industrial firms. *Journal of the Academy of Marketing Science*, 19(1), 43–51.

Downs, GW and LB Mohr (1976). Conceptual issues in the theory of innovation. *Administrative Science Quarterly*, 21, 700–714.

Droge, C, R Calantone and N Harmancioglu (2008). New product success: Is it really controllable by managers in high turbulent environments? *Journal of Product Innovation Management*, 25(3), 272–286.

Duncan, RB (1972). Characteristics of organizational environments and perceived environmental uncertainty. *Administrative Science Quarterly*, 17, 313–327.

Ellonen, R, K Blomqvist and K Puumalainen (2008). The role of trust in organizational innovativeness. *European Journal of Innovation Management*, 11(2), 160–181.

Emery, E and EL Trist (1965). The causal texture of organizational environment. *Human Relations*, 18, 21–32.

Eisenhardt, K and S Brown (1998). *Competing on the Edge: Strategy as Structured Chaos*. Cambridge, MA: Harvard Business School Press.

Ettlie, JE, WP Bridges and RD O'Keefe (1984). Organization strategy and structural differences for radical versus incremental innovation. *Management Science*, 30, 682–695.

Farley, JU, S Hoenig and Z Ismail (2008). Organizational culture, innovativeness, market orientation and firm performance in South Africa: An interdisciplinary perspective. *Journal of African Business*, 9(1), 59–76.

Freel, MS (2005). Perceived environmental uncertainty and innovation in small firms. *Small Business Economics*, 25, 49–64.

Galbraith, J (2002). *Designing Organizations, An Executive Guide to Strategy, Structure, and Process*. San Francisco: Jossey Bass.

Garcia-Morales, VJ, FJ Llorens-Montes and AJ Verdu'-Jover (2006). Antecedents and consequences of organizational innovation and organizational learning in entrepreneurship. *Industrial Management & Data Systems*, 106(1), 21–42.

Germain, R (1996). The role of context and structure in radical and incremental logistics innovation adoption. *Journal of Business Research*, 35, 117–127.

Gibb, J and JM Haar (2010). Risk taking, innovativeness and competitive rivalry: A three-way interaction towards firm performance. *International Journal of Innovation Management*, 14(5), 871–891.

Gopalakrishnan, S and F Damanpour (1997). Innovation research in economics, sociology, and technology management. *Omega*, 25, 15–28.

Grant, RM (1996). Prospering in dynamically competitive environments: Organizational capability as knowledge integration. *Organization Science*, 7(4), 375–387.

Grewal, R and P Tansuhaj (2001). Building organizational capabilities for managing economic crisis: The role of market orientation and strategic flexibility. *Journal of Marketing*, 65(2), 67–80.

Hage, J and M Aiken (1970). *Social Change in Complex Organizations*. New York: Random House.

Hage, JT (1999). Organizational innovation and organizational change. *Annual Review of Sociology*, 25, 597–622.

Hart, S and C Banbury (1994). How strategy-making processes can make a difference. *Strategic Management Journal*, 15(4), 251–269.

Helfat, C, S Finkelstein, W Mitchell, M Peteraf, H Singh, D Teece and S Winter (2007). *Dynamic Capabilities: Understanding Strategic Change in Organisations*. Malden: Blackwell Publishing.

Henderson, RM and KB Clark (1990). Architectural innovation: The reconfiguration of existing product technologies and the failure of established firms. *Administrative Science Quarterly*, 35, 9–30.

Huber, GP, KM Sutcliffe, CC Miller and WH Glick (1993). Understanding and predicting organizational change. In *Organizational Change and Redesign*, GP Huber and WH Glick (Eds.). New York: Oxford University Press.

Hurley, R and GTM Hult (1998). Innovation, market orientation, and organizational learning: An integration and empirical examination. *Journal of Marketing*, 62, 42–54.

Jaworski, BJ and AK Kohli (1993). Market orientation: Antecedents and consequences. *Journal of Marketing*, 57(3), 53–71.

Jimenez-Jimenez, D and R Sanz-Valle (2011). Innovation, organizational learning, and performance. *Journal of Business Research*, 64, 408–417.

Kang, KN and YS Lee (2008). What affects the innovation performance of small and medium sized enterprises (SMEs) in the biotechnology industry? An empirical study on Korean biotech SMEs. *Biotechnology Letters*, 30, 1699–1704.

Kelly, L and R Kumar (2009). Impact of knowledge management on the self-efficacy entrepreneurs: Mexican SME context. *International Journal of Management and Decision Making*, 10(1/2), 111–124.

Ketkar and PK Sett (2010). Environmental dynamism, human resource flexibility, and firm performance: Analysis of a multi-level causal model. *International Journal of Human Resource Management*, 21(8), 1173–1206.

Kimberly, JR and MJ Evanisko (1981). Organizational innovation: The influence of individual, organizational and contextual factors. *Academy of Management Journal*, 24(4), 689–713.

Klein, KJ and JS Sorra (1996). The challenge of innovation implementation. *Academy of Management Review*, 21(4), 1055–1080.

Kumar, R and C Uzkurt (2010). Can innovativeness be impacted by interpersonal influences and do cultural dimensions play a moderating role? An investigation in the Turkish context. *The Journal of Global Business Management*, 6(2).

Kumar, R and L Kelly (2006). Self-efficacy, social and cultural issues in designing online technology skills transfer programs: A Mexican context. *Journal of Information Systems and Technology*.

Levinthal, DA and JG March (1993). The myopia of learning. *Strategic Management Journal*, 14, 95–112.

Lichtenthaler, U (2009). Outbound open innovation and its effect on firm performance: Examining environmental influences. *RD Management*, 39(4), 317–330.

Lin, CH, CH Peng and DT Kao (2008). The innovativeness effect of market orientation and learning orientation on business performance. *International Journal of Manpower*, 29(8), 752–772.

Liu, Y and JS Phillips (2011). Examining the antecedents of knowledge sharing in facilitating team innovativeness from a multilevel perspective. *International Journal of Information Management*, 31, 44–52.

Luk, CL, OHM Yau, LYM Sin, ACB Tse, RPM Chow and JSY Lee (2008). The effects of social capital and organizational innovativeness in different institutional contexts. *Journal of International Business Studies*, 39, 589–612.

Man, MMK (2010). The relationship between distinctive capabilities, innovativeness strategy types and the export performance of small and medium-size enterprises (SMEs) of Malaysian manufacturing sector. *International Journal of Management and Innovation*, 2(1), 15–30.

Martinez-Sanchez, A, MJ Vela-Limenez, M Perez-Perez and P Luis-Carnicer (2009). Interorganizational cooperation and environmental change: Moderating effects between flexibility and innovation performance. *British Journal of Management*, 20, 537–561.

Menguc, B and S Auh (2006). Creating a firm-level dynamic capability through capitalizing on market orientation and innovativeness. *Journal of the Academy of Marketing Science*, 34(1), 63–73.

Mintzberg, H (1990). The design school: Reconsidering the basic premises of strategic management. *Strategic Management Journal*, 11, 171–195.

Montes, FJL, AR Moreno and LMM Fernández (2004). Assessing the organizational climate and contractual relationship for perceptions of support for innovation. *International Journal of Manpower*, 25(2), 167–180.

Morris, MH, A Marks, JA Allen and S Newman (1996). Modeling ethical attitudes and behaviors under conditions of environmental turbulence: The case of South Africa. *Journal of Business Ethics*, 15(10).

Mudrak, T, AV Wagenberg and E Wubben (2005). Innovation process and innovativeness of facility management organizations. *Facilities*, 23(3/4), 103–118.

Musteen, M, VL Barker and VL Baeten (2010). The influence of CEO tenure and attitude toward change on organizational approaches to innovation. *The Journal of Applied Behavioral Science*, 46, 360–387.

Nonaka, I and H Takeuchi (1995). *The Knowledge-Creating Company*. New York: Oxford University Press.

Normann, R (2001). *Reframing Business: When the Map Changes the Landscape*. Chichester: John Wiley and Sons.

Nunnally, J (1978). *Psychometric Theory*, 2nd Ed. New York: McGraw-Hill.

Nystrom, P, K Ramamurthy and A Wilson (2002). Organizational context, climate and innovativeness: Adoption of imaging technology. *Journal of Engineering and Technology Management*, 19(3–4), 221–247.

Olli, K, S Sundqvist, K Puumalainen and JW Cadogan (2004). The effect of environmental turbulence and leader characteristics on international performance: Are knowledge-based firms different? *Canadian Journal of Administrative Sciences*, 21(1), 35–50.

Porter, M (1985). *Competitive Advantage: Creating and Sustaining Superior Performance*. New York: Free Press.

Phromket, C and P Ussahawannitchakit (2009). Effects of organizational learning effectiveness on innovation outcomes and export performance of garments business in Thailand. *International Journal of Business Research*, 9(7), 6–31.

Ravichandran, T (2000). Swiftness and intensity of administrative innovation adoption: An empirical study of TQM in information systems. *Decision Sciences*, 31(3), 691–724.

Read, A (2000). Determinant of successful organizational innovation: A review of current research. *Journal of Management Practice*, 3(1), 95–119.

Robertson, TS and H Gatignon (1986). Competitive effects on technology diffusion. *Journal of Marketing*, 50(3), 1–12.

Rogers, EM (1995). *Diffusion of Innovations*, 4th Ed. New York: Free Press.

Sajeva, S and R Jucevicius (2008). Empowering innovations in organizations: Linking knowledge management and organizational innovativeness. *Social Sciences (Socialiniai Mokslai)*, 1(59), 50–57.

Salavou, H (2004). The concept of innovativeness: Should we need to focus? *European Journal of Innovation Management*, 7(1), 33–44.

Santos-Rodrigues, H, PF Dorrego and CF Jardon (2010). The influence of human capital on the innovativeness of firms. *The International Business & Economics Research Journal*, 9(9), 53–63.

Slater, SF and JC Narver (1994). Does competitive environment moderate the market orientation-performance relationship? *Journal of Marketing*, 58, 46–55.

Song, M, VDH Bij and M Weggeman (2005). Determinants of the level of knowledge application: A knowledge-based and information-processing perspective. *The Journal of Product Innovation Management*, 22, 430–444.

Stapleton, JH (1995). *Linear Statistical Models*. New York: John Wiley & Sons.

Starkey, K and A McKinlay (1988). *Organisational Innovation*. Aldershot: Avebury.

Subramanian, A and S Nilakanta (1996). Organizational innovativeness: Exploring the relationship between organizational determinants of innovation, types of innovations, and measures of organizational performance. *Omega International Journal of Management Science*, 24(6), 631–647.

Thompson, VA (1965). Bureaucracy and innovation. *Administrative Science Quarterly*, 5, 1–20.

Tidd, J, J Bessant and K Pavitt (2005). *Managing Innovation: Integrating Technological, Market, and Organizational Change*, 3rd Ed. New York: John Wiley & Sons.

Tushman, ML, Anderson and CA O'Reilly (1997). *Winning Through Innovation: A Practical Guide to Leading Organizational Change and Renewal*. Boston, MA: Harvard Business School Press.

Verdu-Jover, AJ (2006). Antecedents and consequences of organizational innovation and learning in entrepreneurship. *Industrial Management and Data Systems*, 106(1–2), 21–42.

Wang, CL and PK Ahmed (2004). The development and validation of the organizational innovativeness construct using confirmatory factor analysis. *European Journal of Innovation Management*, 7(4), 303–313.

Xie, XM, SX Zeng and CM Tam (2010). Overcoming barriers to innovation in SMEs in China: A perspective based cooperation network. *Innovation: Management, Policy & Practice*, 12(3), 298–310.

Yang, JT (2007). Knowledge sharing: Investigating appropriate leadership roles and collaborative culture. *Tourism Management*, 28, 530–543.
Zaltman, G, R Duncan and J Holbeck (1973). *Innovations and Organizations*. New York: John Wiley & Sons.

SME Practices for Innovation

Chapter 8

A MODEL OF ORGANIZATIONAL INNOVATION IMPLEMENTATION EFFECTIVENESS IN SMALL TO MEDIUM FIRMS

Sukanlaya Sawang

QUT Business School,
Queensland University of Technology, Australia

Kerrie L. Unsworth

UWA Business School,
University of Western Australia, Australia

The present study aims to validate the current best-practice model of implementation effectiveness in small and mid-size businesses. Data from 135 organizations largely confirm the original model across various types of innovation. In addition, we extended this work by highlighting the importance of human resources in implementation effectiveness and the consequences of innovation effectiveness on future adoption attitudes. We found that the availability of skilled employees was positively related to implementation effectiveness. Furthermore, organizations that perceived a high level of benefits from implemented innovations were likely to have a positive attitude towards future innovation adoption. The implications of our improvements to the original model of implementation effectiveness are discussed.

Keywords: Innovation management; implementation; innovation effectiveness; SMEs.

Introduction

Research on innovation within organizations has focused predominantly on the adoption phase (Drury and Farhoomand, 1999) namely the decision by an organization to make use of an innovation (Rogers, 1995). However, the adoption

This paper was first published in *International Journal of Innovation Management*, Vol. 15, No. 5 (October 2011), pp. 989–1011. © Imperial College Press. DOI: 10.1142/S1363919611003398 decision is only the beginning of the innovation process. The process can only be considered a success when the innovation is accepted and implemented by organizational members and the organization perceives benefits or some improvement as a result (Bhattacherjee, 1998). Very little research has examined the factors that help to successfully implement an innovation (Holahan *et al.*, 2004), so our research was designed to build on and extend this burgeoning domain.

Following previous literature, we define innovation broadly as ideas, systems, technologies, products, processes, services, or policies that are new to the adopting unit (Zaltman *et al.*, 1973). This study focuses on the implementation stage of the innovation and includes activities such as training and support programs for organizational members who are expected to use the innovation. We focus our research on organizational characteristics that are related to successful innovation implementation, rather than system design characteristics related to user acceptance (Davis, 1989).

An Integrated Model of Implementation Effectiveness

A number of researches that were conducted in the area of innovation implementation were generally qualitative case studies with all their attendant limits on generalizable knowledge (Scarbrough, 2003; Storey and Barnett, 2000). Klein *et al.* (2001) proposed the implementation effectiveness model that is based on previous case study literature and suggests that organizational differences in *innovation* effectiveness (perceived benefits from implemented innovation) is related to *implementation* effectiveness (how smoothly the implementation went); and that implementation effectiveness is significantly related to organizational support, financial resource availability, policies and practices, and climate.

Klein *et al.*'s (2001) study has received considerable attention in academic circles including seven empirical papers that applied or modified specific aspects of their final model. These empirical papers integrated and examined some of the posited relationships in the implementation effectiveness model but none have re-examined the full model of implementation effectiveness. Two studies (Holahan *et al.*, 2004; Naveh and Marcus, 2004) examined the path between implementation climate and implementation effectiveness. As with the findings of Klein *et al.* (2001); Holahan *et al.* (2004) found that implementation climate was positively related to implementation effectiveness of computer and telecommunication technologies among 164 K-12 schools in New Jersey, USA. Naveh and Marcus's study (2004) used data from two general hospitals in USA. They also found that implementation climate influenced effective implementation of patient safety practice.

Two other studies (Alexander *et al.*, 2006; Link and Naveh, 2006) have examined the relationship between implementation effectiveness and innovation effectiveness. Alexander *et al.* (2006) conducted a survey among 1,784 community hospitals and, unlike Klein *et al.* (2001), found that successful implementation of quality improvement practice improved financial and cost performance. Link and Naveh (2006) surveyed 40 organizations that implemented ISO 14001 — a standard for environmental management. Similarly, they found that comprehensive implementation of ISO 14001 increased organizational performance and benefits.

Some studies selected the relationship between implementation policies and practices and implementation effectiveness to be their research question. Weiner *et al.* (2007) conducted multiple case studies among six primary care practices in North Carolina, USA. They found that providing policies and practices, such as training, influenced the effective implementation of prevention efforts-diabetes management strategies among healthcare practitioners. Marler *et al.* (2006) surveyed 94 administrative employees and confirmed the relationship. They concluded that implementation policies and practices facilitated the successful implementation of web-based enterprise-wide resource planning software system.

A study from Helfrich *et al.* (2007) tested a slightly trimmed version of Klein *et al.*'s (2001) original model of implementation effectiveness. They conducted interviews with four cancer clinical research networks. Their findings indicated that the original model of implementation effectiveness explained the effective implementation of new programs in cancer prevention and control very well. However, they did not observe a significant relationship between implementation effectiveness and innovation effectiveness.

The intent of the current study is to enhance the original theoretical model from Klein *et al.*'s (2001) study. Although a number of studies applied the original model of implementation effectiveness for their investigation, none of them tested empirically the full original model. Therefore, the first aim in this study is restricted to re-examining the model of implementation effectiveness. The second aim of this study then is to enhance the model of effective implementation, which will be discussed in the later section.

Relationships among variables based on the original model of implementation effectiveness

Financial resources availability and implementation policies and practices

To engage people in the implementation process and using an innovation, organizations should provide supportive schemes, such as training, rewards or technical support. Schuppan (2006) found training had an important role in successful E-governance implementation. Furthermore, supportive factors for elementary teachers' use of computers were identified such as technological accessibility and availability, incentives to use and personnel support (Tranklin, 2007). A quantitative finding in the Australian construction industry indicated that barriers to successful implementation of information communication technology were a lack of training and a difficulty in finding time to participate in the implementation process (Peansupap and Walker, 2006). In the implementation effectiveness model, these supportive schemes (e.g., training) were defined as implementation policies and practices.

Organizations providing supportive policies and practices can incur substantial financial cost. In the absence of slack financial resources, an organization may have considerable difficulty in offering policies and practices for implementation (Ramsey *et al.*, 2008). For example, in the banking industry, innovation implementation was most successful in banks that had sufficient financial resources to offer training, to hire consultants, and to lower organizational performance standards during the implementation effort (Nord and Tucker, 1987). The education sector also faced a similar problem. Schrum and Glassett (2006) reviewed research on the integration of computer technologies by teachers and other educational leaders in the P-12 school environment. They identified barriers that inhibited the successful implementation of technology into classroom instruction and found that limited financial resources inhibited support for using technology within schools. Similarly, empirical findings from small to medium sized firms showed that one of the barriers to providing e-learning training for employees was financial resources (Sambrook, 2003). Helfrich *et al.* (2007) adapted Klein *et al.*'s (2001) hypothesized relationship between financial resources availability and implementation policies and practices to be their research question. Interview results from top management in four cancer clinical research centers indicated that the organizations had adequate funding resource for implementing a new program in cancer prevention and control (CP/C). Interviewees also reported that their clinic established a variety of policies and practices (such as organizing dedicated CP/C research committees) to encourage researchers to participate in CP/C implementation. As such, an adequate budget can improve implementation activities.

Top management support and implementation policies and practices

Financial resources availability may permit an organization to bear the cost of implementation and absorb failure (Rosner, 1968). However, financial resources availability alone may not be sufficient to support implementation policies and practices. Senior management could play a role as facilitators and endorse implementation activities (van der Panne *et al.*, 2003). The support from senior management refers to the degree to which senior management views the implementation activities as a top priority and critical to organizational effectiveness (Jarvenpaa and Ives, 1991). Helfrich *et al.*'s (2007) case studies concluded that senior management signaled their support for CP/C research through specific implementation policies and practices. Findings from an implementation of a client/server computing system at an insurance company in United Kingdom revealed that senior management was highly supportive of the implementation activities (Subramanian and Lacity, 1997). They recognized the successful implementation would enable the business transformation. Furthermore, the senior management was involved with the implementation budget, training approval, and technology maintenance support. As such, top management support could influence implementation policies and practices.

Implementation policies and practices and implementation climate

Climate was initially considered as the general situation that is experienced by individuals in terms of the values or characteristics of the environment (Tagiuri and Litwin, 1968) that influence individual behavior. In the 1980s, the concept of climate was transferred to large units such as organizations, rather than indicating individual emotional reaction. In this context, climate is described as arising from routine organizational practices that influence members' behavior and attitudes (Hoy and Miskel, 1991). Although the terms "climate" and "culture" are often used interchangeably, they are different concepts. While culture in organizations refers to the shared assumptions and values of group members (Moran and Volkwein, 1992), climate refers to shared perceptions of organizational policies, practices, and procedures, both formal and informal (Reichers and Schneider, 1990). Climate is therefore a surface-level indicator of the deeper, more embedded organizational culture.

It is possible for multiple climates to exist concurrently within an organization. Therefore, climate is best defined as a specific construct having a referent

(Schneider *et al.*, 1998). That is, a "climate" is actually a climate *for* something, for example climate for creativity (Ekvall, 1996), climate for workplace safety (Griffin and Neal, 2000), or climate for service (Schneider *et al.*, 1998). In the current study, we examined the climate for implementation. Climate for implementation in this study refers to *managerial perceptions of the extent to which organizational members support the implementation activities*. Given that senior management deliver the importance of the implementation message to organizational members through the endorsement of various policies and practices, the members should perceive the implementation as a top priority.

Implementation climate and implementation effectiveness

Although there is no direct research for this link, related research suggests that organizations that view changes positively are more likely to make those changes smoothly. Martin *et al.* (2005) conducted research in two large public organizations, and found that organizational members' positive perceptions of a restructuring process fostered effective implementation of that restructuring. This suggests that implementation climate may affect implementation. Similar outcomes were found in a study of a successful merger between two non-profit organizations. Giffords and Dina (2003) found that the success of the merger was influenced by organizational climate. On a different but related note, Griffin and Neal (2000) studied the climate for safety in seven Australian manufacturing and mining organizations and found that safety climate was an important predictor of successful safety performance.

Implementation effectiveness and innovation effectiveness

Some innovation research has defined the outcome of innovation implementation as that of a simple, unproblematic process with decreasing resistance among organizational members. For example, Amoako-Gyampah and Salam (2004) defined the implementation outcome as acceptance of ERP systems among target users. They found that training and project communication influenced 571 employees to accept the use of an ERP system. Similarly, Johnston and Linton's (2000) research in the manufacturing industry defined the point of successful implementation of environmentally clean process technology as the time at which firms incorporated the environmental technology into their operation. They found that the inter-firms network facilitated a completed technology implementation.

However, several researchers have distinguished between implementation effectiveness, and innovation effectiveness (e.g., Klein *et al.*, 2001; Klein and

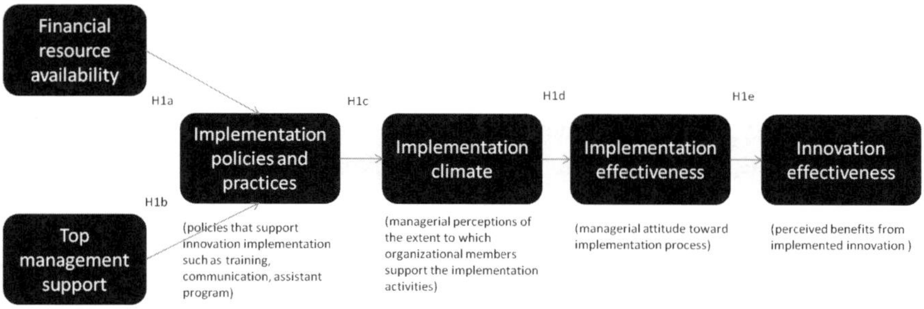

Fig. 1. The proposed hypotheses for effective implementation.

Knights, 2005; Klein and Sorra, 1996). Innovation effectiveness refers to the organizational realization of benefits from the adopted innovation and can be seen as a function of implementation effectiveness, that is a smooth process (e.g., fewer problems during implementation or a less complicated implementation process) and organizational members' acceptance. Accordingly, with a less complicated implementation process and less resistance among organizational members, the greater the perceived benefits of innovation (innovation effectiveness) should be.

According to the previous literature review, Figure 1 displays our proposed hypotheses.

H1a-e: *Availability of general financial resources and top management support for innovation are positively and significantly related to implementation policies and practices; implementation policies and practices are positively and significantly related to implementation climate; implementation climate is positively and significantly related to implementation effectiveness; and implementation effectiveness is positively and significantly related to innovation effectiveness.*

Extending the Model of Implementation Effectiveness

The original model of implementation effectiveness proposed that financial resources availability could indirectly affect innovation implementation effectiveness. However, numerous authors suggest human resource factors may also affect the implementation of innovation. Therefore, there is a potential to enhance the original model of implementation effectiveness by including separate treatment of human resource factors.

Nystrom *et al.* (2002) examined the implementation of the imaging technology among 555 hospitals from Wisconsin, Minnesota, and Illinois states, USA.

The results indicated that organizational resources availability, which defined resources as both financial resources and human resources, influenced the innovation implementation. However, the authors did not distinguish between financial and human resources in their analysis. Although the resources availability affected the effective implementation of the imaging technology, the study did not draw a clear conclusion as to whether financial or human resources would probably have differential impacts on the effective implementation.

Nevertheless, a number of studies have indicated that skilful and competent employees are a key to effective implementation of technological innovation (e.g., Dooley *et al.*, 2002; Snell and Dean, 1992). Implementing technological innovation can improve organizational performance. Effective implementation requires higher average skills from organizational members to manage the implementation process (Spenner, 1983). Furthermore, skilful and talented employees will perhaps adapt themselves to the change process more easily. A study of relocation within a State government department in the Queensland Public Service indicated that competent and confident employees viewed the relocation as an opportunity rather than as a threat, thus they were more willing to participate in the change (Jimmieson *et al.*, 2004). Likewiese, Starkweather (2005) commented that talented and competent K-12 teachers well managed activities that promoted the successful implementation of technology, innovation, design, and engineering curriculum.

Implementing new technologies or practices may enhance work effectiveness, however, it may require more skills and capabilities from organizational members, rather than an unskilled workforce, to deal with the new technologies. For instance, organizations can provide supportive training of how to use computerized bookkeeping. However, if most employees have a low level of computer literacy, the training may be ineffectual and could possibly create resistance to using the technology. On the other hand, computer literate employees may adjust themselves to the new technology more smoothly and have fewer problems. As such, having skilful and talented personnel is likely to lead to a more successful implementation of innovation. Thus, we propose that:

H2: *Human resources availability is positively and significantly related to implementation effectiveness.*

Future Innovation Adoption

Finally, we suggest that innovation effectiveness is not necessarily the end-point for understanding innovation in organizations. We propose a more complex scenario where a positive perception of the benefits of implemented innovations (innovation effectiveness) will have knock-on effects to future innovation adoption. Many innovation researchers (e.g., Damanpour, 1991; Lehman *et al.*, 2002) have identified positive beliefs and motivational readiness as facilitators of adopting new innovations. Furthermore, at an individual level, positive attitudes towards technology are significantly related to system usage (Davis, 1993). We propose that much of this positive attitude will come from past experiences with innovation. This suggestion is supported by a study of the implementation of organizational websites by 288 members of a Chamber of Commerce in the USA (Flanagin, 2000). The research suggested that the perceived benefit from technology was one of the best predictors of future innovation adoption. Likewise, a study of 298 companies in Hong Kong indicated that perceived benefits were positively related to attitudes towards adoption (Au and Enderwick, 2000). On the basis of these findings, we propose that perceiving greater innovation effectiveness with current innovations will correspond to a more positive overall attitude towards future innovation adoption within an organization.

H3: *Innovation effectiveness is positively and significantly related to organizational attitudes towards future innovation adoption.*

Method

Sample

This study is part of a wider program of research on the organizational innovation process. A sample of 750 firms was randomly selected from the Australian Business Register and the contact list from a technology diffusion agency. Our level of analysis was the organizational unit — in our case, the firm itself — and not the innovation, which is similar to the work of Klein *et al.* (2001). This allows us to test the validity of the model for multiple implementations. As the innovation implementation process usually involves top management, senior managers were considered to be knowledgeable about innovation projects within the organization and questionnaires were mailed directly to them.

We received responses from 135 organizations (18% response rate). Although the response rate is relatively low compared to individual-level research, it is comparable to many other surveys conducted at the organizational-level (e.g., Cycyota and Harrison, 2006; Moreno-Luzón and Begoña Lloria, 2008). The definition of an SME differs across countries and has varied over time (McAdam *et al.*, 2004). The most convenient and widely used to categorize SMEs is the number of employees (Jones, 2003). For the purpose of this study, using the

Australian Bureau of Statistics' definition, we defined SMEs as the firms that have fewer than 200 employees or fewer. The majority (104 companies, 77%) had 100 employees or less. Over half (54%) were in the manufacturing industry, while the remainder (46%) were across other industries (e.g., construction and pharmaceuticals).

To ensure that the level of innovation implementation was varied, we asked the organization to identify the number of innovation that they introduced in the last three years. We then also asked the organization to provide the detail of their innovation. We categorized innovation into (a) product innovation;[1] (b) process innovation;[2] and (c) management innovation.[3] The majority of respondents reported that they implemented 15 of product innovation (87.80%), 5 of process innovation (92.53) and 5 of management innovation (97.71%). We also asked the respondents to what extent were these innovations radically different from what the organisation had or did before. On scale of 1 (not at all) to 5 (a great deal), we found that the organizations perceived that their process innovation was relatively radical (Mean = 3.63), and to some extent that their process (Mean = 3.27) and management (Mean = 3.11) innovations were radical.

Measures

As this research was part of Australian Research Council (ARC) industry linkage research project (LP 0455129: Organizational innovation adoption: The effect of external, technology diffusion agencies), there were a number of additional items included in the questionnaire that were constructed by other researchers to gather data designed to address other research questions. A lengthy questionnaire may possibly reduce the response rate, therefore we had selected some items from each construct to reduce the questionnaire length. To confirm the reliabilities and validities of those selected items, the Confirmatory Factor Analysis was employed to ensure adequate item loadings on each construct.

Financial resources availability. Four items from Klein *et al.* (2001) were used to measure financial resource allocation within the company, e.g., "*Money is readily available to pay for special projects in the organization.*" A confirmatory factor analysis of all measures, however, showed that one item from this scale ("*Our organization is performing well relative to our competitors*") cross-loaded.[4]

[1]New product/service development or improvement.
[2]Processing or manufacturing improvement.
[3]New and/or improved business management practices.
[4]Details of the confirmatory factor analysis can be obtained from the first author.

It was therefore deleted from further analysis and the remaining three items displayed an acceptable estimate of internal reliability of 0.76.

Top management support. Three items from Klein *et al.* (2001) measured the extent to which top management supports and is committed to the implementation process. This was referenced to innovation within the organization within the last three years. An example item is "*Innovation implementation is generally carefully planned and costed.*" It had an acceptable level of internal reliability ($\alpha = 0.77$).

Implementation policies and practices. Eight questions from Klein *et al.* (2001) asked individuals to indicate the extent their organization endorsed policies and practices such as training, rewards or incentives, innovation assistance, time for participating in innovation implementation, and communication about innovation implementation (Klein *et al.*, 2001). The confirmatory factor analysis, however, showed that two items from this scale cross-loaded. Those items were therefore deleted from further analysis and the remaining six items displayed an acceptable estimate of internal reliability of – 0.81.

Implementation climate. Three items from Klein *et al.* (2001) explored shared perception of *managerial expectations* of the extent to which employees supported the implementation of innovation. An example item was "*If employees can avoid using the innovation, they do*" ($\alpha = 0.92$).

Implementation effectiveness. Our measure of implementation effectiveness was developed by Klein and colleagues as implementation smoothness (Klein, personal communication). Although Klein *et al.* (2001) focused their paper on extent and quality of MRP use this was too specific for a generalizability study. Therefore, we decided to go with their alternative measure of effectiveness. The scale comprised four adjective pairs: many problems/few problems; employee resistance/employee acceptance; rough/smooth; and complicated/simple. The organizations were asked to describe their experiences with innovation implementation over the past three years. An acceptable estimate of internal reliability was 0.74.

Innovation effectiveness. Because we were investigating the validity of the Klein *et al.* (2001) model across a range of innovations, we needed a broader innovation effectiveness measure than the one used by Klein *et al.* (2001). Thus, we used a measure by Totterdell *et al.* (2002), comprised of 16 items evaluating improvements of accumulative benefits from various aspects, i.e., organizational finances (e.g., cost effectiveness and financial performance), customer issues (e.g., customer satisfaction and customer responsiveness), employee factors (e.g., management–employee relation and employee morale) and quality of life (e.g., health and safety). An index of innovation effectiveness was created by adding the items together. An acceptable estimate of internal reliability was 0.79.

Human resources availability. Two items were adapted from Nystrom *et al.*'s (2002) measure of resources availability. The items included the availability of skilled labor and managerial talent. The other two original items, which were removed from this study, related to financial resources availability, and were similar to Klein *et al.*'s (2001) items as previously described. An acceptable estimate of internal reliability was 0.73.

Organizational attitude toward future innovation adoption. Individuals were asked about their attitude toward innovation adoption in the future. The scale was developed based on the Theory of Planned Behavior (Ajzen, 1991). Questions were represented by five adjective pairs: dislike/like; a bad idea/a good idea; negative/positive; worthless/valuable; bad/good. Respondents were asked to rate their views on 7-point Likert scale (−3 to 3). An acceptable estimate of internal reliability was 0.96.

Organization characteristics. To prevent potential confounding effects on dependent measures, the following organizational characteristics were utilized as statistical controls: company size (determined by employee numbers), and industry types. Due to the small numbers of respondents in each industry, industry type was categorized as "manufacturing" and "non-manufacturing".

Results

The bivariate correlation coefficient for all variables in the current study is given in Table 1. Since the intent is to focus on the causal model assumed to underlie these correlations, they are not discussed in detail here.

Validating the original model of implementation effectiveness

Goodness of fit results indicated that the original model represented a poor fit to the data [$\chi^2(13) = 31.73$, $p < 0.05$; CMIN/DF $= 2.44$; GFI $= 0.94$; TLI $= 0.74$; CFI $= 0.84$; RMSEA $= 0.10$]. A path from top management support to implementation climate was added as this had both a theoretical foundation (it was included in Klein *et al.*'s (2001) final model) and a statistical foundation; this model displayed a significant improvement and an adequate fit to the data [$\chi^2(12) = 24.03$, $p < 0.05$; CMIN/DF $= 2.00$; GFI $= 0.95$; TLI $= 0.89$; CFI $= 0.90$; RMSEA $= 0.08$]. Path loadings shown in Figure 2 indicate that all but one

Table 1. Correlation matrix, means, standard deviations and reliabilties [In bracket] among studied variables.

Variables	Mean	SD	1	2	3	4	5	6	7	8
1. Top management support	4.00	0.72	[0.77]							
2. Financial resources availability	2.70	0.64	0.17*	[0.76]						
3. Implementation policies and practices	3.51	0.72	0.55**	0.04	[0.81]					
4. Implementation climate	4.14	0.83	0.36**	−0.01	0.36**	[0.92]				
5. Implementation effectiveness	3.36	0.57	0.28**	−0.04	0.21*	0.27**	[0.74]			
6. Innovation effectiveness	3.79	0.35	0.34**	−0.03	0.42**	0.18*	0.42**	[0.79]		
7. Human resources availability	2.91	1.00	0.40**	0.15	0.40**	0.30**	0.22**	0.21*	[0.73]	
8. Organizational attitude toward future innovation adoption	2.25	0.81	0.19*	0.06	0.28**	0.14*	0.34**	0.43**	0.12*	[0.96]
9. Organization size	—	—	−0.13	0.06	0.00	−0.05	−0.08	−0.07	0.01	−0.01

Note: ***$p < 0.001$, **$p < 0.01$, *$p < 0.05$. Alpha coefficients are depicted in parentheses along the diagonal. Organization size dummy coded 0 = less than 100 employees, 1 = 100 employees or more.

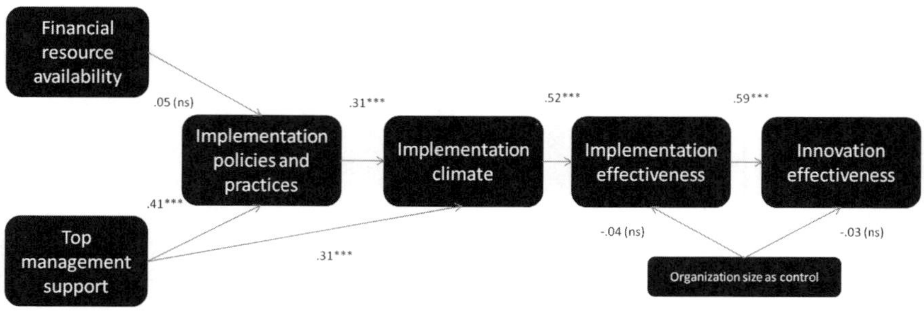

Fig. 2. Results of the model of implementation effectiveness with structural loadings.***$p < 0.001$.

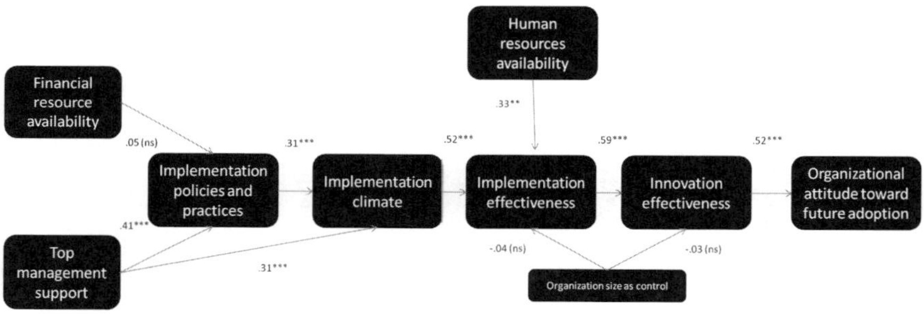

Fig. 3. Results of the enhanced model of implementation effectiveness with structural loadings.***$p < 0.00$; **$p < 0.01$.

of the hypothesized relationships (between financial resources availability and implementation policies) were significant and in the right direction.

Examining the enhanced model of implementation effectiveness

Results indicated that the proposed enhanced model represented an adequate fit to our sample [$\chi^2(23) = 38.77$, $p < 0.05$; CMIN/DF $= 1.68$; GFI $= 0.95$; TLI $= 0.89$; CFI $= 0.93$; RMSEA $= 0.07$] (see Fig. 3). Furthermore, the other fit indices have all been improved from the original model, suggesting that the enhanced model is sound.

To confirm that the enhanced model was fully mediated, we compared a rival, non-mediated model (including additional paths between financial resources and climate and effectiveness, support and climate and effectiveness, human resources and innovation effectiveness, and implementation effectiveness and attitudes). Comparisons between the rival non-mediated model [$\chi^2(9) = 11.51$, *ns*] and the Australian model [$\chi^2(22) = 33.52$, *ns*] indicated no significant differences [$\chi^2_{\text{different}}(13) = 21.93$, *ns*] suggesting that the more parsimonious mediated model be kept. To examine direct and indirect effects, we used a bootstrapping procedure with 95% confidence intervals on 10,000 bootstrap estimates (Efron and Tibshirani, 1993; Yung and Bentler, 1996). We found that top management support had a significant direct effect on implementation policies and practices as well as implementation climate ($\beta = 0.47$, $p < 0.001$; $\beta = 0.36$, $p < 0.01$; respectively). Thus, implementation policies and practices only partially mediated the effect of top management support on implementation climate. Other indirect relationships were as hypothesized and path loadings are detailed in Fig. 3.

Hypotheses two and three were tested by examining the path weights between human resources availability and implementation effectiveness and innovation effectiveness and organizational attitude toward future adoption. Both were significant ($\beta = 0.33$, $p < 0.001$; $\beta = 0.52$, $p < 0.001$; respectively) fully supporting the hypotheses.

Discussion and Implications

Within the literature, the key model for understanding innovation implementation is that of Klein *et al.* (2001), however their model had not been validated nor had it been extended to smaller-sized companies. In our sample of SMEs we were able to replicate most of the hypothesized relationships. Importantly, we found the "missing" relationships that were hypothesized but not found by Klein and colleagues. As Klein *et al.*'s (2001) model has been the yardstick for research in this area, our clarification of their model is an important contribution to understanding innovation in smaller firms.

Given the findings of prior studies, including Klein *et al.* (2001), the lack of a significant association between financial resources availability and implementation policies and practices is surprising. We propose that the different types of innovations studied may be affecting the results. Klein *et al.* (2001) examined a radical innovation producing extensive organizational, operational and managerial changes (Spathis and Ananiadis, 2006; Wu and Wang, 2006). Implementing radical innovations incurs sizeable financial investments in implementation activities (Scupola, 2003). In contrast, incremental innovations, such as upgrades to technology or modifications to existing products or services, are much less likely to need such high levels of resourcing (e.g., Vonortas and Xue, 1997). As our research was designed to maximise generalizability across SMEs we investigated various innovations and it is likely that our data represented both incremental and radical innovations. Unfortunately, the sample size precludes a post hoc analysis; however, we propose that radicalness may moderate the relationship between financial resources availability and implementation policies and practices. We believe that the formulation of this proposition is a significant contribution to the literature and an important next step is to investigate this moderation further. Should this proposed moderation be found, it will have significant implications for managers directing financial resources.

Another key finding is the relationship between top management support and implementation climate. Although this path was not originally specified, Klein *et al.* (2001) found it in their results and the robustness of the relationship is demonstrated by its emergence in our sample also. Two other recent studies on safety and knowledge management innovations lend further credence to its existence (Lee *et al.*, 2006; McFadden *et al.*, 2006). It therefore appears as though this is a robust relationship that can provide additional understanding of the implementation climate. Our research, therefore, adds evidence to the calls for top management to endorse activities that foster implementation effectiveness, such as clarifying communications, providing supportive policies and reducing organizational resistance (Basu *et al.*, 2002).

This study confirms that skilful and capable employees increase the level of implementation effectiveness. This extended finding creates a better understanding of factors influencing the implementation effectiveness. The original model of implementation effectiveness implied that the effective implementation is affected by the climate for implementation. However, only perceiving the importance of the implementation may not be sufficient to drive the successful implementation. Organizations may be required to have a capable staff who can deal with any problem occurring during the implementation as well. Chao and Kozlowski (1986) conducted a study of a robotic manufacturing technology among 461 employees in a plant. The authors concluded that lower skilled employees reported negatively toward the implementation. Although the company provided essential training during the implementation, lower skilled employees lacked problem solving skill during the implementation. As a result, the employees viewed the implementation negatively. Conversely, van Erp *et al.*'s (2007) study indicated that the skilful team members fostered the implementation of the supported employment system in four Dutch hospitals. Further, employees that have better communicative skill were capable to deal with problems during the enterprise resource planning (ERP) implementation

(Wagner and Newell, 2007). Further, Wagner and Newell (2007) concluded that employees who have strong communicative skills tended to negotiate or find out the workable solutions of problems during the implementation.

The current study indicated that when organizations perceive that the innovation is effective in a number of areas, they have a more positive attitude toward future innovation adoption. This finding, while novel in the field of innovation implementation, is consistent with knowledge within social and cognitive psychology regarding attitude formation. Much research over the past decade has shown that attitudes are often based on previous experiences (Anderson *et al.*, 2004; Poortman and Van Tilburg, 2005). In the field of innovation however, the finding has further implications. Based on the theory of planned behavior (Ajzen, 1991), Unsworth *et al.* (2005) developed the theoretical framework called innovation theory of planned behavior (I-TPB) to explain innovation adoption at organizational level. The I-TPB suggests that an organization's positive attitude towards an innovation will enhance the likelihood of further innovation adoption in the future. Thus, those organizations that perceive greater benefits from an implemented innovation are more likely to have positive attitudes towards future innovation adoption, which in turn may lead to an actual innovation adoption.

Generally, published research has developed specific conclusions or models explaining a particular innovation within a single-organizational type. For instance, collective learning (such as learning about others' role, improvising, and adjustability) was a critical predictor of the introduction of minimally invasive cardiac surgery in academic and community hospitals (Edmondson *et al.*, 2001). In a different innovation, just-in-time production, managerial commitment was found to be a key predictor of successful implementation within a manufacturing company (Chong *et al.*, 2001). Without comparative research in various types of organizations and innovations, we can only speculate the generalizability of basics. Without a comprehensive model, it is difficult for managers to design their implementation plan. A model that can be applied to most innovations and contexts is needed to frame new innovation initiatives. The current study thus contributes to the existing theory and integrates key concepts in order to advance understanding of innovation implementation effectiveness.

Implications of the enhanced model of implementation effectiveness

An important contribution of this study is the confirmation that skilful and capable employees increase the level of implementation effectiveness in smaller-sized firms. The previous model of implementation effectiveness implied that implementation was affected only by the climate for implementation (Klein *et al.*, 2001). However, we have shown that a climate for implementation will be of no use if those who are involved are not sufficiently capable of using the innovation or conducting the implementation.

This finding suggests that managers of SMEs need to recognize the importance of human resources, such as skilled staff, for the implementation of innovations. These human resources may help overcome limited finances in some organizations. Organizations that have a high incidence of qualified scientists and engineers are more likely to have a high rate of innovation activities (Hoffman *et al.*, 1998). However, some studies argued that a high incidence of qualified human resources may not be necessary for all SMEs (see Rothwell and Beesley, 1989; Young and Francis, 1991). Although our findings demonstrated that human resource availability assisted the effective implementation, this information should be considered under a circumstance of our SMEs sample, which were mainly in the manufacturing sector and level of innovation was incremental rather than radical. We believe that financial resources should also remain as an important resource for innovation implementation. When SMEs implement radical technological innovation, there is evidence that the availability of finance is a major constraint on the new technology implementation (Hoffman *et al.*, 1998).

A company may have highly skilful employees but how do managers engage these employees to take on the implementation process? We suggest managers should (1) provide clear communication about the implementation process, (2) empower employees to participate in the implementation plan and (3) recognize the employees' contribution to the implementation process. If employees perceive that the implementation is a major organizational priority — promoted, supported, and rewarded by the organization, then employees tend to be more engaged and contribute to the successful implementation.

Our enhanced model of implementation effectiveness also takes the understanding of innovation beyond the implementation stage by incorporating a post-implementation stage. Generally, studies of innovation implementation focus on the implementation's completeness and success. Therefore, this study contributes to the literature by examining relationships among implementation effectiveness, innovation effectiveness and organizations' attitude toward future innovation adoption. We are able to confirm that when an SME realizes the benefits of an innovation they are more likely to have positive attitudes towards adopting an innovation in the future. To increase the innovation adoption rate in the industries, government agencies could provide the additional support to SMEs (such as additional funding or consultation to adopt new technology) assuring their implementation will run smoothly and successfully. As a result, the positive experiences will encourage SMEs to adopt more innovation in the future.

Limitations and Future Research Directions

As with any research, this study has a number of limitations. Common method variance from using single-source data is a potential concern and may possibly account for the high mean for top management support. However we took a number of steps to minimize its effect. First, we guaranteed anonymity to respondents to mitigate the adverse effects of respondents' social desirability and evaluation apprehension. Second, we employed reverse-coding and different scale endpoints to reduce commonalities in scale endpoints and anchoring effects. Third, as suggested by Malhotra *et al.* (2006), our CFA results showed that the items for each construct illustrated good localization and the comparisons across the fully mediated model and the non-mediated model indicated that the non-mediated model did not represent the data any better than the fully-mediated model (see Podsakoff *et al.*, 2003). Although our study minimized a potential bias from single source information via the advance statistics, we strongly recommend the future research to adopt a multi source data collection method to enhance the study.

In addition, with data ranging across multiple industries and multiple innovations, there is the potential for other, confounding differences to occur. Nevertheless, our study attempted to minimize the potential confounding variables by including controls for organizational size and industry into our model testing and we found it did not influence our results. Furthermore, the emergence of the results across different innovations suggests that the model can be generalized across innovations, at least within small to medium-sized businesses — as such, we consider it a strength rather than a limitation.

Conclusion

To date, the benchmark model for understanding innovation implementation has been the research conducted by Klein *et al.* Replication and expansion of their model into different contexts has led to a more developed understanding of innovation, and this study has produced a new and more robust integrative model. The use of the scientific principle of replication and generalizability has allowed us to stand on the shoulders of giants and to confirm and elucidate the view they initially shared with us.

References

Ajzen, I (1991). The theory of planned behavior. *Organizational Behavior and Human Decision Processes*, 50, 179–211.

Alexander, JA, BJ Weiner and J Griffith (2006). Quality improvement and hospital financial performance. *Journal of Organizational Behavior*, 27(7), 1003–1029.

Amoako-Gyampah, K and AF Salam (2004). An extension of the technology acceptance model in an ERP implementation environment. *Information and Management*, 41, 731–745.

Anderson, AG, KP Hodge, D Lavallee and SB Martin (2004). New Zealand athletes' attitudes towards seeking sports psychology consultation. *New Zealand Journal of Psychology*, 33(3), 129–136.

Au, AK and P Enderwick (2000). A cognitive model on attitude towards technology adoption. *Journal of Managerial Psychology*, 15, 266–282.

Basu, V, E Hartono, AL Lederer and V Sethi (2002). The impact of organizational commitment, senior management involvement, and team involvement on strategic information systems planning. *Information and Management*, 39(6), 513–524.

Bhattacherjee, A (1998). Managerial influences on intra-organizational information technology use: A principal-agent model. *Decision Science*, 29, 139–162.

Chao, GT and SW Kozlowski (1986). Employee perceptions on the implementation of robotic manufacturing technology. *Journal of Applied Psychology*, 71(1), 70–76.

Chong, H, RE White and V Prybutok (2001). Relationship among organizational support, JIT implementation, and performance. *Industrial Management and Data Systems*, 101(5/6), 273–281.

Cycyota, CS and DA Harrison (2006). What (not) to expect when surveying executives: A meta-analysis of top manager response rates and techniques over time. *Organizational Research Methods*, 9(2), 133–160.

Damanpour, F (1991). Organizational innovation: A meta-analysis of effects of determinants and moderators. *Academy of management Journal*, 34(3), 555–590.

Davis, FD (1989). Perceived usefulness, perceived ease of use, and user accep. *MIS Quarterly*, 13(3), 319.

Davis, FD (1993). User acceptance of information technology: Systems characteristics, user perceptions and behavioral impacts. *International Journal of Man-Machine Studies*, 38(3), 475–487.

Dooley, KJ, A Subra and J Anderson (2002). Adoption rates and patterns of best practices in new product development. *International Journal of Innovation Management*, 6(1), 85–103.

Drury, DH and A Farhoomand (1999). Innovation diffusion and implementation. *International Journal of Innovation Management*, 3(2), 133–157.

Edmondson, AC, RM Bohmer and GP Pisano (2001). Disrupted routines: Team learning and new technology implementation in hospitals. *Administrative Science Quarterly*, 46(4), 685–716.

Ekvall, G (1996). Organizational climate for creativity and innovation. *European Journal of Work and Organizational Psychology. Special Issue: Innovation in Organizations*, 5(1), 105–123.

Flanagin, AJ (2000). Social pressures on organizational Website adoption. *Human Communication Research*, 26(4), 618–646.

Franklin, C (2007). Factors that influence elementary teachers' use of computers. *Journal of Technology and Teacher Education*, 15(2), 267–293.

Giffords, ED and RP Dina (2003). Changing organizational cultures: The challenge in forging successful mergers. *Administration in Social Work*, 27(1), 69–81.

Griffin, MA and A Neal (2000a). Perceptions of safety at work: A framework for linking safety climate to safety performance, knowledge and motivation. *Journal of Occupational Health Psychology*, 5(3), 347–358.

Helfrich, CD, BJ Weiner, MM McKinney and L Minasian (2007). Determinants of implementation effectiveness — Adapting a framework for complex innovations. *Medical Care Research and Review*, 64(3), 279–303.

Hoffman, K, M Parejo, J Bessant and L Perren (1998). Small firms, R&D, technology and innovation in the UK: A literature review. *Technovation*, 18(1), 39–55.

Holahan, PJ, ZH Aronson, AP Jurkat and FD Schoorman (2004). Implementing computer technology: A multiorganizational test of Klein and Sorra's model. *Journal of Engineering and Technology Management*, 21, 31–50.

Hoy, WK and CG Miskel (1991). *Educational Administration: Theory, Research, and Practice*, 4th edn. New York: McGraw-Hill.

Jarvenpaa, SL and B Ives (1991). Executive involvement and participation in the management of information technology. *MIS Quarterly*, 15(2), 205–227.

Jimmieson, NL, DJ Terry and VJ Callan (2004). A longitudinal study of employee adaptation to organizational change: The role of change-related information and change-related self-efficacy. *Journal of Occupational Health Psychology*, 9(1), 11–27.

Johnston, DA and JD Linton (2000). Social networks and the implementation of environmental technology. *IEEE Transactions on Engineering Management*, 47(4), 465–477.

Jones, O (2003). Competitive advantage in SMEs: Towards a conceptual framework. In Tilley, F and J Tonge (Eds.), *Competitive Advantage in SMEs: Organising for Innovation and Change*. England: John Wiley & Sons.

Klein, KJ, AB Conn and JS Sorra (2001). Implementing computerized technology: An organization analysis. *Journal of Applied Psychology*, 86(5), 811–824.

Klein, KJ and D Knights (2005). Innovation implementation overcoming the challenge. *Current Directions in Psychological Science*, 14(5), 243–246.

Klein, KJ and JS Sorra (1996). The challenge of innovation implementation. *Academy of Management Review*, 21(4), 1055–1079.

Lehman, WEK, JM Greener and DD Simpson (2002). Assessing organizational readiness for change. *Journal of Substance Abuse Treatment*, 22, 197–209.

Link, S and E Naveh (2006). Standardization and discretion: Does the environmental standard ISO 14001 lead to performance benefits? *IEEE Transactions on Engineering Management*, 53(4), 508–519.

Malhotra, NK, SS Kim and A Patil (2006). Common method variance in IS research: A comparison of alternative approaches and a reanalysis of past research. *Management Science*, 52(12), 1865–1883.

Marler, JH, XY Liang and JH Dulebohn (2006). Training and effective employee information technology use. *Journal of Management*, 32(5), 721–743.

Martin, AJ, ES Jones and VJ Callan (2005). The role of psychological climate in facilitating employee adjustment during organizational change. *European Journal of Work and Organizational Psychology*, 14(3), 263–289.

McAdam, R, RS Reid and DA Gibson (2004). Innovation and organizational size in Irish SMEs: An empirical study. *International Journal of Innovation Management*, 8(2), 147–165.

McFadden, KL, GN Stock and CRI Gowen (2006). Implementation of patient safety initiatives in US hospitals. *International Journal of Operations and Production Management*, 26, 326–347.

Moran, T and F Volkwein (1992). The culture approach to the formulation of organizational climate. *Human Relations*, 45, 19–29.

Moreno-Luzón, MD and M Begoña Lloria (2008). The role of non-structural and informal mechanisms of integration and coordination as forces in knowledge creation. *British Journal of Management*, 19, 250–276.

Naveh, E and AA Marcus (2004). When does the ISO 9000 quality assurance standard lead to performance improvement? Assimilation and going beyond. *IEEE Transactions on Engineering Management*, 51(3), 352–363.

Nord, WR and S Tucker (1987). *Implementing Routine and Radical Innovations.* Lexington, MA: Lexington.

Nystrom, PC, K Ramamurthy and AL Wilson (2002). Organizational context, climate and innovativeness: Adoption of imaging technology. *Journal of Engineering and Technology Management*, 19(3,4), 221–247.

Peansupap, V and DHT Walker (2006). Information communication technology (ICT) implementation constraints. *Engineering, Construction and Architectural Management*, 13(4), 364–379.

Podsakoff, PM, SB MacKenzie, JY Lee and NP Podsakoff (2003). Common method biases in behavioral research: A critical review of the literature and recommended remedies. *Journal of Applied Psychology*, 88(5), 879–903.

Poortman, AR and TG Van Tilburg (2005). Past experiences and older adults' attitudes: A life-course perspective. *Ageing and Society*, 25(1), 19–39.

Ramsey, E, P Ibbotson and P McCole (2008). Factors that impact technology innovation adoption among Irish professional service sector SMEs. *International Journal of Innovation Management*, 12(4), 629–654.

Reichers, AE and B Schneider (1990). Climate and culture: An evolution of constructs. In Schneider, B (Ed.), *Organizational Climate and Culture*, San Francisco: Jossey-Bass.

Rogers, EM (1995). *Diffusion of Innovations*, 4th edn. New York: The Free Press.

Rosner, MM (1968). Economic determinants of organizational innovation. *Administrative Science Quarterly*, 12, 614–625.

Rothwell, R and M Beesley (1989). The importance of technology transfer. In Barber, Ja, S Metcalfe and M Proteous (Eds.), *Barriers to Growth in Small Firms*, pp. 87–103. London: Routledge.

Sambrook, S (2003). E-learning in small organisations. *Education and Training*, 45(8/9), 506–516.

Sawang, S and KL Unsworth (In Press). Why adopt now? Multiple case studies and survey studies comparing small, medium and large firms. *Technovation*, DOI: 10.1016/j.technovation.2011.06.002

Scarbrough, H (2003). Knowledge management, HRM and the innovation process. *International Journal of Manpower*, 24(5), 501–516.

Schneider, B, SS White and MC Paul (1998). Linking service climate and customer perceptions of service quality: Test of a causal model. *Journal of Applied Psychology*, 83(2), 150–163.

Schrum, L and KF Glassett (2006). Technology Integration in P-12 Schools: Challenges to implementation and impact of scientifically-based research. *Journal of Thought*, 41–59.

Snell, SA and JW Dean Jr (1992). Integrated manufacturing and human resource management: A human capital perspective. *Academy of Management Journal*, 35(3), 467–504.

Spenner, KI (1983). Deciphering prometheus: Temporal change in the skill level of work. *American Sociological Review*, 48(6), 824–837.

Starkweather, KN (2005). Innovation, TIDE teachers, and the global economy. *Technology Teacher*, 65(2), 28–30.

Storey, J and E Barnett (2000). Knowledge management initiatives: Learning from failure. *Journal of Knowledge Management*, 4(2), 145.

Tagiuri, R and G Litwin (1968). *Organizational Climate*. Boston: Division of Research, Harvard Graduate School of Business.

Totterdell, P, D Leach, K Birdi, C Clegg and T Wall (2002). An investigation of the contents and consequences of major organizational innovations. *International Journal of Innovation Management*, 6(4), 343.

van der Panne, G, C van der Beers and A Kleinknecht (2003). Success and failure of innovation: A literature review. *International Journal of Innovation Management*, 7(3), 309–338.

van Erp, NHJ, FBM Giesen, J van Weeghel, H Kroon, HWC Michon and D Becker (2007). A multi-site study of implementing supported employment in the Netherlands. *Psychiatric Services*, 58(11), 1421–1426.

Vonortas, NS and L Sue (1997). Process innovation in small forms: Case studies on CNC machine tools. *Technovation*, 17(8), 427–438.

Wagner, EL and S Newell (2007). Exploring the importance of participation in the post-implementation period of an ES project: A neglected area. *Journal of the Association for Information Systems*, 8(10), 508–524.

Weiner, BJ, CD Helfrich, LA Savitz and KD Swiger (2007). Adoption and implementation of strategies for diabetes management in primary care practices. *American Journal of Preventive Medicine*, 33(1), S35–S49.

Wu, JH and YM Wang (2006). Measuring ERP success: The ultimate users' view. *International Journal of Operations and Production Management*, 26(8), 882–903.

Young, RC and JD Francis (1991). Entrepreneurship and innovation in small manufacturing firms. *Social Science Quarterly*, 72(1), 149–162.

Zaltman, G, R Duncan and J Holbeck (1973). *Innovations and Organizations*. New York: Wiley.

Chapter 9

A STUDY OF ORGANISATIONAL DETERMINANTS AND INNOVATION PRACTICES IN DUBAI SMEs

Yahya Al-Ansari, Jun Xu and Simon Pervan
Southern Cross Business School
Southern Cross University, Australia

This study explores the relevance of established organisational determinants of innovation practice in small and medium-sized enterprises (SMEs) in the emerging market of Dubai, United Arab Emirates. Drawing upon data from 200 SMEs, the findings suggest that management orientation, technology orientation, alliance and cooperation and market orientation are important antecedents to SMEs' innovative practices, but question the importance of organisational culture. This study is the first to examine the organisational determinants of innovation in both an SME and emerging market context. Implications and suggestions for future research are offered to both scholars and practitioners.

Keywords: Small and medium-sized enterprises; SMEs; innovation practices; organisational determinants; Dubai; United Arab Emirates.

Introduction

The emerging Dubai market in the United Arab Emirates (UAE) is considered to be one of the fastest growing markets in the Middle East and North Africa (WEF, 2010). It is ranked as the most innovation-driven economy in the Arab world and has been moving from an oil-based economy to a knowledge-based economy as it seeks to emerge as a prominent free market that is integrating into the world economy (Grant *et al.*, 2007; Knight, 2011). The UAE attracts businesses and investments from many parts of the world. It has pursued an outward-oriented development strategy by prioritising economic policy reforms and diversifications

This paper was first published in *International Journal of Innovation Management*, Vol. 18, No. 1 (February 2014), 1450003 (28 pages). © Imperial College Press. DOI: 10.1142/S1363919614500030 and streamlining foreign investment regulations (BMI, 2007). The UAE has combined features of developed and developing market economies to become both a commercial and an industrial hub. This hybrid situation offers a mixture of challenges and opportunities for firms of different sizes seeking to transform their current businesses and/or create new ones to meet new market conditions (Rettab *et al.*, 2009; Hertog, 2010).

Of the seven states, *emirates* in Arabic, in the UAE, Dubai has been the first to implement major reforms and establish a strong position in the business world (Grant *et al.*, 2007). In Dubai, small to medium-sized enterprises (SMEs) comprise the majority of firms. For example, more than 90% of firms in the service and manufacturing sectors are SMEs (DCCI, 2010). They contribute to the development of the economy in areas such as entrepreneurship, innovation, productivity, diversification, competition and growth (Massa and Testa, 2008). SMEs establishing in Dubai must have a minimum of 51% UAE national-ownership (though full profit repatriation is permitted), *kafeel in Arabic*, or a local agent and their employees have a three-year work visa cycle (BMI, 2007). These legal requirements and weak regulations in Dubai facilitate fierce competitions for SMEs, heightening the need for innovation and long-term investment (Elewa, 2007; Brik *et al.*, 2011).

Historically, a preoccupation with inward focused head-to-head competition over the buying and selling of existing products and services has decreased the "stickiness" of SMEs' in the local Dubai market, limiting innovation outcomes and contributions to the local economy (Hertog, 2010). The main survival objective in stable markets has been to decrease costs and increase short-term profits while at the same time cautiously moving into new domains when the market condition is dynamic or turbulent (Hertog, 2010; Avci *et al.*, 2011). The outcome is a low margin model that undermines the accumulation of sufficient resources and capabilities to implement new business concepts and innovations.

Change is occurring, however, due to government led incentives for the entry of foreign firms, the opening of new markets, tightening of intellectual property laws and investment in education sectors (Grant *et al.*, 2007). As local and global competitions intensify, SMEs in Dubai are being encouraged to use innovation as a tool to derive competitive advantage and improve their business performance (Forsman and Temel, 2011). Further to this, the UAE has natural advantages in its location between East and West and its many harbours, which have allowed the development of 10 major shipping ports and 6 airports (Grant *et al.*, 2007), meaning *emirates* like Dubai have great potential for becoming innovation hubs for the region (Dutta, 2006). This study examines the innovative practices of

SMEs in the emerging Dubai market, a local economy dominated by SMEs, tasked with surviving in a rapidly changing marketplace.

Firms of all sizes have long recognised the importance of being able to keep innovating and continually providing new and better products and services (Tucker, 2002). Those achieving higher business performances are considered to have competitive advantages because they are likely to possess resources and capabilities that are valuable, unique and difficult to imitate by others, but the sustainability of these competitive advantages depends on the firms' innovative capacities (Porter, 1980; Barney, 1991). Further, innovation is not something that is limited to large firms (Carrier, 1994). Small firms are important engines for innovation and technological advancement (Mulhern, 1995). Many SMEs actively incorporate innovation into their business strategies and activities (Vossen, 1998; Hadjimanolis, 1999), although compared to large firms they are faced with greater challenges such as lack of economies of scale, limited resources, smaller market size and greater susceptibility to market shifts and other environmental shocks (Tether, 1998; Cagliano *et al.*, 2000).

Previous studies have identified key organisational determinants (internal-driven factors) relevant to innovation within large firms in advanced markets. In particular, researchers have emphasised the importance of management orientation (Ghobadian and Gallear, 1997; Aragon-Sanchez and Sanchez-Marin, 2005), organisational culture (Hurley and Hult, 1998; Kenny and Reedy, 2006), technology orientation (Hitt *et al.*, 1990; Humpherys *et al.*, 2005), alliance and co-operation (Baum *et al.*, 2000; Stuart, 2000) and market orientation (Jaworski and Kohli, 1993; Hult *et al.*, 2004) in developing continuous innovation to achieve competitive advantage. There is increasing interest in whether these relationships hold in SMEs as well as in emerging markets (Kim and Lim, 1988; Arnold and Quelch, 1998; Salavou *et al.*, 2004). In view of this, the primary research question addressed in this study is to examine the effect of organisational determinants on innovation practices within SMEs of an emerging economy. Specifically, this research explores the presence of management orientation, organisational culture, technology orientation, alliance and cooperation, and market orientation in SMEs reporting strong innovation practice within SMEs in the Dubai market.

The paper proceeds as follows. The next section lends further precision to background and this is followed by a review of literature leading to the research questions that inform this exploratory study. The research method is then introduced, and the results of the study are presented. Finally, an analysis of the findings is offered and implications, limitations and suggestions for future research are discussed.

Background

SMEs have traditionally relied on local markets but recently have become increasingly involved in the global competitive markets (Cagliano *et al.*, 2000). Ghobadian and Gallear (1997) and Vossen (1998) argue that SMEs need to evaluate their business strategies and consider innovation at the organisational level and within their activities (processes, products and services). SMEs, however, are perhaps more vulnerable than larger firms to market shifts and usually operate with narrower product portfolios, which means that they are at greater risk from industry-related downturns (Shama, 1993). Conversely, they often possess the flexibility to adjust their inputs and outputs (processes, products, services and prices) quickly in response to environmental shocks. The ability to respond quickly in this way is crucial to their business (Berry *et al.*, 2002). Further, SMEs are often more willing than large firms to engage in risky investments and innovative behaviours to improve their competitive advantages and business performances (Latham, 2009).

Innovative practices of SMEs are related to the firm's ability to seek new and better ways to identify, acquire and implement ideas and tasks (processes, products, services, management and administrative systems, organisational structures and marketing methods) in the organisation to cope with market dynamics (North and Smallbone, 2000; Calantone *et al.*, 2002; Blumentritt and Danis, 2006). Having well-established innovation practices can contribute to the willingness of the firm to implement internal and external changes and they can play a key role in the provision of the business solutions and successes of the firm into the future (Hult *et al.*, 2004). Brown (1990, 257) further argues that "the firm gains competitive advantage by performing its activities more cheaply or better than its competitors." SMEs have an advantage due to their entrepreneurial characteristics and the structural flexibility of their short chain decision-making processes for adopting and/or generating innovation practices. SMEs with a record of innovation have more proactive attitudes toward new ideas and tasks and tend to view obstacles as learning opportunities rather than merely as problems (Mahemba and De Bruijn, 2003).

There are two streams of management literature on innovation. The first reflects research on the diffusion of innovation across organisations, industries and market places and the second examines the influence of organisational culture, process and the individual on innovation (Cromer *et al.*, 2011). Research on the diffusion of innovation informs firms looking to apply a new management practice, strategy or technology to carry out a process of restructuring or improvement in order to influence the rate of adoption by customers. It is thus related to the persuasion stage of adoption and implementation of innovation (Tornatzky and Klein, 1982; Rogers, 2003). Research on organisational culture, process and the individual, informs firms looking to create a new and better process, product or service for the market and is therefore centered on the innovation outcomes, which influence the rate of creation and commercialisation (Myers and Marquis, 1969). This study is concern with the second stream. What is more, a resource-based view of the firm is taken whereby an examination of the role of organisational determinants (internal resources and capabilities) and their influences are used to offer a better understanding of innovative potential and continuous innovative practices (Hjalager, 2010; Martinez-Roman *et al.*, 2011).

Studies have extended the resource-based view through applying the concept of dynamic capabilities to explain the ability of the firm to develop and extend resources and competences to adapt to changing business and market environments (Eisenhardt and Martin, 2000). This view can explain how a firm derives competitive advantages by channelling resources into innovation (Hult *et al.*, 2004). The innovation process requires a number of factors such as structures for research and development and capabilities for technological advancement (Mahemba and De Bruijn, 2003; Laforet and Tann, 2006). Studies that have examined these factors almost exclusively deal with large firms in developed markets (Kim and Lim, 1988; Salavou *et al.*, 2004). Yet, there is an acknowledgement that SMEs play important roles (Allocca and Kessler, 2006) and have behavioural advantages (Salavou *et al.*, 2004) in innovation, mainly, in those emerging markets where liberalisation, privatisation and globalisation of the economy is currently happening (Mytelka, 2000; Grant *et al.*, 2007; Rettab *et al.*, 2009).

The organisational determinants under consideration in this study are those internal-driven factors that can facilitate or hinder the ability of the firm to innovate (Avlonitis and Gounaris, 1999). Their importance to innovation practice has proved robust in the relatively homogenous context of large firms operating in developed economies; however, a more dynamic context similar to Dubai is needed to examine these links. The following discussion leads to exploratory research questions for the study.

Research Questions

Management orientation and innovation practices

Management orientation in relation to management characteristics and basic competences has been well studied in the context of large firms and there is general consensus that it is an important indicator of innovation potential (Snow and Hrebiniak, 1980; Aragon-Sanchez and Sanchez-Marin, 2005). However, unlike most large firms, SMEs usually reflect the personalities of their owners and/or managers and these individuals have far greater capacity to influence day to day operations (Hale and Cragg, 1996), which means that they can make decisions rapidly and drive innovation (Ghobadian and Gallear, 1997). Their characteristics and basic competences (leadership and interpersonal skill) have further positive impacts on the progress and success of their firms (Lin, 1998; Yusof and Aspinwall, 1999). Firms with different management strategic directions can differ in how they implement and conduct specific innovative behaviours (Ettlie *et al.*, 1984). Managers might adopt strategies to generate and sustain different interpretations of environmental information, which can further influence the innovative behaviours and activities of their firms (McGinnis and Ackelsberg, 1983; Blumentritt and Danis, 2006). Successful innovation requires strong managerial support and resource commitment (Cromer *et al.*, 2011). Management orientation appears to play an important role in determining and supporting the firm's decision to adopt and/or generate innovations regardless of its size (Baldridge and Burnham, 1975). In Dubai, the proxy indicators that may lend insight to management orientation toward innovation like entry speed into new markets and/or dynamic environments and develop new products (Hertog, 2010; Avci *et al.*, 2011) tend to indicate a conservative orientation. However, despite this, the Dubai work environment, more so than other markets in the Middle East and North Africa, is characterised by relatively open communication between managers and subordinates (Suliman, 2006). Further, local SMEs have high exposure to foreign management practices given the UAE reliance on foreign labour sources. There is likely to be recognition of the importance of management orientation. However, Dubai has also embarked on a process of *emiratisation* of the labour force since the early nineties whereby locals are prioritised for senior management roles (Grant *et al.*, 2007). There are likely to be a number of different forms of management orientation that may influence the motivation to develop new initiatives and to innovate (Gulati, 2000; Abubakr, 2006):

RQ1: *Is management orientation a characteristic of firm's with strong innovation practices in Dubai?*

Organisational culture and innovation practices

Organisational culture and learning within the firm can inspire innovation and give individuals plenty of space to take risks, make mistakes and create opportunities and successful solutions (Martins and Terblanche, 2003; Peebles, 2003).

A supportive culture can further drive value creation and encourage innovative ways of representing problems and searching for solutions (Buckler, 1997). In achieving long-term success, Tushman and O'Reilly (1997) argue that firms need to integrate innovation into their management process and organisational culture. Organisational learning enhances the creativity of firms and their ability to identify innovative opportunities, which involves different types of learning by providing incremental and discontinuous skills and encouraging the institutionalisation of innovation (McKee, 2003). In order to encourage innovation, knowledge already available in a firm has to be further captured, made explicit, properly understood and adopted into specific outcomes (Hjalager, 2010). In SMEs, an orientation towards learning and continuous improvement can generate new ideas and enable them to possess new technologies, which can encourage innovation and renew competitive advantage (Hurley and Hult, 1998; Keskin, 2006). With other advantages that include more streamlined decision-making process, faster adaption capability and quicker response rate, organisational design and flexibility can distinguish SMEs from large firms (Feigenbaum and Karnani, 1991). Despite evidence of relatively open communication between personnel at different levels within firms in Dubai, many firms still follow directive and paternalistic management styles with centralised organisational structures and top-down and formal communication (Sinangil, 2004; Abubakr, 2006). This may impact the flexibility typically associated with an organisational culture considered optimal for innovation practice. Further, the cultural tradition of *wasta*, whereby family members and personal contacts are favoured for senior roles within the firm, makes the connection between a strategically developed organisational culture and innovation difficult to predict:

RQ2: *Is organisational culture a characteristic of firm's with strong innovation practices in Dubai?*

Technology orientation and innovation practices

Technology is considered to be a success factor of the firm in achieving competitive advantage (Simon, 1996). For firms to explore new and superior technologies, and to get these technologies effectively launched, depends on the existence of niche and fringe markets, or experimental users or both (Malerba *et al.*, 2007). Technological capabilities of the firm in creating products and services, according to its consumer and market needs, can produce successful innovation (Berkhout *et al.*, 2010). Accepting and exploring new technology enables a firm to gain quality and productivity and to generate new products and services

(Hjalager, 2010). Being technology-oriented is viewed as a source of competitive advantage and, in time, innovative capability (Humpherys *et al.*, 2005). A firm having technological resources (knowledge and new technology adoption) can support its innovation process and product development (Hadjimanolis, 1999). Internal technology policy reflects the commitment of the firm to innovation (Ettlie and Bridges, 1982; Wilson *et al.*, 1999) and can enable it to achieve competitive advantage (Hitt *et al.*, 1990). Firms that proactively acquire new technologies might be more innovative owing to their emphasis on employing these technologies for developing new products and services (Cooper, 1994). SMEs are being challenged in their abilities to afford the heavy investments needed for new technologies that may lead them to seek technology through outsourcings and/or value-chain activities (Mahemba and De Bruijn, 2003; Salavou *et al.*, 2004). Technological adoption and/or development in Dubai vary for different SMEs depending on dynamism of the marketplace. The relatively short, three year, visa cycle for the foreign workers, who run many of these firms, does not encourage long-term investment in technological resources and capabilities (Srinivasan *et al.*, 2002; Grant *et al.*, 2007; Brik *et al.*, 2011). However, there appears to be a general recognition of the importance of technology, with the recent development of the Dubai Internet City and Techno Park among other initiatives, creating substantial employment in sectors like information and computer technology (Grant *et al.*, 2007):

RQ3: *Is technology orientation a characteristic of firm's with strong innovation practices in Dubai?*

Alliance and cooperation and innovation practices

Having a well-developed alliance at the business network level is becoming an increasingly important mechanism for driving success (Batonda and Perry, 2003), which can benefit the firm in many ways such as enhancing learning, developing a market focus, transferring technology and expertise, sharing cost and risk, increasing production efficiency and encouraging firm and industry collaboration (Allocca and Kessler, 2006). Strategic alliances and external networks of the firm offer access to information, knowledge, resources, capabilities and technologies that enable it to compete more effectively in the marketplace (Gulati *et al.*, 2000; Ireland *et al.*, 2002), which is mainly important for SMEs with limited resources looking to enhance innovation (Baum *et al.*, 2000). It is important for SMEs in Dubai to establish trade relationships, network with business partners and link with customers (Grant *et al.*, 2007) that can further enable them to access resources and competencies without merging with others, maintain flexibility and adapt to environmental changes. These arrangements can offer innovations that are complete solutions and not just isolated products and services (Glaister and Buckley, 1996). As one of seven *emirates* in a small country, Dubai is acutely aware of its dependence on the outside world. To this end, a number of free trade zones have been set up allowing full foreign ownership of companies and significant tax benefits. Outside of these zones, legal requirements of the ownership law limits foreign investments to 49% may present obstacles to the formation of alliances. Further, firms that choose instead to set up "representative" office or franchises must employ a UAE national as agent distributor (Grant *et al.*, 2007). Further complicating the potential for collaboration is the *wasta* network, which can lead to business behaviours based on personal connections and social networks rather than strategic alliance and cooperation (Hutchings and Weir, 2006). Again though this is not necessarily a limitation as placement in the network may either promote or inhibit collaboration:

RQ4: *Is alliance and cooperation a characteristic of firm's with strong innovation practices in Dubai?*

Market orientation and innovation practices

Firms with market orientations have an outward focus on customers and competitors to increase their knowledge in order to generate actions and encourage innovation (Gray *et al.*, 1998; Varadarajan and Jayachandran, 1999). Jaworski and Kohli (1993, 56) argue that market orientation essentially "involves doing something new or different in response to market conditions, it may be viewed as a form of innovative behaviour." While innovation has not been presented in models of market orientation it is considered an important outcome (Hult *et al.*, 2004; Salavou *et al.*, 2004). SMEs with strong market orientations are considered more innovative because of a better understanding of both customer needs and competitive situations (Lin, 1998; Henard and Szymanski, 2001). The lack of market research and information in Dubai is a limiting factor to SMEs seeking to understand customer and competitor behaviours (Rettab *et al.*, 2009; Hertog, 2010; Brik *et al.*, 2011). There is evidence to suggest that emerging economies have a higher proportion of demand uncertainty where customers want to be led rather than listened to, as they explore product and service categories (Gao *et al.*, 2007). However, the Dubai market is small and quickly maturing and SMEs compete fiercely to survive and often need to look offshore for new markets where customer certainty may be higher (Rettab and Rao, 2008). Further, SMEs tend to operate in environments intimate enough for customers to be directly accessed to provide feedback and to work through problems with (Herb *et al.*, 2001):

RQ5: *Is market orientation a characteristic of firm's with strong innovation practices in Dubai?*

Method

Sample design

This study draws on a sample of SMEs from a broad range of industries including information and communication technology, healthcare services, entertainment and media, manufacturing and industrial engineering, construction and architecture and retail and repairing services (see Table 1). Different manufacturing and service industries were included so as to be able to generalise beyond particular industries, to the population of SMEs, and to cater for the exploratory nature of the study (Dawes, 2000; Scozzi *et al.*, 2005). The definition of SMEs is based on a commonly used criterion of number of employees (Freel, 1999). The Dubai Chamber of Commerce and Industry (DCCI, 2010) classification is

Table 1: Sample composition.

Industry	Firm size			Total[b,c]
	1–9 Employees	10–99 Employees	100–199 Employees	
Information and Communication Technology	18	38	7	63 (31.5)
Healthcare Services	9	17	1	27 (13.5)
Entertainment and Media	19	4	4	27 (13.5)
Manufacturing and Industrial Engineering	7	13	6	26 (13)
Construction and Architecture	5	6	4	15 (7.5)
Retail and Repairing Services	5	8	1	14 (7)
Other Services[a]	10	15	3	28 (14)
Total	73	101	26	200 (100)

[a]Other services include financial intermediation and insurance, education, training and consultancy, tourism and hospitality, and real estate and renting services.
[b]Eight surveys are incomplete. Three are from information and communication technology and five are from other services.
[c]Percentage figures are given in parentheses.

adopted: Small (1–9 employees), medium (10–199 employees) and large (more than 200 employees). A sample frame of 600 SMEs was drawn from the DCCI commercial database. The population was stratified and segmented by firm size to better represent each element of the population and to provide enough variance with respect to constructs under study (Homburg and Jensen, 2007). Firms were divided by size into three groups: 1 to 9, 10 to 199 and greater than 200 employees. In accordance with the DCCI database classification, only the groups of 1 to 9 and 10 to 199 employees were selected according for the current study. Within each group, selection was conducted through simple random sampling. Beyond the information contained in Table 1, it is important to note that 97% (194/200) of the respondents were the owner-managers (27%) or the manager (70%) of the firm while the remaining 3% (6/200) held senior management positions.

Measures and instrument development

The items used to measure the constructs are based on an extensive review of the related literature and are listed in Appendix A. The independent and dependent variables were measured using a seven-point Likert scale with anchors for the organisational determinants "1 — strongly disagree" and "7 — strongly agree" and innovation practices "1 — much below" and "7 — much above."

Organisational determinants

Management orientation was measured using a five-item scale, in responses to questions about management characteristics and strategic direction, adapted from studies by Aragon-Sanchez and Sanchez-Marin (2005), Blumentritt and Danis (2006) and Laforet and Tann (2006). Organisational culture was measured using a five-item scale, in responses to questions about learning processes, design and flexible practice, adapted from studies by Hurley and Hult (1998), Kenny and Reedy (2006) and Martensen *et al.* (2007). Technology orientation was measured using a five-item scale, in responses to questions about technological policy, position and adoption, adapted from studies by Salavou *et al.* (2004), Aragon-Sanchez and Sanchez-Marin (2005) and Allocca and Kessler (2006). Alliance and cooperation was measured using a five-item scale, in responses to questions about collaborative agreements and networks with business groups and supporting industries, adapted from studies by Hoffmann and Schlosser (2001), Aragon-Sanchez and Sanchez-Marin (2005) and Martensen *et al.* (2007). Market orientation was measured using a five-item scale, in responses to questions about customers, competitors and inter-functional market information sharing, adapted from studies by Narver and Slater (1990), Kohli *et al.* (1993) and Pelham and Wilson (1996).

Innovation practices

Innovation practices were measured using a 10-item scale, in responses to questions about innovative practices within a three-year time period, adopted from studies by Calantone *et al.* (2002), Aragon-Sanchez and Sanchez-Marin (2005) and Blumentritt and Danis (2006).

Before the actual data collection procedure was conducted, the face validity of the survey items was established through consultation ($n = 20$) with four academic researchers, eight industry and market experts and eight owners and/or managers of SMEs. The survey was then pretested ($n = 30$) using 14 owners and/or managers of SMEs, 10 academic researchers and six industry and market experts in order to review the survey for structure, readability, ambiguity and completeness (Dillman, 2007). To assess the reliability and validity (internal consistency) among constructs and items in the scales, the coefficient alpha, average item-total correlations and inter-item correlations were used. The coefficient alphas for all constructs were above the cutoff value of 0.60 and the internal consistencies (item-to-total and inter-item correlations) for all items were above the minimum levels of 0.50 and 0.30 respectively (Hair *et al.*, 2006).

Data collection

A cross-sectional structured survey was utilised for data collection. The targeted respondents were owners and/or managers of SMEs operating in Dubai. The rationale for selecting individuals with senior-level responsibilities was that their values and philosophies influence the strategic choices and decisions of their firms (Covin and Slevin, 1990). This makes them key informants who have control over most of the activities (Kumar *et al.*, 1993). In an effort to increase the response rate, a modified version of Dillman's total design method was followed (Dillman, 2007). All respondents received a folder that contained a cover letter, the survey and a postage-paid return envelope. The method of successive waves was applied (Armstrong and Overton, 1977). Three weeks after the first delivery of surveys and introductory letters, reminder letters were sent out to non-respondents. Of the 600 surveys (see Table 1), 208 surveys were returned, resulting in a response rate of 34% (208/200). A total of eight surveys were discarded due to incomplete information, resulting in an effective response rate of 33% (200/600).

Non-response bias was evaluated by using the extrapolation technique to compare the mean difference between the early-respondents ($n = 150$) and late-respondents ($n = 50$). Using an independent-samples *t*-test ($p < 0.05$), early and late respondents were compared on a number of key demographic indicators such as number of employees and number of established years to check for any significant differences. The *t*-tests performed revealed that non-response bias did not appear to be a significant issue between the sampled and targeted populations (Armstrong and Overton, 1977). As this study collected data from a single respondent in each firm, a test for common method bias was also conducted by using the Harman's single factor test. An un-rotated factor analysis, with an eigenvalue-greater-than-one criterion, revealed that six distinct factors accounted for 65.76% of the variance with the first factor capturing only 40.27%. As a single factor did not emerge from the factor analysis and the first factor did not account for most of the common variance, common method bias did not appear to be a problem (Doty and Glick, 1998).

Results

The research questions were tested using the Partial Least Squares (PLS) modelling technique (Smart-PLS 2.0 M3) with component-variance matrices for the estimation of the measurement and structural models (Henseler *et al.*, 2009). The PLS technique has a different requirement for data distribution characteristics (for example normality, skew and kurtosis) that avoids many of the restrictive assumptions underlying covariance-based technique such as the need for multivariate normality (Chin, 1998). It can be described by two models: a measurement model (outer model) relating items to their specific latent variables and a structural model (inner model) relating latent variables to other latent variables (Vinzi *et al.*, 2010). The PLS technique is able to compute the scores of latent variables through a process that iteratively computes first an outer estimation (measurement model) and an inner estimation (structural model) and to examine a series of dependence relationships simultaneously (Barclay *et al.*, 1995; Tenenhaus *et al.*, 2005). It should be further noted that this study applies a reflective, as opposed to formative, construct model where items are affected by their latent variables (Vinzi *et al.*, 2010).

Measurement model

The measurement model consists of 35 measured items and six latent variables. To estimate the parameters in the outer model, a centroid weighting scheme (300 maximum iterations) and an abort criterion for the iterative estimation process were selected as a change of the estimated values of just 0.1–5% between two iterations (Henseler *et al.*, 2009).

The reliability and validity of the latent variables and their items in the outer model were established first through the individual item reliability, Cronbach's alpha and composite reliability both of which were above the recommended 0.60/ 0.707 criterion (see Table 2). This indicated that items were performing well in capturing their latent variables (Barclay *et al.*, 1995; Chin, 1998; Hair *et al.*, 2006). Second, the Average Variance Extracted (AVE) was above the recommended 0.50 threshold (see Table 2), which indicated that, for each latent variable, more than 50% of the variance was explained by their respective items (Chin, 1998; Vinzi *et al.*, 2010). Third, the convergent and discriminant validities were above the recommended 0.707 minimum loadings value (see Table 2 and Appendix A), which indicates that all latent variables were strongly correlated with their items and at least 50% of the variance of all items was shared with their respective latent variables (Barclay *et al.*, 1995).

Table 2: Construct reliability, AVE, cross-loadings, and correlations.

Latent variables	Outer-loadings[a]	Cronbach's α	Composite reliability	AVE	Cross-loadings[a]	Correlations matrix[b]
Management Orientation	0.718–0.848	0.852	0.894	0.629	0.718–0.848	0.793 (0.413–0.676)
Organisational Culture	0.768–0.848	0.870	0.906	0.658	0.768–0.848	0.811 (0.474–0.626)
Technology Orientation	0.809–0.844	0.886	0.916	0.686	0.809–0.844	0.828 (0.514–0.636)
Alliance and Cooperation	0.737–0.826	0.842	0.887	0.611	0.737–0.826	0.782 (0.413–0.549)
Market Orientation	0.684–0.839	0.842	0.888	0.614	0.684–0.839	0.784 (0.504–0.626)
Innovation Practices	0.736–0.802	0.922	0.934	0.587	0.736–0.802	0.766 (0.549–0.676)

[a]Outer-loadings and cross-loadings columns include the lowest and highest loadings.

[b]Correlation matrix column includes squared root AVE and the lowest and highest correlations in bracket.

Discriminant validity was further evaluated at two levels (see Table 2). Firstly, at the item level, the factor analysis of each item's cross-loadings revealed that the items loaded onto their respective latent variable to the highest degree and that no item loaded higher onto other latent variables (Gefen and Straub, 2005). For example, the item "flexible organisational culture" loaded more on to the organisational culture latent variable (0.790) than to the market orientation latent variable (0.439) or any other latent variable. Secondly, at the latent variable level, the square root of the AVE for each construct exceeded correlations with any others in the outer model, what is more these inter-correlations were low (Chin, 1998). For example, the square root of the AVE of the management orientation latent variable (0.793) exceeded the correlations (0.413–0.676) with other latent variables. With these results, it was determined that reliability, convergent and discriminant validities were acceptable to continue with the analysis.

Structural model

The structural model consists of six latent variables. To estimate the parameters in the inner model, a bootstrapping procedure with 200 and 500 re-samples consisting of the same number of cases as in the original sample (that is, 200) was applied (Barclay *et al.*, 1995; Vinzi *et al.*, 2010). Potential sign changes were treated by means of the option of individual sign changes. The goodness of the inner model was established by the strength of each structural path and the combined productiveness of its latent variables (Vinzi *et al.*, 2010). Blindfolding (communality and redundancy) on the latent variable innovation practices with 7 and 21 omission distances was applied.

First, the variance explained by the inner model in terms of the Squared Multiple Correlations (R^2) was examined (see Table 3). The R^2 value was between 1 and 0 and higher than the cut-offs of 0.67, 0.33 or 0.19 levels that are taken to indicate substantial, moderate or weak effect sizes, in that order (Chin, 1998; Vinzi *et al.*, 2010). The R^2 value for innovation practices was 0.631. This was satisfactory and indicated that 63.1% of the variation this construct was explained by organisational determinants latent variables. This result indicated an association of moderate to high strength.

Second, the goodness-of-fit (GoF) was calculated by the square root of the geometric mean of the average of R^2 (0.631) multiplied by the cross-validated communality (0.589), (Tenenhaus *et al.*, 2005). GoF values of 0.36, 0.25 or 0.10 levels are considered to indicate large, medium and small model fit in that order (Wetzels *et al.*, 2009). The GoF was 0.608 indicating good model fit. To further assess the predictive relevance and stability of the inner model, the Stone–Geisser test (Q^2) was determined by the latent variable innovation practices score from the cross-validated communality and redundancy, calculated by the blindfolding procedure of 7 and 21 omission distances (Vinzi *et al.*, 2010). The Q^2 value exceeding 0 means a good predictive relevance and a higher positive value is considered to have more predictive relevance (Vinzi *et al.*, 2010). The Q^2 value (0.589) was higher than zero indicating that the prerequisites of predictive relevance for the structural model are fulfilled (Chin, 1998).

Third, the path coefficients and *t*-values ($p < 0.05$) were calculated (see Table 3). The research questions and their relationships were examined by determining the magnitude of the standardised parameter estimates between latent variables together with the corresponding *t*-values (>1.96, $p < 0.05$). Results indicated support for the relationships in RQ1, RQ3, RQ4 and RQ5 but did not in RQ2. Management orientation (RQ1: 0.353, 5.49), technology orientation (RQ3: 0.230, 3.87), market orientation (RQ4: 0.224, 3.83) and alliance and cooperation (RQ5: 0.152, 2.31) have significant positive correlations with innovation practices. However, organisational culture (RQ2: 0.040, 0.80) had non-significant positive correlations with innovation practices. The path coefficients estimates and *t*-ratios were further recalculated using the bootstrap routine of 200 and 500 re-samples on the latent variable innovation practices (see Table 4). With these results, the path coefficients were very stable with respect to the original sample estimates and four paths were revealed as significant, yielding identical results to those reported for the original sample above.

The relative path coefficient strengths at the structural level were further evaluated by the size effects (see Table 4). This was done using Cohen's (1988)

Table 3: Structural Model R^2, path coefficients, and *t*-values measures.

Latent variables	R^2	Path coefficients	*t*-Values[a]
RQ1: Management Orientation → Innovation Practices	0.631	0.353	5.49***
RQ2: Organisational Culture → Innovation Practices	0.631	0.040	0.80
RQ3: Technology Orientation → Innovation Practices	0.631	0.230	3.87***
RQ4: Alliance and Cooperation → Innovation Practices	0.631	0.152	2.31*
RQ5: Market Orientation → Innovation Practices	0.631	0.224	3.83***

[a]*t*-values are significant at $p < 0.05$*, 0.01**, and 0.001***.

Table 4: Path coefficients and total effects of 200 and 500 bootstrapping procedures measures.

Latent variables	Mean of original sample	Mean of resample[a]	*t*-Ratios[b]	Supported
Total Effects on Innovation Practices ($R^2 = 0.631$)				
RQ1: Management Orientation (+ Sign Expected)	0.353	0.352 (0.351)	5.49*** (5.64***)	Yes
RQ2: Organisational Culture (+ Sign Expected)	0.040	0.042 (0.045)	0.56 (0.54)	No
RQ3: Technology Orientation (+ Sign Expected)	0.230	0.236 (0.231)	3.87*** (3.58***)	Yes
RQ4: Alliance and Cooperation (+ Sign Expected)	0.152	0.153 (0.153)	2.23* (2.39*)	Yes
RQ5: Market Orientation (+ Sign Expected)	0.224	0.224 (0.223)	3.83*** (3.67***)	Yes

[a]500 re-sample values are in bracket.
[b]*t*-ratios are significant at $p < 0.05$*, 0.01**, and 0.001***.

recommended gauges of > 0.02 for weak, > 0.15 for medium and > 0.35 for strong effects. Management orientation was found to have a strong effect (just > 0.35), technology orientation, alliance and cooperation and market orientation were found to have medium effects (all > 0.15 but < 0.02) while organisational culture was found to have a negligible effect (> 0.02) on innovation practices. Based on these results, there is evidence to suggest that all of the firm characteristics identified in the research questions, except for organisational culture, have a positive effect on innovation practice.

Discussion

This study examines the presence of established organisational determinants in SMEs reporting strong innovation practice in the emerging Dubai market in the UAE. The findings revealed that management orientation, technology orientation, market orientation and alliance and cooperation are important characteristics of these firms thus confirming extant literature. However, contrary to previous research, a firm's organisational culture did not emerge as an important factor. Based on the results of this study, a number of implications can be offered to both scholars and practitioners.

First, management orientation appears to be important to firms reporting strong innovation practices within SMEs, confirming the finding of previous studies (Covin and Slevin, 1991; Kumar *et al.*, 2008). The results imply management prioritisation and strategic direction towards innovation play important roles in shaping and supporting decisions to adopt and/or generate innovation; for example, incorporating innovations as strategic goals and future ambitions, focusing on long-term objectives, exploring new opportunities, involving proactively in new initiatives, and allocating resources for research and development activities.

Second, organisational culture, thought to be conducive to innovative practice, was not strongly evident in SMEs in Dubai, a finding inconsistent with previous studies in other countries (Tushman and O'Reilly, 1997; Kenny and Reedy, 2006) and in need of further research. The finding may be explained by studies that suggest SMEs in Dubai have a centralised organisational structure, formal procedures and process and high administrative intensity which inhibit the free flow of information that an organisational culture optimal for innovation suggests (Iseri and Demirbag, 1999; Kabasakal and Bodur, 2002). Further, it may be typical of SMEs in general. Scott *et al.* (1996) suggests that manufacturing SMEs have difficulties with knowledge absorption, learning attitude and individuals' contributions to new ideas, with the managing director taking many of the decision in a relatively autocratic manner.

Third, technology orientation seems to be moderately important to SMEs and their innovation practices thus confirming extant literature (Wilson *et al.*, 1999; Humpherys *et al.*, 2005). The results indicate that SMEs in Dubai realise that their technology policies and adaptation of new and emerging technology play important roles toward improving internal processes and allocating resources to support innovation. Cumming (1998) argues that the use of technologies by SMEs is considered to be innovative in its own right. Investment in new technology is rare due to resource constraints and capabilities, with most seeking to keep abreast of leading-edge technologies and acquiring available technologies (Hassink, 1996).

Fourth, alliance and cooperation look to be potentially important to innovation practices again confirming extant literature (Stuart, 2000; Hoffmann and

Schlosser, 2001). The results show that SMEs in Dubai utilise their networks and business linkages to collaborate more effectively. The low medium effect of this finding may indicate the impact of *wasta* and foreign ownership restrictions both of which are in need of future research. However, the diversity of resources in any alliance and cooperation portfolios appears to benefit innovation when resources and capabilities are shared across partners (Cui and O'Connor, 2012).

Finally, market orientation is evident in firms with strong innovation practices thus supporting the literature (Han *et al.*, 1998; Verhees and Meulenberg, 2004). The results imply that SMEs in Dubai realise the importance of their customer needs, their competitive situations, and their internal sharing of market information within the firm. The moderate effect may indicate the need to lead customers in this emerging market (Gao *et al.*, 2007) and/or the limited market research available (Rettab *et al.*, 2009). However, overall the findings point to recognition of the importance of a market orientation to a firm's innovative capacity (Von Hippel, 1988).

Managerial implications

This study has implications for policymakers, managements and scholars alike. First, the findings are generally supportive of current market environment in Dubai as findings largely produced in developed Western economies have also born out in Dubai SMEs. The notable exception of organisational culture points to the need to consider the impact of cultural norms like *wasta* or a directive or a paternalistic management style, however, it may equally be indicative of the autocratic style of many SMEs.

The importance of technology, management and market orientations vindicate the significant investment in the education and technology sectors in the UAE and this focus should be maintained. The investment in education seems particularly important, in parallel with the *emiratisation* programme, to help to ensure that young local managers have an understanding of the resource, capability and process requirements needed to instill a culture of innovation.

Policymakers also need to take further steps to encourage alliance and collaboration, for example, the relaxing of legal and regulatory conditions around ownership and governance requirements. Effective collaborations appears critical to the development of a country so dependent on outside resources for its development beyond the oil sector.

Limitations and future research

There are a number of limitations to this study. First, the surveys were answered by a single informant within each firm. Tests did not indicate a problem, but common method bias cannot be ruled out. Second, the generalisability of the findings is limited to those industries covered in the Dubai *emirate*. Third, a cross-sectional survey design was applied and as such any causality suggested is tentative. Fourth, participants in the study were single respondents and senior managers. Using multiple respondents (from different managerial levels or functional units) in each firm would provide deeper insight. Fifth, items under each constructs were modified and evaluated, but this study cannot claim a definite non-ambiguous link between items and constructs. Finally, self-report measures of innovation practice were relied upon. An objective measure of innovation outcomes would heighten the external validity of findings.

Future research should also extend to the other *emirates* of the UAE, as well as SMEs in comparable emerging markets like Singapore and other countries in the Middle East, Africa, Asia, Latin America and the Caribbean. Further, collection of longitudinal data should be conducted to support the casual relationships suggested in this study. The conceptualisation of organisational determinants and innovation practices could also be extended. While this study examines the most frequently mentioned dimensions, cognitive orientations such as quantitative versus qualitative and analytical versus intuitive, are also worthy of examination.

The lack of evidence, reported in this study, of an organisational culture considered optimal for innovation practices by previous studies, is perhaps the most interesting area of future research. If confirmed in new studies then researchers need to assess the theoretical and managerial implications. For example, is it possible that the Dubai business environment compensates in other ways?

Conclusion

This study examines the presence of established organisational determinants of innovation practice in SMEs in the Dubai market of the UAE. Applying a resource-based view of the firm; it lends support to previous studies, which were undertaken in developed markets. In doing so, this study confirms the importance of management orientation, technology orientation, alliance and cooperation and market orientation in an emerging economy context but highlights a notable inconsistency in the lack of evidence of a supportive organisational culture.

Overall, it appears that SMEs in Dubai already apply most of the procedures and processes recommended for innovation practice thus confirming the validity of findings in large firms in developed economies. However, an important anomaly has been highlighted and future research in varied markets is required.

Appendix A. Construct Measurement Items

Latent variables	Loadings[a]
Management Orientation	0.718–0.848
Item 1: Our management considers innovation to be part of our firm's strategic goals and future ambitions.	0.848
Item 2: Our management focuses on long-term objectives with interests in adopting innovations.	0.786
Item 3: Our management favors high-risk projects with attitudes in exploring new opportunities.	0.718
Item 4: Our management is involved in new initiatives and innovative programmes.	0.845
Item 5: Our management allocates resources to support research and development into innovative products or services.	0.762
Organisational Culture	0.768–0.848
Item 1: Our firm is open to different perspectives that accepts and adopts new and external ideas.	0.767
Item 2: Staff within our firm obtain and exchange new knowledge and skills in fair and collegial ways.	0.805
Item 3: Our firm accurately shares important information such as success and failure and customer feedback with all relevant staff as part of our internal learning processes.	0.843
Item 4: Our firm has a flexible organisational structure.	0.790
Item 5: Our firm encourages staff to think freely, generate ideas, follow-up on ideas, learn by experience, and take risks.	0.848
Technology Orientation	0.809–0.844
Item 1: Our firm's policy is to adopt up-to-date technologies.	0.809
Item 2: Our firm purchases and uses technologies to position itself ahead of competitors.	0.837
Item 3: Our firm is often to be first to try out new methods and technologies.	0.844
Item 4: Our firm frequently improves internal processes such as speed, reliability, and information management.	0.817

Appendix A. (*Continued*)

Latent variables	Loadings[a]
Item 5: Our firm allocates resources for investments in latest technologies and future forecasted technological changes.	0.834
Alliance and Cooperation	0.737–0.826
Item 1: Our firm has proactive relationships with partners, suppliers, and sub-contractors.	0.740
Item 2: Our firm has memberships in local and/or international business and industry associations.	0.790
Item 3: Our firm systematically identifies possible strategic partners to explore new knowledge and technology in order to improve resources and capabilities.	0.826
Item 4: Our firm has collaborative agreements with other firms to in/outsource research and development activities.	0.737
Item 5: Our firm has proactive networks with other firms to share innovation benefits and risks.	0.812
Market Orientation	0.684–0.839
Item 1: Our firm has proactive dialogues and mutual relationships with customers.	0.805
Item 2: Our firm is geared toward quality customer service culture.	0.839
Item 3: Our firm implements immediate responses when our customers are targeted by other firms.	0.782
Item 4: Our firm frequently takes advantage of targeted opportunities to benefit from competitors' weaknesses.	0.684
Item 5: Our firm encourages internal sharing of market information to understand consumer/competitor behaviours.	0.799
Innovation Practices	0.736–0.802
Item 1: Our firm frequently tries out new ideas.	0.739
Item 2: Our firm introduces number of new products, services, processes, or organisation/management systems.	0.736
Item 3: Our firm is first to market with new products or services.	0.796
Item 4: Our management seeks out new ways to do things.	0.785
Item 5: Our firm is creative in its methods of operation.	0.802
Item 6: Our firm uses up-to-date technologies.	0.758
Item 7: Our firm develops new market segments.	0.783
Item 8: Our firm uses new marketing methods.	0.771

Appendix A. (*Continued*)

Latent variables	Loadings[a]
Item 9: Our firm develops new ways of establishing relationships with customers.	0.746
Item 10: Our firm spends resources on research and development for new products, services, or processes.	0.738

[a]Outer-loadings and cross-loadings are included.

References

Abubakr, M (2006). *Managing Human Resource in the Middle East: Human Resource Management in the United Arab Emirates*. London: Taylor & Francis.

Allocca, MA and EH Kessler (2006). Innovation speed in small and medium-sized enterprises. *Creativity and Innovation Management*, 15, 279–295.

Aragon-Sanchez, A and G Sanchez-Marin (2005). Strategic orientation, management characteristics, and performance: A study of Spanish SMEs. *Journal of Small Business Management*, 43, 287–308.

Armstrong, J and T Overton (1977). Estimating nonresponse bias in mail surveys. *Journal of Marketing Research*, 14, 396–402.

Avci, U, M Madanoglu and F Okumus (2011). Strategic orientation and performance of tourism firms: Evidence from a developing country. *Tourism Management*, 32, 147–157.

Avlonitis, G and S Gounaris (1999). Marketing orientation and its determinants: An empirical analysis. *European Journal of Marketing*, 33, 1003–1037.

Baldridge, JV and RA Burnham (1975). Organisational innovation: Individual, organisational and environmental impacts. *Administrative Science Quarterly*, 20, 165–76.

Barclay, D, C Higgins and R Thompson (1995). The partial least squares (PLS) approach to causal modeling: Personal computer adoption and use as an illustration. *Technology Studies*, 2, 285–309.

Barney, JB (1991). Firm resources and sustained competitive advantage. *Journal of Management*, 17, 99–120.

Batonda, G and C Perry (2003). Approaches to relationship development processes in inter-firm networks. *European Journal of Marketing*, 37, 1457–1484.

Baum, JA, T Calabrese and BS Silverman (2000). Don't go it alone: Alliance network composition and start-ups' performance in Canadian biotechnology. *Strategic Management Journal*, 21, 267–294.

Berkhout, G, D Hartmann and P Trott (2010). Connecting technological capabilities with market needs using a cyclic innovation model. *R&D Management*, 40, 474–490.

Berry, A, E Rodriquez and H Sandee (2002). Firm and group dynamics in the small and medium enterprises sector in Indonesia. *Small Business Economics*, 18, 141–161.

Blumentritt, T and WM Danis (2006). Business strategy types and innovative practices. *Journal of Management Issues*, 18, 274–291.

BMI (2007). *The UAE Business Forecast Report*. UAE: Business Monitor International Ltd.

Brik, AB, B Rettab and K Mellahi (2011). Market orientation, corporate social responsibility, and business performance. *Journal of Business Ethics*, 99, 307–324.

Brown, L (1990). *Competitive Marketing Strategy, Developing, Maintaining, and Defending Competitive Position*. Melbourne, Victoria: Thomas Nelson Inc.

Buckler, SA (1997). The spiritual nature of innovation. *Research-Technology Management*, 40, 43–47.

Cagliano, R, K Blackmon and C Voss (2000). Small firms under the microscope: International differences in production/operation management practices and performance. *Integrated Manufacturing Systems*, 12, 469–482.

Calantone, R, S Cavusgil and Y Zhao (2002). Learning orientation, firm innovation capability, and firm performance. *Industrial Marketing Management*, 31, 515–524.

Carrier, C (1994). Intrapreneurship in large firms and SMEs: A comparative study. *International Small Business Journal*, 12, 54–61.

Chin, WW (1998). The partial least square approach for structural equation modeling. In Marcoulides GA (ed.), *Modern Methods for Business Research*, pp. 295–336. Mahwah, NJ: Lawrence Erlbaum Associates.

Cohen, J (1988). *Statistical Power Analysis for the Behavioral Sciences*. Hillsdale, IL: Lawrence Erlbaum.

Cooper, RG (1994). New products: The factors that drive success. *International Marketing Review*, 11, 60–76.

Covin, JG and DP Slevin (1990). New venture strategic posture, structure, and performance: An industry life cycle analysis. *Journal of Business Venturing*, 5, 123–135.

Covin, JG and D Slevin (1991). A conceptual model of entrepreneurship as firm behavior. *Entreprenurship: Theory and Practice*, 16, 7–25.

Cromer, CT, C Dibrell and JB Craig (2011). A study of Schumpterian (radical) vs. Kirzerian (incremental) innovation in knowledge intensive industries. *Journal of Strategic Innovation and Sustainability*, 7, 28–42.

Cui, AS and G O'Connor (2012). Alliance portfolio resource diversity and firm innovation. *Journal of Marketing*.

Cumming, BS (1998). Innovation overview and future challenges. *European Journal of Innovation Management*, 1, 21–29.

Dawes, J (2000). Market orientation and company profitability: Further evidence incorporating longitudinal data. *Australian Journal of Management*, 25, 173–199.

DCCI (2010). *Dubai Chamber of Commerce and Industry*. Dubai: The Information Centre.

Dillman, DA (2007). *Mail and Internet Survey: The Tailored Design Method*. Hoboken: John Wiley & Sons Inc.

Doty, DH and WH Glick (1998). Common methods bias: Does common methods variance really bias results? *Organizational Research Methods*, 1, 374–406.

Dutta, S (2006). *The Arab Innovation Index*. France: INSEAD.

Eisenhardt, KM and JA Martin (2000). Dynamic capabilities: What are they? *Strategic Management Journal*, 21, 1105–1121.

Elewa, AA (2007). New UAE commercial companies law soon. *Gulf News*, January 26.

Ettlie, JE and WP Bridges (1982). Environmental uncertainty and organizational technology. *IEEE Transactions on Engineering Management*, EM-29, 2–10.

Ettlie, JE, WP Bridges and RD O'Keefe (1984). Organization strategy and structural differences for radical versus incremental innovation. *Management Science*, 30, 682–695.

Feigenbaum, A and A Karnani (1991). Output flexibility: A competitive advantage for small firms. *Strategic Management Journal*, 12, 101–114.

Forsman, H and S Temel (2011). Innovation and business performance in small enterprises. An enterprise-level analysis. *International Journal of Innovation Management*, 15, 641–665.

Freel, MS (1999). Where are the skills gaps in innovative small firms? *International Journal of Entrepreneurial Behaviour and Research*, 5, 144–145.

Gao, GY, KZ Zhou and CK Yim (2007). On what should firms focus in transitional economies? A study of the contingent value of strategic orientations in China. *International Journal of Research in Marketing*, 24, 3–15.

Gefen, D and DW Straub (2005). A practical guide to factorial validity using PLS-Graph: Tutorial and annotated example. *Communications of the AIS*, 16, 91–109.

Ghobadian, A and D Gallear (1997). TQM and organisation size. *International Journal of Operations and Production Management*, 17, 121–163.

Glaister, KW and PJ Buckley (1996). Strategic motives for international alliance formation. *Journal of Management Studies*, 33, 301–332.

Granell, E (2000). Culture and globalisation: A Latin American challenge. *Industrial and Commercial Training*, 32, 89–93.

Grant, J, FS Golawala and DS McKechnie (2007). The United Arab Emirates: The twenty-first century beckons. *Thunderbird International Business Review*, 49, 507–533.

Gray, B, S Matears, C Boshoff and P Matheson (1998). Developing a better measure of market orientation. *European Journal of Marketing*, 32, 884–903.

Gulati, R, N Nohria and A Zaheer (2000). Strategic networks. *Strategic Management Journal*, 21, 203–215.

Hadjimanolis, A (1999). Barriers to innovation for SMEs in a small less developed country (Cyprus). *Technovation*, 19, 561–570.

Hair, J, WC Black, BJ Babin, RE Anderson and RL Tatham (2006). *Multivariate Data Analysis*. Upper Saddle River, NJ: Pearson Education Inc.

Hale, A and P Cragg (1996). Business process reengineering in the small firm: A case study. *INFOR*, 34, 15–29.

Han, JK, N Kim and RK Srivastava (1998). Market orientation and organizational performance: Is innovation a missing link? *Journal of Marketing*, 62, 30–45.

Hassink, R (1996). Technology transfer agencies and regional economic development. *European Planning Studies*, 14, 167–184.

Henard, DH and DM Szymanski (2001). Why some new products are more successful than others. *Journal of Marketing Research*, 38, 362–375.

Henseler, J, CM Ringle and RR Sinkovics (2009). The use of partial least squares path modeling in international marketing. In Sinkovics RR and Ghauri PN (eds.), *Advances in International Marketing*, pp. 277–320. Bingley: Emerald.

Herb, E, K Leslie and C Price (2001). Teamwork at the top. *The McKinsey Quarterly*, 2, May.

Hertog, S (2010). *Benchmarking SME Policies in the GCC: A Survey of Challenges and Opportunities*. Brussels: Euro Chamber.

Hitt, M, R Hoskisson and R Ireland (1990). Mergers and acquisitions and managerial commitment to innovation in M-form firms. *Strategic Management Journal*, 11, 29–47.

Hjalager, A-M (2010). A review of innovation research in tourism. *Tourism Management*, 31, 1–12.

Hoffmann, WH and R Schlosser (2001). Success factors of strategic alliances in small and medium-sized enterprises: An empirical study. *Long Range Planning*, 34, 357–381.

Homburg, C and O Jensen (2007). The thought worlds of marketing and sales: Which differences make a difference? *Journal of Marketing*, 71, 124–142.

Hult, GT, RF Hurley and GA Knight (2004). Innovativeness: Its antecedents and impact on business performance. *Industrial Marketing Management*, 33, 429–438.

Humpherys, P, R McAdam and J Leckey (2005). Longitudinal evaluation of innovation implementation in SMEs. *European Journal of Innovation Management*, 8, 283–304.

Hurley, RF and GT Hult (1998). Innovation, market orientation, and organizational learning: An integration and empirical examination. *Journal of Marketing*, 62, 42–54.

Hutchings, K and D Weir (2006). Guanxi and wasta: A comparison. *Thunderbird International Business Review*, 48, 141–156.

Ireland, R, M Hitt and D Vaidyanath (2002). Alliance management as a source of competitive advantage. *Journal of Management*, 28, 413–446.

Iseri, A and M Demirbag (1999). Overcoming stereotyping: Beyond cultural approach. *Middle East Business Review*, 3, 1–21.

Jaworski, BJ and AK Kohli (1993). Market orientation: Antecedents and consequences. *Journal of Marketing*, 57, 53–70.

Kabasakal, H and M Bodur (2002). Arabic cluster: A bridge between East and West. *Journal of World Business*, 37, 40–54.

Kenny, B and E Reedy (2006). The impact of organisational culture factors on innovation levels in SMEs: An empirical investigation. *Irish Journal of Management*, 27, 119–142.

Keskin, H (2006). Market orientation, learning orientation, and innovation capabilities in SMEs: An extended model. *European Journal of Innovation Management*, 9, 396–417.

Kim, L and Y Lim (1988). Environment, generic strategies and performance in a rapidly developing country: A taxonomic approach. *Academy of Management Journal*, 31, 802–827.

Knight, J (2011). Education hubs: A fad, a brand, an innovation? *Journal of Studies in International Education*, 15, 221–240.

Kohli, AK, BJ Jaworski and A Kumar (1993). MARKOR: A measure of market orientation. *Journal of Marketing Research*, 30, 467–477.

Kumar, N, LW Stern and LC Anderson (1993). Conducting interorganizational research using key informants. *Academic Management Journal*, 36, 1633–1651.

Kumar, U, V Kumar and D Grosbois (2008). Development of technological capability by Cuban hospitality organizations. *International Journal of Hospitality Management*, 27, 12–22.

Laforet, S and J Tann (2006). Innovative characteristics of small manufacturing firms. *Journal of Small Business and Enterprise Development*, 13, 363–380.

Latham, S (2009). Contrasting strategic response to economic recession in start-up versus established software firms. *Journal of Small Business Management*, 47, 180–201.

Lin, C (1998). Success factors of small and medium-sized enterprises in Taiwan: An analysis of cases. *Journal of Small Business Management*, 36, 43–56.

Mahemba, CM and EJ De Bruijn (2003). Innovation activities by small and medium-sized manufacturing enterprises in Tanzania. *Creativity and Innovation Management*, 12, 162–173.

Malerba, F, R Nelson, L Orsenigo and S Winter (2007). Demand, innovation, and the dynamics of market structure: The role of experimental users and diverse preferences. *Journal of Evolutionary Economics*, 17, 371–399.

Martensen, A, JJ Dahlgaard, SM Park-Dahlgaard and L Gronholdt (2007). Measuring and diagnosing innovation excellence — Simple contra advanced approaches: A Danish study. *Measuring Business Excellence*, 11, 51–65.

Martinez-Roman, JA, J Gamero and JA Tamayo (2011). Analysis of innovation in SMEs using an innovative capability-based non-linear model: A study in the province of Seville (Spain). *Technovation*, 31, 459–475.

Martins, EC and F Terblanche (2003). Building organizational culture that stimulates creativity and innovation. *European Journal of Innovation Management*, 6, 64–74.

Massa, S and S Testa (2008). Innovation and SMEs: Misaligned perspectives and goals among entrepreneurs, academics, and policy makers. *Technovation*, 28, 393–407.

McGinnis, MA and RM Ackelsberg (1983). Effective innovation management: Missing link in strategic management. *The Journal of Business Strategy*, 4, 59–66.

McKee, D (2003). An organizational learning approach to product innovation. *Journal of Product Innovation Management*, 9, 232–245.

Mulhern, A (1995). The SMEs sector in Europe: A broad perspective. *Journal of Small Business Management*, 33, 83–89.

Myers, S and DG Marquis (1969). *Successful Industrial Innovation: A Study of Factors Underlying Innovation in Selected Firms*. Washington: National Science Foundation.

Mytelka, LK (2000). Local systems of innovation in a globalised world economy. *Industry and Innovation*, 7, 15–32.

Narver, JC and SF Slater (1990). The effect of a market orientation on business profitability. *Journal of Marketing*, 54, 20–35.

North, D and D Smallbone (2000). The innovativeness and growth of rural SMEs during the 1990s. *Regional Studies*, 34, 145–157.

Peebles, E (2003). Inspiring innovation. In *Harvard Business Review on the Innovative Enterprise*, pp. 89–110. Cambridge: Harvard Business School Press.

Pelham, AM and DT Wilson (1996). A Longitudinal study of the impact of market structure, firm structure, strategy, and market orientation culture dimensions of small-firm performance. *Journal of the Academy of Marketing Science*, 24, 27–43.

Porter, ME (1980). *Competitive Strategy: Techniques for Analyzing Industries and Competitors*. New York: Free Press.

Rettab, B, AB Brik and K Mellahi (2009). A study of management perceptions of the impact of corporate social responsibility on organisational performance in emerging economies: The case of Dubai. *Journal of Business Ethics*, 89, 371–390.

Rettab, B and A Rao (2008). Analyzing performance of exporters in Dubai an emerging economy. *International Review of Business Research Papers*, 4, 278–301.

Rogers, E (2003). *Diffusion of Innovations*. New York: Free Press.

Salavou, H, G Baltas and S Lioukas (2004). Organisational innovation in SMEs: The important of strategic orientation and competitive structure. *European Journal of Marketing*, 38, 1091–1112.

Scott, P, B Jones, A Bramley and B Bolton (1996). Enhancing technology and skills in small and medium-sized manufacturing firms: Problems and prospects. *International Small Business Journal*, 14, 85–97.

Scozzi, B, C Garavelli and K Crowston (2005). Methods for modelling and supporting innovation processes in SMEs. *European Journal of Innovation Management*, 8, 120–137.

Shama, A (1993). Marketing strategies during recession: A comparison of small and large firms. *Journal of Small Business Management*, 31, 62–72.

Simon, H (1996). *Hidden Champions, Lessons from 500 of the World's Best Unknown Companies*. Boston, MA: Harvard Business Press.

Sinangil, HK (2004). Globalisation and managing organisational culture change: The case of Turkey. *Psychology Developing Societies*, 16, 27–40.

Snow, CC and LG Hrebiniak (1980). Strategy, distinctive competence, and organizational performance. *Administrative Science Quarterly*, 25, 317–336.

Srinivasan, R, GL Lilien and A Rangaswamy (2002). Technological opportunism and radical technology adoption: An application to e-business. *Journal of Marketing*, 66, 47–60.

Stuart, TE (2000). Interorganizational alliances and the performance of firms: A study of growth and innovation rates in a high-technology industry. *Strategic Management Journal*, 21, 791–811.

Suliman, AMT (2006). Human resource management in the United Arab Emirates. In *Managing Human Resources in the Middle East*, PS Budhwar and K Mellahi (eds.), pp. 59–78. London: Routledge.

Tenenhaus, M, VE Vinzi, Y-M Chatelin and C Lauro (2005). PLS path modeling. *Computational Statistics and Data Analysis*, 48, 159–205.

Tether, B (1998). Small and large firms: Sources of unequal innovations? *Research Policy*, 27, 725–745.

Tornatzky, LG and KJ Klein (1982). Innovation characteristics and innovation adoption implementation: A meta-analysis of findings. *IEEE Transactions on Engineering Management*, 29, 28–45.

Tucker, RB (2002). *Driving Growth Through Innovations*. San Francisco, CA: Berrett-Koehler Publishers.

Tushman, ML and CA O'Reilly (1997). *Winning Through Innovation: A Practical Guide to Leading Organisational Change and Renewal*. Cambridge, UK: Harvard Business School Press.

Varadarajan, PR and S Jayachandran (1999). Marketing strategy: An assessment of the state of the field and outlook. *Journal of the Academy of Marketing Science*, 27, 120–143.

Verhees, F and M Meulenberg (2004). Market orientation, innovativeness, product innovation, and performance in small firms. *Journal of Small Business Management*, 42, 134–154.

Vinzi, VE, WW Chin, J Henseler and H Wang (2010). *Handbook of Partial Least Squares: Concepts, Methods and Applications*. Heidelberg, Baden-Württemberg: Springer-Verlag.

Von Hippel, E (1988). *The Sources of Innovation*. New York, NY: Oxford University Press.

Vossen, R (1998). Relative strengths and weaknesses of small firms in innovation. *International Small Business Journal*, 16, 88–94.

WEF (2010). Arab world competitiveness review. World Economic Forum.

Wetzels, M, G Odekerken-Schröder and C van Oppen (2009). Using PLS path modeling for assessing hierarchal construct models: Guidelines and empirical illustration. *MIS Quarterly*, 33, 177–195.

Wilson, AL, K Ramamurthy and PC Nystrom (1999). A multi-attribute measure for innovation adoption: The context of imaging technology. *IEEE Transactions on Engineering Management*, 46, 311–321.

Yusof, S and E Aspinwall (1999). Critical success factors for total quality management implementation in small and medium enterprises. *Total Quality Management*, 10, 803–809.

Chapter 10

PERSPECTIVES ON INNOVATION MANAGEMENT OF ECUADORIAN COMPANIES: EMPIRICAL EVIDENCE

Guillermo Javier Diaz-Villavicencio
Universidade Federal da Integração Latino-Americana UNILA
Instituto Latino-Americano de Economia, Sociedade e Política

Simone Didonet
Department of Business Administration
Federal University of Paraná UFPR, Av. Lothario Meissner

Alexander Dodd
Universidad Técnica Particular de Loja UTPL
Faculty of Business Administration, San Cayetano Alto

This paper examines the relationship of 40 practices of innovation management lifted from Tidd *et al.* (2013, p. 634) in 477 Ecuadorian companies. The sample is from the period 2013–2014 and are divided into micro, small, medium and large enterprises. The management practices are divided a priori into 5 constructs (strategy, processes, organisation, relationships and learning). We then carry out a descriptive analysis, where the degree of the implementation of practices in the entirety of the businesses are stipulated. Later, we test a model with inferential analysis (multiple regression analysis) where the degrees of significance of the practices are determined within each construct and their relation to the size of the companies. The innovation management potential in the Ecuadorian business environment is also discussed. The results show, that companies are far from displaying constant management practices in innovation, as such, general recommendations for companies in the study are given.

Keywords: Innovation; strategy; management.

This paper was first published in *International Journal of Innovation Management*, Vol. 20, No. 6 (August 2016), 1650048 (20 pages). © World Scientific Publishing Europe Ltd. DOI: 10.1142/S1363919616500481

Introduction

The competitiveness of a nation depends on the ability of its industries to innovate and improve their activities (Porter, 1998). In the same vein, Scott and Lodge (1995) reason that competitiveness is increasingly a matter of structures and strategies, and the reverse is increasingly a consequence of the natural resource endowment of a country. Meanwhile, Ezeala-Harrison (1999) shows that competitiveness in international terms could be defined as the relative ability of companies from one country to produce and market products of superior quality at lower prices.

As such, nations seek competitive advantages that differentiate and empower their business in international markets as this produces the expected benefits for the population (Porter, 1998). At the same time, each organisation knows *a priori* that innovation is an important part of the industrial engine of an economy. Heunks (1998) reinforces this fact by noting that innovation tends to increase growth and business efficiency due to the investment incurred. Therefore, the extent to which knowledge and financial resources are effectively managed is linked to the advantages obtained from innovation (Darroch, 2005). In this sense, we see that companies have had to reallocate their factors of production — as demonstrated by a study with Asian companies (van der Zee, 2006) — to build a competitive advantage through the discovery, access, mobilisation and utilisation of knowledge worldwide (Doz *et al.*, 2001). Innovative practices should be an integral part of everyday operations for organisations as they generate potential economic and social changes for maintaining high levels of profitability (Drucker, 1998; Christensen, 1997). In itself, innovation becomes the expression of the dynamic capabilities of organisations (Teece *et al.*, 1997). In this regard, it is important to mention the Competitiveness Index measured by the The Global Competitiveness Report 2011–2012, which stresses Ecuador as one of the countries with progress on the issue of competitiveness (compared to the previous year). In fact, Ecuador jumped four places from 105 to 101. In terms of innovation, Ecuador is ranked 103rd with a score of 3.76 with 5.79 (Switzerland) scoring highest. Furthermore, the progress of Ecuador in the general index of basic citizen requirements in 82nd position, highlights progress in macroeconomic policies (40th position), health and primary education (70th position) followed by rates in infrastructure and institutional development (94th and 125th positions, respectively). The report highlights Ecuador as one of the three Latin American countries that have experienced sustained growth in the competitive factors and environmental indicators.

This overall increase in productivity for Ecuador is due to trade liberalisation which has had a positive and significant effect on various economic agents which have focused mainly on industrial manufacturing exports in recent years (Wong, 2009). This helped with the commercial liberalisation of production factors, aimed at developing the country, theoretically helping the country's poorest (Winters *et al.*, 2004).

As such, the aim of this paper is to analyse the context of innovation management in Ecuadorian companies based on questionnaire proposed by Tidd *et al.* (2005). The paper is structured as follows: in the next section we present the theoretical framework of innovation. The methodology employed follows by proposing a basic multi-variable regression model. Finally, we present the discussion and conclusions sections.

Theoretical Framework

The search for the "creative destruction" of Schumpeter (1950) directs us towards a constant exploration to create something new, which simultaneously destroys the old rules and sets new ones, altering the economic balance of a given market and generating extraordinary profits above the market average. This is what we call *innovation*. Innovation is divided into radical and/or incremental innovation, which generates something new for the world in the first case and one that generates continuous changes in processes in the incremental case, with both mechanisms having positive effects on organisational performance (Ettlie, 1999: Lin and Chen, 2007; Bhaskaran, 2006; Auken *et al.*, 2008). The above guides the organisation to stay in the market depending on the spaces and navigating between the different types of innovation known as the four Ps. (Process, Paradigm, Product, Position) (Francis and Bessant, 2005).

In the same vein, we know that manufacturing sectors maintain high levels of organisational practices and management for new product development (Miles, 2000; Tidd *et al.*, 2005). We also note that innovation in products/processes depend on the volumes of resources (internal and/or external) that an organisation can acquire, as such to be able to stay in the market in which it operates (Oerlemans *et al.*, 2001b; Freel, 2005). It can be also seen that the service sector makes a contribution to the development of management practices in terms of Organisation, Processes, Tools, Technologies and Systems (OPTS) analysed in Tidd and Hull (2006). Likewise, we understand that there are dynamic innovation models and not all companies have equal capacities for innovation in their branches as is the case of multi-nationals (Richardson and Khan, 2002). This perspective goes along with the complexity of acquiring and developing new technologies for human and business networks and capabilities to maintain high levels of continuous training of their professional staff (Tidd, 1997).

The organisations take information from the competitive environment in which they are involved in order to develop new values (technological solutions) to maintain or expand their market position. From this, we understand that innovation is based on knowledge as an ongoing process that has a pattern to follow (routines and/or organisational practices) to thus reduce the uncertainty or lack of resources required to maintain continuous growth and/or strategic maintenance, generating competitive advantages that position us in the market (Porter, 1980).

Continuous innovation, as an incremental strategy, very often occurs *a priori* within a set of rules and/or established relationships, where actors (entrepreneurs of all sizes) tend to play an active role in the search for greater productivity, efficiency and market dominance. Discontinuous innovation, as a management strategy, aims to achieve a radical innovation in products and/or processes, which requires a huge effort for any organisation, as it will need large investments in R&D. This would practically make normal development go awry which should be organised (a lack of rules of the game), leaving much of the decision making to chance (random), without guaranteeing a product and/or process being truly radical (Tidd *et al.*, 2005).

Based on the above, it can be argued that innovation depends not only on a strategy, but on several components that generate a knowledge-based architecture with different mechanisms in different types of innovation (Tidd *et al.*, 2005). Therefore, is difficult to associate innovation to a particular type of technology or market as such (Henderson and Clark, 1990).

Our research argues that the evaluation of innovation management is not such an easy task, and there have been several attempts to generate kits, models and tools to measure innovation management (Ayhan and Oztemel, 2014; Francis, 2001; Utterback and Abernathy 1975; By, 2005; By *et al.*, 2011; Bessant and Caffyn, 1997; Didonet and Díaz, 2012). Our particular study focuses on evaluating five important areas in innovation management suggested by Tidd *et al.* (2005): organisation; processes; strategy; relationship; learning. We specified these areas in the questionnaire that has been applied to the studied sample of Ecuadorian firms.

According to Tidd *et al.* (2005), "organisation" allows us to know whether an organisation has adequate methods of developing new technological processes, if there is the flexibility to implement new practices that lead us to develop new products/processes, and to see if the organisation generates spaces to develop new ideas in the workers, etc. "Processes" indicate how to generate radical innovation in business, to see if we allocate specific resources for the exploitation of innovative options, and analyse if we are afraid to change everything we do, etc. "Strategy" indicates whether we employ approaches to research and learning in order to explore new directions in technology and markets, where we see if we work as peripheral users to develop new products and services, etc. "Relationships" allow us to see whether we focus on innovative practices and improve practices, to see if we practice open innovation, networking outside our company from which we get a constant stream of challenging ideas, etc. Finally, "learning" lets us know whether we use formal techniques to manage and learn from outside of our industry, we use technology that helps us to be more agile and quick in awareness and response to emerging threats and peripheral opportunities etc.

The above leads us to ask whether there are institutionalised practices that lead to sustained growth of the innovation variable within a scheme of "innovation management" within productive organisations in Ecuador. Otherwise, everything could be random, and down to the circumstances of an organisation (of its own struggle to survive). In this regard, we discuss the following hypothesis:

H1: *Ecuadorian firms emphasize different practices of innovation management according to their size.*

To discuss this proposition, we reviewed the literature and found that the measurement model of innovation management presented by Tidd *et al.* (2005) is a suitable model and has been tested by the authors in various countries with different types of organisations. The measurement model proposed by the authors contains 40 questions that refer to management practices regarding continuous and/or incremental innovation pattern, and discontinuous innovation patterns. The practices are enclosed in five constructs: strategy, processes, organization, relationships and learning.

Methodology and Model

The research is descriptive with a quantitative emphasis. In this regard, our work is based on the implementation of the questionnaire proposed by Tidd *et al.* (2005) and consisted of 40 questions divided into the 5 factors: strategy, processes, organisation, relationships and learning. The questionnaire was translated into Spanish for application in Ecuador.

The questionnaire was designed to answer the question "How well do we manage innovation?" The survey was a self-assessment tool that consists of assertions and denials, and was measured via the Likert Scale (1932) in which 1 is "not true at all" and 7 is "Very true" (see the questionnaire in Appendix A). The results helped us see if there was a behavioural pattern of the organisations when approaching the subject of continuous and discontinuous innovation in different time periods.

In order to classify the micro, small, medium and large enterprises of Ecuador, we have taken as reference the company size by the number of employees, as set forth in Resolution 1260 issued by the Secretariat of the Andean Community, Lima (Peru). This resolution has been accepted by the Organic Code of Production, Trade and Investment of Ecuador. In this regard, we considered micro enterprises as having between 1 to 9 employees, small enterprises as having 10 to 49 employees, medium enterprises as having 50 to 199 employees, and large enterprises as having 200 or more employees. According to the National Economic Census of 2010, there are 496.708 SMEs exist in Ecuador. For our study we split the companies according to their most profitable activity, dividing them into production companies, services and trade.

The first stage of the study was to apply the questionnaire outlined above adding to it basic data of the companies surveyed (company name, number of workers, educational level manager or senior person, indicate whether the company's $R+D$ and how much percentage spent on R&D) and then generate a descriptive and inferential analysis using a second step of the basic econometric model of multiple linear regression. As such, we can consider that and y^*_{it} represents a censored sample (see Novales, 1993, p. 552) in a linear function of K explanatory variables (x_k where $k = 1, 2, 3, \ldots, k$):

$$y^*_{it} = \beta_0 + \beta_1 x_{1it} + \beta_2 x_{2it} + \cdots + \beta_k x_{kit} + u_{it} \tag{3.1}$$

simpler:

$$y^*_{it} = \beta_0 + \sum_{k=1}^{k} \beta_k x_{kit} + u_{it}, \tag{3.2}$$

where $i = 1, \ldots, \mathrm{N}$ units and $t = 1, \ldots, \mathrm{T}$ observations in the period; in matrix notation would be our econometric model:

$$y^*_{it} = \beta'_k x_{kit} + u_{it}. \tag{3.3}$$

Furthermore, u_{it} is also the error term that represents the effects of all other variables omitted from the model. That is, it is the observed variation of the dependent variable, and cannot get explained by the variation observed in the k independent variables. Expressions (3.1)–(3.3) are identical. The letters β_0 β_1 $\beta_2, \ldots, \beta_k$

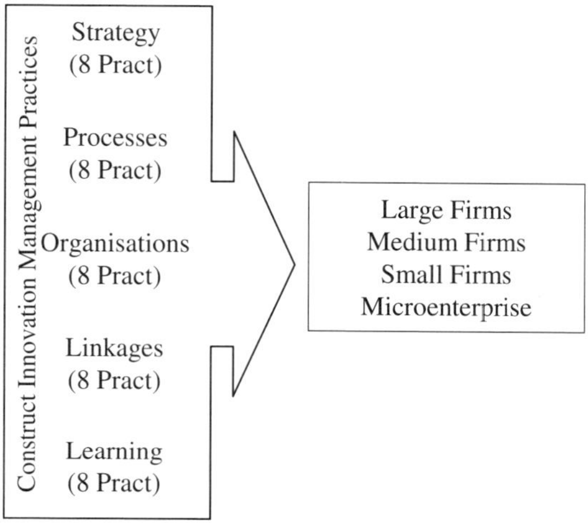

Fig. 1. Conceptual model.

are the parameters that need to be measured. β_0 is the intercept, the constant term; while the other parameters are the slopes of y^*_{it} with respect to each of K independent variables.

For the purpose of this investigation, the formulation (3.3) was used to test the 40 questions within each construct (see Appendix 1) and its relation to the size of the company (see Table 3) in this way we can determine the significance of each question with your construct.

To do this, we generated four separate databases by company size (with information from the 477 surveys and 40 practices evaluated in two years) and generate a regression for each construct (Strategy, Processes, Organisation, Relationships and Learning). In this regard, 24 models that validate practices regarding its construct and firm size were performed. The conceptual model is shown in Fig. 1.

Data Sampling

The sample consists of 477 micro, small, medium and large companies surveyed in Ecuador. The questionnaire was answered in two periods: second half of 2013 and first half 2014. The survey was administered by students of the Master of Business Administration, The Technical University of Loja, Ecuador, as part of the module of Research, Development and Business Innovation. The students themselves selected the companies surveyed, as they were generally companies in which they worked.

Results

Descriptive analysis of the sample

The scope of the sample reaches 477 companies, with 46.8% surveyed in 2013 and 53.2% companies in 2014. The division based on Ecuadorian law resulted in over the two years the following groups of enterprises in terms of size: 3.4% of microenterprises, 40.7% of small businesses with fewer than 50 employees, 23.9% of companies having between 50 and 199 employees (medium) and 32.1% are considered "large Ecuadorian companies" as can be observed in Table 1.

The sectors of production and services bring together 82.3% of the sample, with 17.6% in the commercial sector.

We sampled a large number of professionals across all business levels (63.7% of the total), concentrating mostly on small business, and draw attention to the low number of technicians across the board in all categories of business (3.1%). Similarly, a sharp increase in the number of graduates according to company size can be observed. As the company grows the number of professionals with post graduate qualifications grows too, with it being evident the number of post graduate in large companies.

Regarding R&D for just 2014, when the questions of investment in R&D and the use of a R&D department are included, the data shows clearly that there are low levels of investment in R&D by the 254 companies surveyed in 2014. A fewer than 80% of companies state that they invest less than 1% in R&D, whereas only 42.9% of companies have a R&D department. Also, we can see that only the small, medium and large companies feel as though they are capable of generating some sort of real long-term innovations. What is remarkable is that, as the firm size increases, so too does the creation of R&D departments, with one being able to assume that there is greater awareness for the need to have a department of R&D.

Table 1. Sample size.

Sample size (%)	
Micro	3.4
Small	40.6
Medium	23.9
Large	32.1
Total	100

Descriptive analysis of the survey

The questionnaire titled "How well do we manage innovation?" was applied in consecutive years in 2013 and 2014.

The strongest reply for the 2013 questionnaire was in relation to the section concerning strategy. This returned results demonstrating a mean high of 5.5, slightly higher than learning and organisation. With a mean of 5.9, Question 26 was the most strongly answered, claiming, that there is the support and agreement from higher levels of management for innovation within the company.

Three questions (1, 6 and 32) tied with a mean low of just 5.3 for year 2013 in strategy variable, including, that the employees have a clear understanding of how innovation can make a company more competitive, that the goals and strategy of innovation are clearly explained to the staff so that they know what it is they are looking to improve and that they have processes to examine new technological or market developments and determine what they mean to their business strategy.

The results of "strategy" construct demonstrate that the interviewees claim for to be support for innovation from higher management, however this support is not transmitted to the workers via explanations of innovation nor do they have processes to examine new technological and market developments and the outcomes of these for their business strategy.

Moving on to the results of "learning", the overall findings for the topic retrieved a mean of 5.4. Tellingly, the strongest answered question (number 15), yielding a mean of 5.9 was that the companies studied innovated most frequently when learning from their mistakes. The weakest answered question returned with a mean of just 4.5 was concerning whether companies innovate when meeting and discussing issues with other firms as a method of learning through others (question number 25).

In relation to "organisation", we were returned results with an overall mean of 5.4, a similar level to learning. A lofty 5.8 was found in reply to Question 38 showing that the companies work well in internal teams. However, the companies did not appear to agree that their reward and recognition system supported innovation, with a mean of 4.9 for question number 28 for the 224 firms researched and interviewed. These results appear to show that whilst the companies work well together in search of learning and innovation, they do not have a system of reward or recognition when this is achieved.

"Processes" was the next topic analysed in questionnaire one, having an overall mean of 5.3 for the 224 firms in the eight questions aimed at processes. The interviewees answered strongly (with a mean of 5.6) that the companies had in place appropriate processes that help the management develop new products in an effective way, from the start of the process to the product launch (Question 2). However, they did not agree so strongly with the statement that they had a clear structure for choosing innovation, with a mean of just 4.9 returned for question number 32. This is a similar sort of result that we found for organisation — with the company's management appearing to be very hierarchical in that the management is in place to decide what is done and where with little incentive for the workers in general — but without definite structures in place for the choosing of their innovations.

Finally, we arrive at "relationship" construct with a mean of 5.2. This is the most curious of the results because we actually have not only the lowest averaged answered question but also the highest. With a mean of 6.0, in question number 5, the interviewees stated that the companies have good relationships with the suppliers in which both parties are winning. However, with a mean of 4.3, in question number 14, the companies do not appear to have good ties with the universities or other centres of research to help develop understanding of innovation. This is interesting because it begs the question as to why, if they have good relationships with their suppliers, do they not work with them to discuss innovations (as shown previously with the results of learning). Perhaps it is an innate protection from the companies in wanting to internalise procedures and not use outside sources such as other companies, universities, or research institutions to consult in methods of innovation or change.

In regards to the 2014 questionnaire it is shown that there has been a decrease in all five dimensions (Strategy, Processes, Organisation, Relationships and Learning) as opposed to the results from the 2013 questionnaire.

The strongest reply for the 2014 questionnaire (similarly to 2013) was in relation to the sections concerning strategy and learning. These returned results demonstrating a mean high of 5.3, slightly higher than organisation. With a mean of 6, question 5 was the most strongly answered, claiming, that the company has good relations with the suppliers, in which both companies come out winning.

The weakest reply for the 2014 questionnaire was in regards to relationships (the same as the 2013 questionnaire). However, the results of this construct revealed a lower mean of 4.9 as opposed to 5.2 a year earlier. Question 14 (belonging to the relationships section) had the lowest score with a mean of just 3.8, suggesting that the companies do not have good contacts with universities or research institutions helping them to work together and improve knowledge and understanding.

It can be observed in Table 2, that the mean of each dimension and its relation with size differs in all types of companies which lead us to accept H_1. The dimension *organisation* is the one with the highest average for the years (2013,

Table 2. Relationship size/years and constructs of innovation.

	2013				2014			
	Micro	Small	Medium	Large	Micro	Small	Medium	Large
Strategy	5.3	5.5	5.5	5.5	5.4	5.2	5.3	5.4
Processes	5.3	5.3	5.2	5.4	5.2	5	5.1	5.2
Organisation	5.7	5.4	5.5	5.3	5.6	5.3	5.3	5.1
Linkages	4.7	5.3	5.2	5.3	5.4	4.9	5	5.1
Learning	5	5	5	5	5	5	5	5

2014) with 5.7 and 5.6, respectively in microenterprises. The lowest average is from the year 2013 in the relationships dimension with 4.7, increasing in the year 2014 to 5.4 points for microenterprises.

These results showed quite a dip in regards to the results from 2013. Figure 2 (radar graph) shows these findings in graph form. Strategy, learning and organisation are the strongest ways of demonstrating how the companies manage innovation, whilst processes and relationships are the weakest.

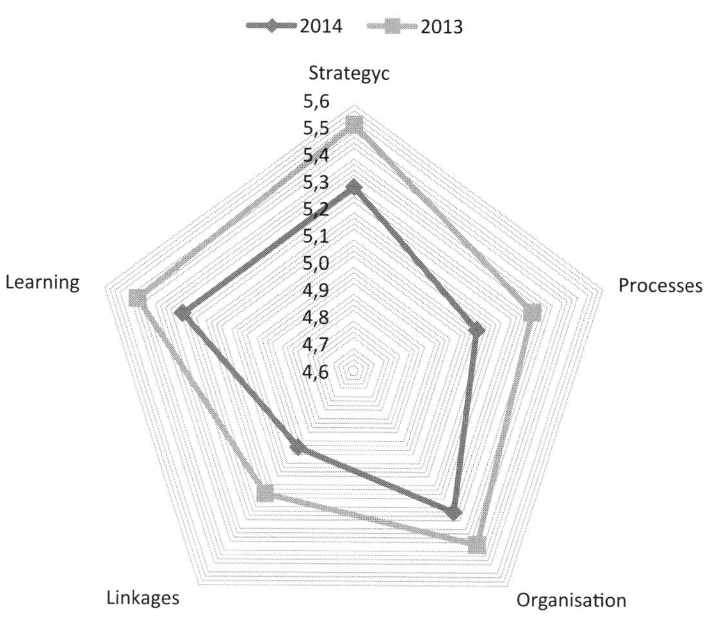

Fig. 2. Radar graph.

Table 3. Result of linear regression model: including questions of each construct given (dependent variables) and firm size (independent variables).

Companies	Strateg*		Proces*		Organ*		Linkag.*		Learn.*	
	Quet.	Sig.	Quet.	Sig.	Quet.	Sig.	Quet.	Sig.	Quet.	Sig.
**Large	Q6	0.018	Q7	0.001	Q13	0.031	Q19	0.018	Q30	0.05
					Q38	0.001				
**Medium	Q1	0.007	Q7	0.005	Q3	0.018			Q25	0.043
	Q16	0.008					No Sig.		Q15	0.047
	Q21	0.035								
**Small	Q31	0.016	Q2	0.001	Q33	0.001	Q9	0.034	Q15	0.028
			Q27	0.043	Q38	0.022	Q14	0.026	Q20	0.000
							Q24	0.002	Q25	0.000
							Q34	0.007		
							Q39	0.000		
***Micro	No Sig.		No Sig.		No Sig.		No Sig.		No Sig.	

*$N = 477$ (2013–2014).
**Lineal regression models $p < 0.05$.
***Lineal regression models $p > 0.05$.

Inferential analysis of the survey

For the inferential analysis, we have generated a Pearson correlation analysis with 40 variables. Only two variables can be observed with a relatively high level of correlation (0.70). More, the results of the Pearson correlation for the independent variables did not show problems of multicollinearity once the highest was 0.70, which did not represent a problem for subsequent analysis that implies a relationship of dependence (Lin and Chen, 2005). Following the methodology of Tidd *et al.* (2005), there are eight questions related to each construct studied.

Generating a detail of the questions that were significant in the sample we note that, in the construct *strategy*, results of Question 6 of the questionnaire reveals that all employees of large companies have clarity on the innovation goals of the company, i.e., there would be a reinforcement of teamwork and a vertical communication theme of innovation.

Regarding the segment of medium companies Questions 1, 16 and 21, that relate to the knowledge of employees in general terms are significant. This idea is reinforced by the fact of using tools and forecasting techniques to overcome the threats and enhance opportunities (Question 16 of the questionnaire).

For the construct *process*, the results of Question 7 are repeated in the large and medium enterprises of the sample, hinting to an internal order of organisations regarding deadlines and operating costs of innovation projects. This would lead us to think that there is a good drive between both companies (large and medium enterprises), given that there is a more structured relationship between the two groups. In the case of small businesses, the results of Question 2 indicates that the organisations have adequate processes that help them manage or develop new products (from idea to launch) and results of Question 27 indicates that companies have adequate mechanisms to ensure the involvement of all departments of the company in developing new products/processes.

In the construct *organisation*, the results of Question 38 revealed greater significance in large and small businesses. We argue, therefore, that organisations teamwork is essential for Ecuadorian large and small companies. In the case of the large firms, the results of Question 13 reinforce the fact that in a high level business, the company staff suggests ideas to improve products and processes, which can also denote the fact that much R&D departments is concentrated in large firms. In the case of medium-sized enterprises, the results of Question 3 indicate that organisational structure does not stifle innovation, in fact, it helps generate innovation.

The construct *relationships* becomes the most dispersed variable from the sample, since it only has significant questions for the large and small enterprises. The questions associated with analysis revealed that have no business relationship between the two groups (not significant between groups were repeated questions). In large companies, only Question 19 is significant and indicates the importance of working close to customers in developing new concepts in products and processes. These levels of work close by and for customers in large companies is something that we could call normal, which in the case of Ecuador, would reflect an accommodation of business universal ideals. For small businesses and their level of relationship, we can see that are in line with all matters relating to generate actions that positioned at a place in the market because. In this case, the results of Question 9 indicates that small enterprises revise their projects to improve future performance. The results of Question 14 showed that working with universities and research centres to develop new knowledge is a practice in small firms. The results of Question 24 indicate the effort to collaborate with other companies to develop new processes and products. The results of Question 34 exhibit the fact of being close to the formal education systems to communicate their training needs and/or training.

It can be seen for the construct *learning*, that Questions 15 and 25 are repeated with a higher level of significance. In the sample, large and small businesses have an important significance for Question 15, which makes it clear that learning from mistakes in each organisation (beyond size) is something that generates significant thought and attention to the particular. The Question 25 is repeated in medium and small enterprises, indicating a strong interest in generating ties with other companies to learn. Both questions 15 and 25 show us that Ecuadorian companies are willing to learn as much from their mistakes and finding new ways to make them grow with their peers, which would mean a clear focus on the business associativeness.

Conclusion and Recommendations

Although, we know that companies do not need to be highly innovative in order to have good results (Yamin *et al.*, 1999), we cannot stop thinking about the results of global innovation index (Global Competitiveness Report-GCR 2011–2012) of Ecuador. In that sense, perhaps the increased rate is improved by 12 pillars of global competitiveness which measures the GCR (mainly macroeconomic pillars) and not for a strong change within Ecuadorian companies. In this respect, the results of our study point out that any micro manages to have significant responses to the regressioned variables. If so, the issue of the management of innovative practices in Ecuador, will be presented only in small, medium and large enterprises with mixed results.

Having implemented Tidd and Bessant's (2013) survey in 477 Ecuadorian companies it can be argued that only 22 of the 40 practices (55%) were significant in relation to the proposed econometric model, which revealed that the number of responses is low, since we must consider that of the 22 variables 4 are repeated in different econometric models. This shows a concentration of responses in different business levels except for microenterprises, which could be interpreted as a repetition of practices in specific topics depending on the size and activities of the company. For example, Question 25 has the same significance for small and medium enterprises (sharing experiences with other companies). This shows that they are making efforts together (without discrimination even if they are private sector efforts as such or the same economic sector) through the support of business associations or government support (through special public policy).

Out of the five constructs studied, the strategy construct was the highest rated by respondents, with 5.4 points on average, followed by learning (5.35), organisation (5.30), processes (5.3) and linkages (5.0). This shows that the Ecuadorian companies have a strong focus on teamwork, demonstrating clarity on issues of vision, mission and values that surround an organisation. Worth noting that this symmetry is found strongly marked in medium and large enterprises.

Finally, our study has generated a basic radiography of innovation management in Ecuador. Overall, companies present different emphasis on innovation management practices depending on their size. Therefore, our study hypothesis (H_1) was accepted. The businesses are still far from generating sustained innovation practices. There are no indications that companies maintain a register of ongoing and/or incremental innovation or maintain a pattern of discontinuous innovation.

In this regard, we generally recommend that the Ecuadorian companies generate incentive programs for the management of innovation and next to it, exploit synergies that could be achieved with strategic agreements with universities and research centres, eliminating the academic distance today clearly reflects an organisational disadvantage. At the same time, we suggest to companies to maximise the good relationships they have with suppliers and customers in order to strengthen the internal supply chain. This could help enterprises to expand the search for open innovation as a way to grow in the various sectors of international markets (Chesbrough, 2005) and achieve a level of performance with multinational companies as evidenced by the study of Williams and Vossen (2014).

We recommend that companies generate programs and practices of teamwork oriented to the management of innovative practices that enhance the participation of employees at all levels. From the above, companies must generate effective communication systems (from top to bottom and from bottom to top) and generate economic and social incentives to support innovation and/or form specialised teams in the field, stimulating a project-based organisation, bringing together different elements into an integrated whole. This usually includes several companies, long periods and high levels of technological hazards that may help in the long term (Gann and Salter, 2000).

Regarding business relationships, it is observed that most of the practices are exercised, but more concentrated in small businesses, especially those related to improving systems training and staff development, this helps to strengthen networking with other firms and markets. Therefore, it should be recommended that companies should strengthen relations with its suppliers, better understand the needs of their customers (through market surveys and other means of approachment), learn from the experience of competition organisations, and constantly evaluating the innovation management of the organisation.

Lines of future research should focus on the similarities and differences between business sizes, to explore whether there is a universal model for innovation as noted by Mazzarol *et al.* (2014) in a study of OECD countries, more approaching the concepts and comparisons within the Ecuadorian context.

Appendix A. Questionnaire

How Well Do We Manage Innovation?

This simple self-assessment tool focuses attention on some of the important areas of innovation management. Below you will find statements which describe 'the way we do things around here' — the pattern of behaviour which describes how the organization handles the question of innovation. For each statement simply put a score between 1 (= not true at all) to 7 (= very true).

	Statement	Score 1 = Not true at all to 7 = Very true
1	People have a clear idea of how innovation can help us compete	
2	We have processes in place to help us manage new product development effectively from idea to launch	
3	Our organization structure does not stifle innovation but helps it to happen	
4	There is a strong commitment to training and development of people	
5	We have good 'win-win' relationships with our suppliers	
6	Our innovation strategy is clearly communicated so everyone knows the targets for improvement	
7	Our innovation projects are usually completed on time and within budget	
8	People work well together across departmental boundaries	
9	We take time to review our projects to improve our projects to improve our performance next time	
10	We are good at understanding the needs of our customers/end-users	
11	People know what our distinctive competence is — what gives us a competitive edge	

Appendix A. (*Continued*)

	Statement	Score 1 = Not true at all to 7 = Very true
12	We have effective mechanisms to make sure everyone (not just marketing) understands customer needs	
13	People are involved in suggesting ideas for improvements to products or processes	
14	We work well with universities and other research centres to help us develop our knowledge	
15	We learn from our mistakes	
16	We look ahead in a structured way (using forecasting tools and techniques) to try and imagine future threats and opportunities	
17	We have effective mechanisms for managing process change from idea through to successful implementation	
18	Our structure helps us to take decisions rapidly	
19	We work closely with our customers in exploring and developing new concepts	
20	We systematically compare our products and processes with other firms	
21	Our top team have a shared vision of how the company will develop through innovation	
22	We systematically search for new product ideas	
23	Communication is effective and works top-down, bottom-up and across the organization	
24	We collaborate with other firms to develop new products or processes	
25	We meet and share experiences with other firms to help us learn	
26	There is top management commitment and support for innovation	
27	We have mechanisms in place to ensure early involvement of all departments in developing new products/processes	

Appendix A. (*Continued*)

	Statement	Score 1 = Not true at all to 7 = Very true
28	Our reward and recognition system supports innovation	
29	We try to develop external networks of people who can help us — for example, with specialist knowledge	
30	We are good at capturing what we have learned so that others in the organization can make use of it	
31	We have processes in place to review new technological or market developments and what they mean for our firm's strategy	
32	We have a clear system for choosing innovation projects	
33	We have a supportive climate for new ideas — people don't have to leave the organization to make them happen	
34	We work closely with the local and national education system to communicate our needs for skills	
35	We are good at learning from other organizations	
36	There is a clear link between the innovation projects we carry out and the overall strategy of the business	
37	There is sufficient flexibility in our system for product development to allow small 'fast-track' projects to happen	
38	We work well in teams	
39	We work closely with 'lead users' develop innovative new products and services	
40	We use measurements to help identify where and when we can improve our innovation management	

References

Auken, H (2008). Innovation and performance in Spanish manufacturing SMEs. *International Journal of Technology Management*, 8(1), 36–56.

Ayhan, BM and Oztemel, E (2014). A methodology to measure the degree of managerial innovation. *Journal of Industrial Engineering and Management*, 7(1), 153–173. doi: http://www.jiem.org/index.php/jiem/article/view/927.

Bessant, J and S Caffyn (1997). High involvement innovation. *International Journal of Technology Management*, 14(1), 7–28.

Bhaskaran, S (2006). Incremental innovation and business performance: Small and medium-size food enterprises in a concentrated industry environment. *Journal of Small Business Management*, 44(1), 64–80.

By, RT (2005). Organizational change management: A critical review. *Journal of Change Management*, 5(4), 369–380.

By, RT, B Burnes and Coswick (2011). Change management: The road ahead. *Journal of Change Management*, 11(1), 1–6.

Chesbrough, H (2005). *Open Innovation: The New Imperative for Creating And Profiting from Technology*. US: Harvard Business Review Press.

Christensen, CM (1997). *The Innovator's Dilemma: When New Technologies Cause Great Firms to Fail*. Boston, MA: Presidents and Fellows of Harvard College.

Darroch, J (2005). Knowledge management, innovation and firm performance. *Journal of Knowledge Management*, 9(3), 101–115.

Didonet, S and G Díaz (2012). Supply chain management practices as a support to innovation in SMEs. *Journal of Technology Management & Innovation*, 7(3), 91–109. doi:http://dx.doi.org/10.4067/S0718-27242012000300009.

Doz, Y, J Santos and P Williamson (2001). *From Global to Metanational: How Companies Win in the Knowledge Economy*. Boston, Massachusets: Harvard Business Press.

Drucker, P (1998). The discipline of innovation. *Harvard Business Review*, Reprint 98604, November–December 1998, 03–08.

Ettlie, J (1999). *Managing Innovation*. New York: John Wiley & Sons, Inc.

Ezeala-Harrison, F (1999). *Theory and Policy of International Competitiveness*. London: Praeger.

Francis, D (2001). Developing innovative capability, PhD Thesis, University of Brighton, Brighton.

Francis, D and J Bessant (2005). Targeting innovation and implications for capability development. *Technovation*, 25(3), 171–183.

Freel, M (2005). Patterns of innovation and skills in small firms. *Technovation*, 25, 123–134.

Gann, D and J Salter (2000). Innovation in project-based, servce-enhanced firms: The construction of complex product and systems. *Research Policy*, 29, 955–972.

Henderson, R and K Clark (1990). Architectural innovation: The reconfiguration of existing product technologies and the failure of established firms. *Administrative Science Quarterly*, 35, 9–30.

Heunks, F (1998). Innovation, creativity and success. *Small Business Economics*, 10(3), 263–272.

Mazzarol, T, D Clark, S Reboud, N Gouch and P Olson (2014). Perceptions of innovation climate and the influence of others: A multi-country study of smes. *International Journal of Innovation Management*, 18(1), 1–24.

Novales, A (1993). *Econometría*. 2nd Edition. Interamericana de España, Madrid: McGraw-Hill.

Oerlemans, L, M Meeus and F Boekema (2001b). Firm clustering and innovation: Determinants and effects. *Papers in Regional Science*, 80, 337–356.

Porter, M (1980). *Competitive Strategy*. New York: Free Press.

Porter, M (1990). *Competitive Advantage of Nations*, The Free Press Review, pp. 73–91.

Richardson, Z and A Khan (2002). The implementation of innovation by a multinational operating in two different environments: A comparative study. *Creativity and Innovation Management*, 11(2), 98–106.

Schumpeter, J (1950). *Capitalism. Socialism and Democracy*. 3rd Edition. New York: Harper&Row.

Scott, BR and GC Lodge (1995). *US Competitiveness in the World Economy*. Boston, MA: Harvard University Press.

Teece, DJ, G Pisano and A Shuen (1997). Dynamic capabilities and strategic management. *Strategic Management Journal*, 18(7), 509–533.

Tidd, J (1997). Complexity, networks & learning: Integrative themes for research on innovation management. *International Journal of Innovation Management*, 1(1), 1–21.

Tidd, J (2006). *A Review of Innovation Models*. Imperial College London. Tanaka Business School.

Tidd, J and F Hull (2006). Managing service innovation: The need for selectivity rather than 'best practice'. *New Technology, Work and Employment*, 21(2), 139.

Tidd, J, J Bessant and K Pavitt (2005). *Managing Innovation: Integrating Technological, Market and Organizational Change*. 3rd Edition. New York: Wiley John & Sons, Ltd.

Utterback, J and W Abernathy (1975). A dynamic model of process and product innovation. *Omega*, 3(6).

van der Zee, FA (2006). The New Asian Growth Dynamics in a Context of Globalisation. TNO, Delft. Working Paper.

Williams, C and J Vossen (2014). How open MNCs need to be extract value in open innovation?. *International Journal of Innovation Management*, 18(5), 1–27.

Winters, LA, N Mcculloch and A Mckay (2004). Trade liberalization and poverty: The evidence so far. *Journal of Economic Literature*, 42, 72–115.

Wong, SA (2009). Productivity and trade openness in Ecuador's manufacturing industries. *Journal of Business Research*, 62, 868–875.

World Economic Forum (2011). *The Global Competitiveness Report 2011–2012*. Geneva, Switzerland. http://www3.weforum.org/docs/WEF_GCR_Report_2011-12.pdf.

Yamin, S, A Gunasekaran and FT Mavondo (1991). Innovation index and its implication on organisational performance: A study of Australian manufacturing companies. *International Journal of Technology Management*, 17(5), 495–503.

Chapter 11

THE RELATIONSHIP BETWEEN TOP MANAGEMENT TEAM INNOVATION ORIENTATION AND FIRM GROWTH: THE MEDIATING ROLE OF FIRM INNOVATIVENESS

Nils D. Kraiczy and Andreas Hack

Institute of Organization and Human Resource Management
University of Bern

Franz W. Kellermanns

Department of Management
University of North Carolina

Institute of Family Businesses
WHU – Otto Beisheim School of Management

Upper echelon theory and research on innovation have considered top management teams and their behaviour and characteristics as important factors that positively influence innovativeness and organizational outcomes. Yet, innovation research has mostly focused on individual new product projects, and their performance and impact on firm performance. Recent research has started to apply a more holistic view in terms of innovation, by considering firm-wide innovation instead of single new products. Upper echelon research has concentrated on direct relationships between top management team characteristics and organizational outcomes. But recent research calls for mediating effects of the relationship between top management team characteristics and organizational outcomes. Hence, this study introduces firm innovativeness as a mediator between top management team innovation orientation and firm growth. Focusing on small and medium-sized firms, which often represent highly innovative firms, results show that firm innovativeness fully mediates the

This paper was first published in *International Journal of Innovation Management*, Vol. 19, No. 1 (February 2015), 1550005 (24 pages). © Imperial College Press. DOI: 10.1142/S136391961550005X relationship between top management team innovation orientation and firm growth. Implications and future research are discussed.

Keywords: Top management team; innovation orientation; firm innovativeness; firm growth; small and medium-sized firms; upper echelon theory.

Introduction

Decisions made by the top management team (TMT), in particular in small and medium-sized firms, are essential for the long-term survival of the firm (Hambrick, 2007; Hambrick and Mason, 1984). Consequently, research has focused on this group to understand its impact on organisational outcomes (for an overview, see Certo *et al.*, 2006; Fernández-Ortiz and Lombardo, 2009). In particular, upper echelon theory highlights the influence of TMT characteristics on organizational outcomes. In the context of innovation, TMTs have also been identified as important success factor (e.g., Ernst, 2002; Henard and Szymanski, 2001; Montoya-Weiss and Calantone, 1994). Combining these two research streams, we introduce TMT innovation orientation as an innovation fostering TMT behaviour and firm innovativeness as a mediating variable between TMT innovation orientation and firm growth.

Previous research on innovation has mostly focused on the project level and the impact of the TMT on individual innovation projects (for an overview, see Ernst, 2002). Upper echelon research has focused on direct relationships between TMT characteristics and organisational outcomes by neglecting indirect effects (Certo *et al.*, 2006; Hambrick, 2007). Although recent research has started to investigate mediating effects (e.g., Cho and Hambrick, 2006; Talke *et al.*, 2011), the black box between TMT characteristics and organisation outcomes is under-researched.

By introducing firm innovativeness as a mediator between TMT innovation orientation and firm growth, we extend research that analyses the relationship between TMT characteristics and organisational outcomes. We apply upper echelon theory to hypothesise a positive effect of TMT innovation orientation on firm growth and use findings of innovation literature that suggest a positive relationship between firm innovativeness and firm growth.

Our study makes several important contributions to TMT research and innovation literature. First, we introduce TMT innovation orientation as an innovation fostering TMT behaviour, which has a positive impact on firm growth. Second, we extend upper echelon research on mediating effects between TMT characteristics and organisational outcomes by integrating firm innovativeness as a mediator between TMT innovation orientation and firm growth. Third, we contribute to innovation literature that focuses on the firm-level impact of TMTs on firm innovativeness by changing the view from the product-/project level to the firm level.

In the following section, we develop hypotheses based on upper echelon theory that introduce TMT innovation orientation as a TMT behaviour that fosters firm growth. Next, we integrate firm innovativeness as a mediator between TMT innovation orientation and firm growth, followed by the presentation of our research results. We conclude by discussing theoretical and practical implications and suggestions for future research.

Theoretical Framework and Hypotheses

Top management team characteristics and organisational outcomes

In the context of TMTs, upper echelon theory analyses the relationship between TMT characteristics and organisational outcomes. Based on the concept of bounded rationality (Cyert and March, 1992), upper echelon theory is an information processing theory which states that in situations where TMT members are faced with more information than they can process, they base their decisions on their experiences, preferences, and other biases (Cho and Hambrick, 2006; Hambrick and Mason, 1984). Additionally, decision making of TMT members is often influenced by accustomed procedures and rules of thumb heuristics (Cyert and March, 1992). Upper echelon theory highlights the importance of TMTs for an organization by using observable demographic factors as proxies of unobservable psychological factors of TMT members (Hambrick, 2005) and suggests that these characteristics (e.g., functional background, experience, education, age, and tenure) influence decision-making and, thus, organisational outcomes such as firm performance (for an overview see Certo *et al.*, 2006) and firm growth (e.g., Boeker, 1997; Kor, 2003).

Extensions of upper echelon research focus not only on traditional TMT characteristics but also on several additional facets of personality such as self-esteem, locus of control, emotional stability, hubris, overconfidence, and narcissism (e.g., Chatterjee and Hambrick, 2007; Hayward and Hambrick, 1997); and behaviours such as TMT advice seeking (Alexiev *et al.*, 2010), TMT behavioural integration (collaborative behaviour, quantity and quality of information exchange, joint decision-making) (Simsek *et al.*, 2005), TMT entrepreneurial drive (Wood and Michalisin, 2010), and TMT risk taking (Gilley *et al.*, 2002).

Results support a strong influence of personality and TMT behaviours on organisational outcomes.

TMT innovation orientation and firm growth

Introducing TMT innovation orientation as an innovation fostering TMT behaviour, this study extends recent research. On the firm level, research demonstrates that innovation orientation both directly and indirectly affects new product performance and thus firm performance (Calantone *et al.*, 2003; Hurley and Hult, 1998; Narver *et al.*, 2004). For example, Narver *et al.* (2004) show a direct and positive impact of innovation orientation on new product performance. Similarly, Rosenbusch *et al.* (2011, p. 1) conclude "fostering an innovation orientation has more positive effects on firm performance than creating innovation process outcomes such as patents or innovative products or services." The literature provides various constructs such as (product) innovation/innovativeness (Deshpandé *et al.*, 2000; Miller, 1983; Miller and Friesen, 1982), entrepreneurial style (Naman and Slevin, 1993), entrepreneurial orientation (George *et al.*, 2001; Jantunen *et al.*, 2005), innovative capacity (Eddleston *et al.*, 2008), or innovation orientation (Manu, 1992; Olson *et al.*, 2005; Worren *et al.*, 2002) that tend to measure similar phenomena. Most of these constructs are measured on the firm level. Thereby, the terms "innovation orientation" and "innovativeness" appear synonymously in literature, although only a few studies focus on the innovation orientation of TMTs (Zahra and Fescina, 1991; Kraiczy *et al.*, 2014). In order to analyse innovation orientation on the TMT level, this study defines TMT innovation orientation as innovation fostering behaviour of TMT members as a group. It describes the willingness of the TMT to engage in innovative activities.

Focusing on the TMT-level, TMT innovation orientation may have three important effects that positively influence firm growth. First, a higher TMT innovation orientation indicates a higher number of more innovative ideas, which may result in more incremental and radical innovations that have been identified as driver of firm performance (Bhaskaran, 2006; Kleinschmidt and Cooper, 1991). Although radical innovations are riskier than incremental improvements, most of the literature suggests that radical innovations are more valuable because they create more opportunities for firms in terms of differentiation and competitive advantage (Chaney *et al.*, 1991; Sorescu *et al.*, 2003; Sorescu and Spanjol, 2008). But without incremental innovations, focusing only on radical innovations cannot increase short term firm performance. Hence, a balanced mix of incremental and radical innovations may yield highest firm performance and reduce risks by higher diversity of product innovativeness levels in the new product portfolio (Kang and Montoya, 2014). Second, as the willingness to be innovative is higher, the number of new product projects and their probability of implementation increase. Although not all projects will succeed, a wider and more balanced portfolio of new product projects enhances the probability of successful projects and decreases the impact of failed projects, because the portfolio contains a sufficient number of successful new product projects (Cooper *et al.*, 1999). Third, particularly in small and medium-sized firms, the positive effect of TMT innovation orientation may be stronger, because organisational structures are less complex (Levy and Powell, 1998) and TMT members are more actively involved in operational tasks such as innovation projects (Gupta and Cawthon, 1996). Accordingly, we hypothesise:

H1: *TMT innovation orientation will be positively related to firm growth.*

TMT innovation orientation, firm innovativeness, and firm growth

In the previous section, we argued for a direct effect between TMT innovation orientation and firm growth. However, recent research on TMTs has started to open the black box between TMT characteristics and organisational outcomes by integrating mediating variables (Cho and Hambrick, 2006; de Jong *et al.*, 2013; Pelled *et al.*, 1999; Smith *et al.*, 1994; Talke *et al.*, 2011). Indeed, Certo *et al.* (2006, p. 835) conclude in their meta-analysis "that further investigation of the intermediate mechanisms between the TMT and firm performance may provide valuable insights into the ways in which TMTs can effectively enhance firm performance." Accordingly, we follow the call in the literature by investigating a potential mediator of the aforementioned relationship, by focusing on firm innovativeness. Following the definition by Lumpkin and Dess (1996, p. 142), we define innovativeness as "a firm's tendency to engage in and support new ideas, novelty, experimentation, and creative processes that may result in new products, services, or technological processes."

TMT innovation orientation, particularly in small and medium-sized firms, which are the focus of our study, should have a stronger and perhaps more critical relationship with firm innovativeness. In small and medium-sized firms, organisational management structures are less complex and rather flat compared to large organizations (Levy and Powell, 1998), TMT members are more strongly involved in operational tasks (Gupta and Cawthon, 1996), and resources that can be invested in innovation projects are limited (Acs *et al.*, 1999). Hence, in small and medium-sized firms, TMT behaviours and decisions become even more reflected in innovation activities and in turn in firm innovativeness.

The importance of innovation for firm performance and resulting competitive advantages has been identified by numerous studies (e.g., Dutta *et al.*, 2005; Geroski *et al.*, 1993; Hall, 2000). Although not all studies support an explicit, positive impact of firm innovativeness on firm performance (e.g. Baum *et al.*, 2000; Menguc and Auh, 2006), firm innovativeness is widely accepted as a critical firm characteristic providing competitive advantages (e.g., Dougherty and Hardy, 1996) that result in increased firm performance (e.g., Srinivasan and Hanssens, 2009; Tellis *et al.*, 2009). Particularly for small and medium-sized firms, research has identified investments in innovation as the most important strategy for expansion (Hay and Kamshad, 1994). Accordingly, higher levels of firm innovativeness should be positively related to growth in small and medium-sized firms.

Taken together, we argue that the relationship between TMT innovation orientation and firm growth in small and medium-sized firms is mediated by firm innovativeness. The beneficial effect of TMT innovation orientation on firm growth will be facilitated by firm innovativeness. If TMT innovation orientation matches a firm's tendency to actively pursue innovation, facilitated implementation should generate higher growth for the organisation. Accordingly, we hypothesise:

H2: *Firm innovativeness mediates the relationship between TMT innovation orientation and firm growth.*

Figure 1 presents the mediation model of our study.

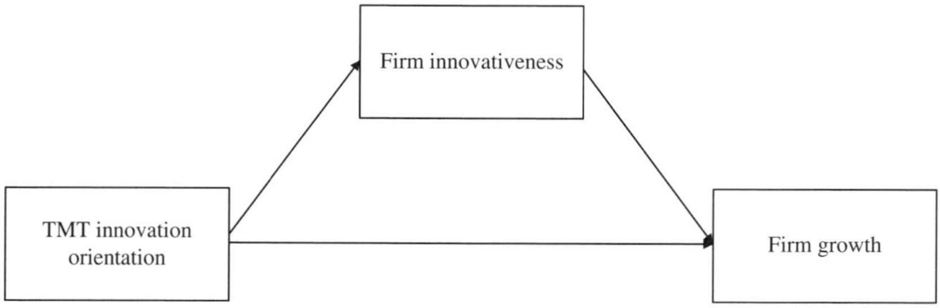

Fig. 1. Mediation model.

Methodology

Sample

As part of a larger study (Kraiczy *et al.*, 2014; Kraiczy *et al.*, forthcoming), we collected data from TMTs of German small and medium-sized firms in different manufacturing industries; these industries were chosen based on their high potential for innovation in order to test our hypotheses. We draw firm contacts from a database of the Southwestfalian Chamber of Commerce and Industry (Hagen, Germany) and chose a random sample of 1000 firms with 50–499 employees. We specifically focused on small and medium-sized firms because in these firms the influence of TMTs affects firm outcomes more strongly than in large companies with complex organisational structures (Wright and Kellermanns, 2011). We identified the CEO of each firm and their personal email addresses by scanning firm websites and the Markus database, a product of the German credit agency Creditreform that contains comprehensive information on over 1.3 million companies in Germany, Austria, and Luxembourg.

To collect data for our study, we sent a personalised email to the CEO of each firm that included a request to participate in the study and distribute the email link to key TMT members involved in decisions regarding innovations in the firm. The multiple respondents approach reduces key informant bias. Each link contained a code to assign members to a team and to aggregate team-specific data; aggregated data enhances performance assessments (Simons and Peterson, 2000).

Two weeks after the first contact, we sent out reminder emails followed by personal phone calls. In total, 127 TMT members completed the survey, which yielded 77 TMTs of small and medium-sized firms. The response rate of 7.7% may be due to the privacy concerns and inward orientation of small and medium-sized firms (Upton *et al.*, 2001; Zahra *et al.*, 2004). Additionally, we assume that TMT members of small and medium-sized firms are more involved in operational tasks and have less time to respond to surveys. The final response rate is consistent with recent studies focusing on TMTs and small and medium-sized firms (Ensley and Pearson, 2005; Hambrick *et al.*, 1993; Simsek, 2007). The mean firm size is 168.61 employees, the mean firm age is 66.76 years.

We include both multiple respondents (51 firms) and single respondents (26 firms) in our analysis. All single respondents are CEOs of the respective firms. As CEOs have been identified as reliable key informants, they should be able to provide valid information (Kumar *et al.*, 1993; Seidler, 1974). To justify the aggregation of multiple respondents to one TMT-level score, we used r_{wg}, which represents a measure of agreement among TMT members' perceptions (James *et al.*, 1984, 1993). The average r_{wg} estimates are equal to or larger than 0.80, thus exceeding the general 0.70 rule-of-thumb cut off (Klein and Kozlowski, 2000). High levels of agreement among TMT members support the decision to include single respondent firms in our analysis (see also Eddleston *et al.*, 2008).

We tested for multicollinearity issues by assessing the correlation matrix, the variance inflation factor (VIF), and the condition index (CI). Levels of correlations between the variables are modest. All VIFs (< 2.448) show acceptable values well below the suggested critical limit (Hair *et al.*, 2010). The CIs (< 21.993) are also below the critical values suggested by Hair *et al.* (2010). Hence, we assume that multicollinearity is not a significant concern.

As our data was collected via a single questionnaire, we tested for potential common method bias, as suggested by Podsakoff and Organ (1986). We entered all items of the control, mediator, independent, and dependent variable into a one-factor analysis and identified five factors with eigenvalues > 1.0, which account for 66.15% of the variance. The first factor accounted for 31.77% of the variance, whereas the remaining factors accounted for 34.38%. Because no single method factor emerged and the factors separated cleanly, we believe that a common method bias is not a concern.

We assessed a potential non-response bias by comparing early and late respondents via a one-way ANOVA (analysis of variance), which is an approach that assumes that non-respondents tend to be more similar to late respondents (Kanuk and Berenson, 1975; Oppenheim, 1966). Because there are no significant differences between the two groups, we conclude that non-response bias is not a concern.

Measures

When possible we used validated and reliable measurement scales based on measures of related constructs in the literature.

TMT innovation orientation

We measured TMT innovation orientation on a Likert-type scale with six items and a five-point response format anchored by "not at all" to "to an exceptional degree". This scale has originally been developed by Scott and Bruce (1994) to rate the innovative behaviour of employees by their managers and has recently been applied by Kraiczy *et al.* (2014). The construct demonstrates acceptable reliability with a Cronbach's alpha of 0.91. All items appear in the Appendix.

Firm growth

We indicated firm growth with a two-item construct that measures sales and employee growth compared to competitors. Respondents were asked to rate the sales and employee growth of their own firm on a Likert-type scale with a five-point response format anchored by "much worse than the competitors" to "much better than the competitors" (Wiklund and Shepherd, 2003). Reliability analysis shows an acceptable internal consistency of the measure with a Cronbach's alpha of 0.79. All items appear in the Appendix.

Firm innovativeness

We assessed firm innovativeness by using two items: "Our firm has emphasised taking bold, wide-ranging actions in positioning itself and its products or services over the past 3 years" and "Our firm has shown a strong commitment to research and development, technological leadership and innovation." This construct is a part of a wider corporate entrepreneurship construct adapted from Miller (1983), which has recently been applied by other studies investigating innovativeness (e.g., Kellermanns *et al.*, 2012). The construct demonstrates acceptable reliability with a Cronbach's alpha of 0.83. All items appear in the Appendix.

Control variables

Consistent with prior research, we utilized a number of variables shown to influence innovation and, in turn, firm growth (Chen and Huang, 2009; Tsai, 2009). We controlled for three different manufacturing industries because previous research stresses that industry can affect innovation intensity and firm growth. The representation of firms by industry is as follows: manufacturing of machinery and vehicles (Industry 1, 29.82%); manufacturing of chemicals, chemical, rubber, and plastic products (Industry 2, 18.18%); and manufacturing of basic metals (Industry 3, 36%). The remaining firms operate in other industries (16%). We also controlled for firm size because larger firms may possess more slack that they could invest in capital-intensive radical innovations (George, 2005; Nohria and Gulati, 1996). Firm size is measured by the number of employees according to the Markus database. We used the ln of the number of employees to control for firm size. A transformation was necessary to achieve normal distribution. To control for liability of newness concerns, as well as to address the potential for higher innovativeness and firm growth in younger organisations, we also controlled for firm age (Stinchcombe, 1965). Further, TMT size might affect dynamics in decision-making processes and growth intentions (Carpenter *et al.*, 2004; Siegel and

Hambrick, 2005). Hence, we controlled for TMT size by measuring the total number of TMT members (Ling and Kellermanns, 2010). The mean TMT size is 2.57 members. Last, we controlled for CEO tenure, which is calculated as the number of years that the CEO has occupied the position as CEO in the firm (Wallace *et al.*, 2010; Zhang and Rajagopalan, 2010). CEOs with long tenures are less likely to pursue strategies of innovation and growth, and more likely to focus on stability and efficiency (Barker and Mueller, 2002). The mean CEO tenure is 11.21 years.

In order to show the independence of the two innovation constructs, TMT innovation orientation and firm innovativeness as well as the difference between these, we conducted an additional confirmatory factor analysis. The normed fit index (0.946), comparative fit index (0.988), and the root mean square error approximation (0.059) showed a good fit (Bentler and Bonett, 1980; Browne and Cudeck, 1993; Hu and Bentler, 1995) indicating that TMT innovation orientation and firm innovativeness measure different constructs.

Data analysis

For testing the two hypotheses, we applied multiple regression analysis (Cohen *et al.*, 2002). In addition, to test hypothesis 2, we assessed the presence of a mediator by following Baron and Kenny (1986) and Preacher and Hayes (2004). To support a mediating effect, the mediator must account for the relationship between an independent and a dependent variable. In total and in order to determine a mediating effect, literature uses four criteria that must be met. First, the independent variable must have a significant impact on the dependent variable when the mediator is not included in the model. Second, the mediating variable must have a significant impact on the dependent variable when the independent variable is not included in the model. Third, the independent variable must significantly affect the mediating variable when the dependent variable is not included in the model. Fourth, including all variables in a model, the significant effect of the independent variable on the dependent variable must diminish. While full mediation exists when the impact of the independent variable on the dependent variable is reduced to zero, partial mediation occurs when the impact of the independent variable on the dependent variable shrinks, but not to zero.

Results

Table 1 displays the correlations, means, and standard deviations. In order to test the two hypotheses, five models were estimated using multiple regression analysis

Table 1. Descriptive statistics and correlations.

	Mean	SD	1	2	3	4	5	6	7	8	9
1. Firm age	66.76	44.90									
2. Firm size (employees)	168.61	306.20	0.19								
3. Industry 1 (manufacturing of machinery and vehicles)	0.30	0.46	0.15	−0.05							
4. Industry 2 (manufacturing of chemicals, chemical rubber, and plastic products)	0.18	0.39	−0.30**	−0.03	−0.31**						
5. Industry 3 (manufacturing of basic metals)	0.36	0.48	0.01	0.14	−0.49**	−0.36**					
6. CEO tenure	11.21	10.53	−0.01	−0.18	0.21	0.05	−0.25*				
7. TMT size	2.57	0.85	0.39**	0.02	−0.04	−0.16	0.00	−0.19			
8. TMT innovation orientation	3.14	0.65	0.02	0.28*	−0.14	−0.12	−0.03	−0.12	0.04		
9. Firm innovativeness	4.14	1.35	0.06	0.20	−0.01	0.03	−0.19	0.13	0.10	0.63**	
10. Firm growth	3.40	0.57	−0.24*	−0.01	−0.15	0.23*	−0.07	−0.18	−0.04	0.37**	0.43**

Note: $N = 77$, $^{\dagger}p < 0.10$: $^{*}p < 0.05$; $^{**}p < 0.01$.

(Cohen *et al.*, 2002). Table 2 contains the results. In Model 1, where the control variables are entered, firm age shows a moderate significant effect ($\beta = -0.24, p < 0.10$). This control variable has a negative impact on firm growth. Hence, older firms grow less compared to their counterparts. Results appear in Table 2.

Model 2 tests for hypothesis 1 and the first criterion of the mediation analysis. Hypothesis 1, which suggests a positive effect of TMT innovation orientation on firm growth, is supported. Results support a significant positive effect ($\beta = 0.31$, $p < 0.01$). The observed R^2 change is significant ($\Delta R^2 = 0.09$, $p < 0.01$). The first criterion of the mediation analysis is met/satisfied.

In Model 3, the second criterion of the mediation analysis is tested. The effect of the mediating variable, firm innovativeness, on firm growth is significant positive ($\beta = 0.47$, $p < 0.001$). The significant change in R^2 ($\Delta R^2 = 0.19$, $p < 0.001$) supports the second criterion of the mediation analysis.

In order to test the third criterion of the mediation analysis, Model 4 analyses the relationship between TMT innovation orientation and firm innovativeness. TMT innovation orientation is significantly positively related to firm innovativeness ($\beta = 0.62, p < 0.001$). The observed R^2 change is significant ($\Delta R^2 = 0.34, p < 0.001$), and the third criterion of the mediation analysis is satisfied.

To test hypothesis 2, TMT innovation orientation and firm innovativeness were entered in Model 5. In order to meet/satisfy the fourth criterion and support hypothesis 2, the significant effect of TMT innovation orientation on firm growth must reduce to zero (full mediation) or shrink (partial mediation) in comparison to Model 2. Results show that the impact of TMT innovation orientation on firm growth is not significant, which supports hypothesis 2 and a full mediation effect of firm innovativeness ($\beta = 0.44$, $p < 0.01$).

Discussion

The positive impact of TMTs characteristics and behaviours on organisational outcomes has been theorized by upper echelon theory (Hambrick, 2007; Hambrick and Mason, 1984) and supported by empirical research (e.g., Alexiev *et al.*, 2010; Gilley *et al.*, 2002). Our results support the importance of TMT behaviours by showing a positive effect of TMT innovation orientation on firm growth (hypothesis 1). More precisely, a higher innovation orientation of the TMT has a positive influence on sales and employee growth compared to the firms' competitors. Our results show that the positive impact of innovation orientation that

Table 2. Multiple regression analysis.

	Firm growth Model 1	Firm growth Model 2	Firm growth Model 3	Firm innovativeness Model 4	Firm growth Model 5
Step 1: Controls					
Firm age	−0.24†	−0.23†	−0.21†	−0.03	−0.22†
Firm size (Ln)	0.17	0.07	0.04	0.07	0.03
Industry 1	−0.13	−0.07	−0.00	−0.14	0.00
Industry 2	0.06	0.07	0.16	−0.19	0.16
Industry 3	−0.16	−0.10	0.02	−0.27*	0.02
CEO tenure	−0.15	−0.14	−0.24*	0.20*	−0.23*
TMT size	−0.01	0.01	−0.02	0.07	−0.02
Step 2: Main effect					
TMT innovation orientation		0.31**		0.62***	0.04
Step 3: Mediator					
Firm innovativeness			0.47***		0.44**
R^2	0.15	0.24	0.34	0.48	0.34
Adjusted R^2	0.07	0.15	0.26	0.42	0.25
ΔR^2	0.15	0.09**	0.19***	0.34***	0.19***
F	1.76	2.64*	4.33***	7.95***	3.81***

Note: Regression coefficients are reported as β values. $N = 77$, $^{\dagger}p < 0.10$; $^{*}p < 0.05$; $^{**}p < 0.01$, $^{***}p < 0.001$.

has been identified on the firm level (e.g, Calantone *et al.*, 2003; Hurley and Hult, 1998; Narver *et al.*, 2004) can also be found on the TMT level. The strong influence of the investigated TMT behaviour on organisational outcomes is in line with recent extensions of upper echelon research that focuses on the effect of other TMT behaviours like advice seeking (Alexiev *et al.*, 2010), behavioural integration (Simsek *et al.*, 2005), entrepreneurial drive (Wood and Michalisin, 2010), and TMT risk taking (Gilley *et al.*, 2002) on organizational outcomes.

Although not hypothesised but relevant for the mediation analysis, our results also show that TMT innovation orientation positively affect firm innovativeness. In particular on the project level, existing innovation literature has identified the TMT as one driver of innovation (Henard and Szymanski, 2001; Pattikawa *et al.*, 2006). Our results support this positive effect on the firm level by applying a firm-level approach.

Finding support for hypothesis 2, which introduces firm innovativeness as a mediator of the relationship between TMT innovation orientation and firm growth, shows that the relationship between TMT behaviours and characteristics and organisational outcomes is not exclusively direct. Our results show that TMT innovation orientation positively influences new firm innovativeness, which in turn has a positive effect on firm growth. Results indicate a full mediation. Recent research has introduced other mediating effects such as attention (Cho and Hambrick, 2006), strategic innovation orientation (Talke *et al.*, 2011), process-oriented variables (Pelled *et al.*, 1999; Smith *et al.*, 1994), or conflict (de Jong *et al.*, 2013) and showed that TMT characteristics and behaviours have an indirect effect on organisational outcomes via the investigated mediating variables.

Our study is one of the first to consider TMT innovation orientation as an innovation-fostering TMT behaviour. Following recent research that investigates TMT behaviours and its effect on organizational outcomes (Alexiev *et al.*, 2010; Wood and Michalisin, 2010), our study extends this field of research. By introducing firm innovativeness as an indirect effect between TMT innovation orientation and firm growth, we contribute to existing research that opens the black box between TMT characteristics and organisational outcomes (Cho and Hambrick, 2006; de Jong *et al.*, 2013; Talke *et al.*, 2010). Additionally, we change the level of analysis in three ways. First, we focus on all innovation activities on the firm level (firm innovativeness) instead of individual innovation projects (product-/project level). Second, we change the TMT influence from the project level to the firm level. Third, we transfer innovation orientation from the firm level (firm characteristic) to the TMT level. These new perspectives acknowledge the actual role and importance of the TMT, which comprises responsibilities on the firm level (Talke *et al.*, 2010).

Limitations and Future Research

This study has several limitations, which offer routes for future research. First, because our study design is cross-sectional, causal relationships cannot be deduced and results should be interpreted carefully. For future research, a longitudinal research design would further improve our understanding of the relationship between TMT innovation orientation, firm innovativeness, and firm growth because effects of the independent and mediating variable may first become reflected in firm growth after a certain time.

Second, the use of cross-sectional survey data creates a risk of common method bias. Hence, a common method bias may also be a limitation of our study; however, as our test suggests, this concern is diminished (Podsakoff and Organ, 1986). Indeed, the literature also suggests that potential existing effects should not significantly affect the results (Doty and Glick, 1998; Spector and Brannick, 1995).

Third, because the sample size is limited, concerns about statistical power might arise. Small sample sizes always increase the probability of a Type II error (i.e., failure to reject the null hypothesis) (Mazen *et al.*, 1987a,b); however, a bias does not appear to be present (Barrick *et al.*, 2007; Eddleston and Kellermanns, 2007) because effect size and power are in line with many studies in the field of strategic and management research (Mazen *et al.*, 1987a,b).

Fourth, focusing only on German small and medium-sized firms raises the question of generalisability. Therefore, a multi-country study design that considers cultural contingencies may increase our understanding of the relationships under investigation (Miller *et al.*, 2009). Future research should also focus on large companies with complex organisational structures where the effect of TMTs on organisational outcomes may be different. For example, managerial discretion is lower compared to small and medium-sized firms, which may reduce the effect of TMT innovation orientation on firm innovativeness and in turn firm growth.

Fifth, focusing on the manufacturing industry was intended to control for industry effects. Future research could explore the effects of TMT innovation orientation on firm innovativeness and firm growth in other industries. For example, in high-tech industries, TMT innovation orientation may have an even stronger impact on firm innovativeness and firm growth because the implementation of more complex and risky new products may be fostered by the TMT. In this context, Chandler *et al.* (2000) show that an innovative-supportive organisational culture enhances firm earnings only in dynamic environments but not in static ones. Our results may also vary depending on the business strategy of the firm. A study by Olson *et al.* (2005) indicates that innovation orientation has only a positive effect on performance for prospectors but not for analysers and low-cost defenders. Hence, future research might test our research model for different business strategies and environments to find more detailed and sophisticated effects.

Future research could also consider other innovation related TMT characteristics that may have a strong impact on firm innovativeness and firm growth. For example, recent research shows that TMT advice seeking has a positive effect on exploratory innovation (Alexiev *et al.*, 2010). Further, future research could integrate other antecedents of firm innovativeness such as new product strategy and organisational culture (Henard and Szymanski, 2001; Montoya-Weiss and Calantone, 1994), which could help to increase the explained variance of our research model. Another fruitful path for future research is the integration of different indirect effects in the relationship between TMT characteristics and organisational outcomes as suggested by Hambrick (2007). Looking inside the black box may help to explain existing inconsistent findings in this area of research. In the context of innovation, possible mediating variables are for example R&D intensity, R&D investment or new product portfolio performance. Focusing on new product portfolio related research might also be interesting for future research. In particular, new product portfolio management offers unexplored research questions.

Practical Implications

Our results also provide implications for practice purposes. In particular, we suggest that the findings of our study may help governmental authorities to improve firm growth programs and family firms to facilitate firm growth.

Governmental authorities often try to foster the economic development of their national economy by supporting specific innovation projects of firms. For example, the German Federal Ministry of Economics and Technology offers the Central Innovation Program and the European Recovery Innovation Program to foster innovations, and, in turn, growth in small and medium-sized firms (German Federal Ministry of Economics and Technology, 2013). These programmes provide funding and long-term loans for single company, cooperative, network innovation projects. Our results support the idea behind the governmental programs to foster innovation by providing financial resources as firm innovativeness increases firm growth. Governments that have not implemented innovation-fostering programs yet should follow the German approach to achieve comparable growth rates. The innovation leadership of German industries reflects the success of these programmes (European Commission, 2013).

For family firms, firm growth is important to financially satisfy the increasing group of family members over generations. Hence, it is important for family firms to grow sustainably and remain innovative. Our study shows that TMT innovation orientation is one important antecedent of firm innovativeness and, in turn, firm growth. Recent research also shows that TMT innovation orientation has a positive impact on new product portfolio performance in family firms (Kraiczy *et al.*, 2014). When selecting new executives, these firms may use innovation orientation as an additional selection criterion. At the same time, monitoring and incentive systems need to be implemented to ensure that TMT members do not jeopardies family firm existence by focusing only on high-risk innovation projects. A balanced mix of incremental and radical innovations may provide acceptable risk levels for family firms to grow sustainably.

Conclusion

The present study introduces TMT innovation orientation as innovation fostering TMT behaviour and indicates a positive effect on firm growth. Additionally, in order to extend upper echelon research on indirect effects between TMT characteristics and organisational outcomes, we introduced firm innovativeness as a mediator between TMT innovation orientation and firm growth. Results demonstrate that firm innovativeness mediates the investigated relationship.

Appendix

Scale items and reliabilities

Construct	Items	Individual α	Firm α^a	r_{wg}^a
TMT innovation orientation	My colleagues within the TMT…	0.90	0.91	0.91
	… search out new technologies, processes, techniques, and/or product ideas			
	… generate creative ideas			
	… promote and champion ideas to others			
	… investigate and secure funds needed to implement new ideas			
	… develop adequate plans and schedules for the implementation of new ideas			
	… are innovative			
Firm growth	Please rate your company regarding the following key performance indicators in comparison to other companies in your industry!	0.78	0.79	0.93
	Employee growth			
	Sales growth			
Firm innovativeness	Our firm has emphasised taking bold, wide-ranging action in positioning itself and its products or services over the past 3 years	0.83	0.83	0.81
	Our firm has shown a strong commitment to research and development, technological leadership and innovation			

[a]Reported for firms with multiple respondents only.

References

Acs, ZJ, B Carlsson and C Karlsson (1999). The linkages among entrepreneurship, SMEs and the macroeconomy. In *Entrepreneurship, Small and Medium-Sized Enterprises and the Macroeconomy*. ZJ Acs, B. Carlsson and C. Karlsson (Eds.), pp. 3–42, Cambridge: Cambridge University Press.

Alexiev, AS, JJP Jansen, FAJ Van den Bosch and HW Volberda (2010). Top management team advice seeking and exploratory innovation: The moderating role of TMT heterogeneity. *Journal of Management Studies*, 47(7), 1343–1364.

Barker, VL and GC Mueller (2002). CEO characteristics and firm R&D spending. *Management Science*, 48(6), 782–801.

Baron, RM and DA Kenny (1986). The moderator-mediator variable distinction in social psychological research: Conceptual, strategic, and statistical considerations. *Journal of Personality and Social Psychology*, 51(6), 1173–1182.

Barrick, MR, BH Bradley, AL Kristof-Brown and AE Colbert (2007). The moderating role of top management team interdependence: Implications for real teams and working groups. *Academy of Management Journal*, 50(3), 544–557.

Baum, JAC, T Calabrese and BS Silverman (2000). Don't go it alone: Alliance network composition and startups' performance in Canadian biotechnology. *Strategic Management Journal*, 21(3), 267–294.

Bentler, PM and DG Bonett (1980). Significance tests and goodness of fit in the analysis of covariance structures. *Psychological Bulletin*, 88(3), 588–606.

Bhaskaran, S (2006). Incremental innovation and business performance: Small and medium-size food enterprises in a concentrated industry environment. *Journal of Small Business Management*, 44(1), 64–80.

Boeker, W (1997). Strategic change: The influence of managerial characteristics and organizational growth. *Academy of Management Journal*, 40(1), 152–170.

Browne, MW and R Cudeck (1993). Alternative ways of assessing model fit. In *Testing Structural Equation Models*, Bollen KA and SJ Long (Eds.), pp. 136–162, Newbury Park: Sage Publications.

Calantone, RJ, R Garcia and C Dröge (2003). The effects of environmental turbulence on new product development strategy planning. *Journal of Product Innovation Management*, 20(2), 90–103.

Carpenter, MA, MA Geletkanycz and WG Sanders (2004). Upper echelons research revisited: Antecedents, elements, and consequences of top management team composition. *Journal of Management*, 30(6), 749–778.

Certo, ST, RH Lester, CM Dalton and DR Dalton (2006). Top management teams, strategy and financial performance: A meta-analytic examination. *Journal of Management Studies*, 43(4), 813–839.

Chandler, GN, C Keller and DW Lyon (2000). Unraveling the determinants and consequences of an innovation-supportive organizational culture. *Entrepreneurship: Theory and Practice*, 25(1), 59–76.

Chaney, PK, TM Devinney and RS Winer (1991). The impact of new product introductions on the market value of firms. *Journal of Business*, 64(4), 573–610.

Chatterjee, A and DC Hambrick (2007). It's all about me: Narcissistic chief executive officers and their effects on company strategy and performance. *Administrative Science Quarterly*, 52(3), 351–386.

Chen, C-J and J-W Huang (2009). Strategic human resource practices and innovation performance — The mediating role of knowledge management capacity. *Journal of Business Research*, 62(1), 104–114.

Cho, TS and DC Hambrick (2006). Attention as the mediator between top management team characteristics and strategic change: The case of airline deregulation. *Organization Science*, 17(4), 453–469.

Cohen, J, P Cohen, SG West and LS Aiken (2002). *Applied Multiple Regression/Correlation Analysis for the Behavioral Sciences.* Mahwah, NJ: Lawrence Erlbaum Associates.

Cooper, RG, SJ Edgett and EJ Kleinschmidt (1999). New product portfolio management: Practices and performance. *Journal of Product Innovation Management*, 16(4), 333–351.

Cyert, RM and JG March (1992). *A Behavioral Theory of the Firm.* Malden, MA: Blackwell Publishers.

de Jong, A, M Song and LZ Song (2013). How lead founder personality affects new venture performance: The mediating role of team conflict. *Journal of Management*, 39(7), 1825–1854.

Deshpandé, R, JU Farley and FE Jr Webster (2000). Triad lessons: Generalizing results on high performance firms in five business-to-business markets. *International Journal of Research in Marketing*, 17(4), 353–362.

Doty, DH and WH Glick (1998). Common methods bias: Does common methods variance really bias results? *Organizational Research Methods*, 1(4), 374–406.

Dougherty, D, and C Hardy (1996). Sustained product innovation in large, mature organizations: Overcoming innovation-to-organization problems. *Academy of Management Journal*, 39(5), 1120–1153.

Dutta, S, O Narasimhan and S Rajiv (2005). Conceptualizing and measuring capabilities: Methodology and empirical application. *Strategic Management Journal*, 26(3), 277–285.

Eddleston, KA and FW Kellermanns (2007). Destructive and productive family relationships: A stewardship theory perspective. *Journal of Business Venturing*, 22(4), 545–565.

Eddleston, KA, FW Kellermanns and R Sarathy (2008). Resource configuration in family firms: Linking resources, strategic planning and technological opportunities to performance. *Journal of Management Studies*, 45(1), 26–50.

Ensley, MD and AW Pearson (2005). An exploratory comparison of the behavioral dynamics of top management teams in family and nonfamily new ventures: Cohesion, conflict, potency, and consensus. *Entrepreneurship Theory and Practice*, 29(3), 267–284.

Ernst, H (2002). Success factors of new product development: A review of the empirical literature. *International Journal of Management Reviews*, 4(1), 1–40.

European Commission (2013). Innovation Union Scoreboard 2013. http://ec.europa.eu/enterprise/policies/innovation/files/ius-2013en.pdf.

Fernández-Ortiz, R and GF Lombardo (2009). Influence of the capacities of top management on the internationalization of SMEs. *Entrepreneurship & Regional Development*, 21(2), 131–154.

George, G (2005). Slack resources and the performance of privately held firms. *Academy of Management Journal*, 48(4), 661–676.

George, G, DR Wood and R Khan (2001). Networking strategy of boards: implications for small and medium-sized enterprises. *Entrepreneurship & Regional Development*, 13(3), 269–285.

German Federal Ministry of Economics and Technology (2013). Promoting innovation by SMEs. http://www.bmwi.de/EN/Topics/Technology/Technology-policy/promoting-innovation-by-smes.

Geroski, P, S Machin and JV Reenen (1993). The profitability of innovating firms. *The RAND Journal of Economics*, 24(2), 198–211.

Gilley, KM, BA Walters and BJ Olson (2002). Top management team risk taking propensities and firm performance: Direct and moderating effects. *Journal of Business Strategies*, 19(2), 95–114.

Gupta, M and G Cawthon (1996). Managerial implications of flexible manufacturing for small/mediumm-sized enterprises. *Technovation*, 16(2), 77–94.

Hair, JF, WC Black, BJ Babin and RE Anderson (2010). *Multivariate data analysis.* Upper Saddle River, NJ: Prentice Hall.

Hall, B (2000). Innovation and market value. In *Productivity, Innovation and Economic Performance*, Barro R, G Mason and M O'Mahony (Eds.), pp. 177–199, Cambridge: Cambridge University Press.

Hambrick, DC (2005). Upper echelons theory: Origins, twists and turns, and lessons learned. In *Great Minds in Management: The Process of Theory Development*, Smith KG and MA Hitt (Eds.), pp. 109–127, New York: Oxford University Press.

Hambrick, DC (2007). Upper echelons theory: An update. *Academy of Management Review*, 32(2), 334–343.

Hambrick, DC, MA Geletkanycz and JW Fredrickson (1993). Top executive commitment to the status-quo: Some tests of its determinants. *Strategic Management Journal*, 14(6), 401–418.

Hambrick, DC and PA Mason (1984). Upper echelons: The organization as a reflection of its top managers. *Academy of Management Review*, 9(2), 193–206.

Hay, M and K Kamshad (1994). Small firm growth: Intentions, implementation and impediments. *Business Strategy Review*, 5(3), 49–68.

Hayward, MLA and DC Hambrick (1997). Explaining the premiums paid for large acquisitions: Evidence of CEO hubris. *Administrative Science Quarterly*, 42(1), 103–127.

Henard, DH and DM Szymanski (2001). Why some new products are more successful than others. *Journal of Marketing Research*, 38(3), 362–375.

Hu, L-T and PM Bentler (1995). Evaluating model fit. Structural equation modeling: Concepts, issues, and applications. In *Structural Equation Modeling: Concepts, Issues, and Applications*, Hoyle RH (Eds.), pp. 76–99. Thousand Oaks: Sage Publications.

Hurley, RF and GTM Hult (1998). Innovation, market orientation, and organizational learning: An integration and empirical examination. *Journal of Marketing*, 62(3), 42–54.

James, LR, RG Demaree and G Wolf (1984). Estimating within-group interrater reliability with and without response bias. *Journal of Applied Psychology*, 69(1), 85–98.

James, LR, RG Demaree and G Wolf (1993). rwg: An assessment of within-group interrater agreement. *Journal of Applied Psychology*, 78(2), 306–309.

Jantunen, A, K Puumalainen, S Saarenketo and K Kyläheiko (2005). Entrepreneurial orientation, dynamic capabilities and international performance. *Journal of International Entrepreneurship*, 3(3), 223–243.

Kang, W and M Montoya (2014). The impact of product portfolio strategy on financial performance: The roles of product development and market entry decisions. *Journal of Product Innovation Management*, 31(3), 516–534.

Kanuk, L and C Berenson (1975). Mail Surveys and response rates: A literature review. *Journal of Marketing Research*, 12(4), 440–453.

Kellermanns, FW, K Eddleston, R Sarathy and F Murphy (2012). Innovativeness in family firms: A family influence perspective. *Small Business Economics*, 38(1), 85–101.

Klein, KJ and SWJ Kozlowski (2000). From micro to meso: Critical steps in conceptualizing and conducting multilevel research. *Organizational Research Methods*, 3(3), 211–236.

Kleinschmidt, EJ and RG Cooper (1991). The impact of product innovativeness on performance. *Journal of Product Innovation Management*, 8(4), 240–251.

Kor, YY (2003). Experience-based top management team competence and sustained growth. *Organization Science*, 14(6), 707–719.

Kraiczy, ND, A Hack and FW Kellermanns (2014). New product portfolio performance in family firms. *Journal of Business Research*, 67(6), 1065–1073.

Kraiczy, ND, A Hack and FW Kellermanns (forthcoming). What makes a family firm innovative? CEO risk-taking propensity and the organizational context of family firms. *Journal of Product Innovation Management.*

Kumar, N, LW Stern and JC Anderson (1993). Conducting interorganizational research using key informants. *Academy of Management Journal*, 36(6), 1633–1651.

Levy, M and P Powell (1998). SME flexibility and the role of information systems. *Small Business Economics*, 11(2), 183–196.

Ling, Y and FW Kellermanns (2010). The effects of family firm specific sources of TMT diversity: The moderating role of information exchange frequency. *Journal of Management Studies*, 47(2), 322–344.

Lumpkin, GT and GG Dess (1996). Clarifying the entrepreneurial orientation construct and linking it to performance. *Academy of Management Review*, 21(1), 135–172.

Manu, FA (1992). Innovation orientation, environment and performance: A comparison of U.S. and European markets. *Journal of International Business Studies*, 23(2), 333–359.

Mazen, AM, LA Graf, CE Kellogg and M Hemmasi (1987a). Statistical power in contemporary management research. *Academy of Management Journal*, 30(2), 369–380.

Mazen, AM, M Hemmasi and MF Lewis (1987b). Assessment of statistical power in contemporary strategy research. *Strategic Management Journal*, 8(4), 403–410.

Menguc, B and S Auh (2006). Creating a firm-level dynamic capability through capitalizing on market orientation and innovativeness. *Journal of the Academy of Marketing Science*, 34(1), 63–73.

Miller, D (1983). The correlates of entrepreneurship in three types of firms. *Management Science*, 29(7), 770–791.

Miller, D and PH Friesen (1982). Innovation in conservative and entrepreneurial firms: Two models of strategic momentum. *Strategic Management Journal*, 3(1), 1–25.

Miller, D, J Lee, S Chang and I Le Breton-Miller (2009). Filling the institutional void: The social behavior and performance of family vs non-family technology firms in emerging markets. *Journal of International Business Studies*, 40(5), 802–817.

Montoya-Weiss, MM and R Calantone (1994). Determinants of new product performance: A review and meta-analysis. *Journal of Product Innovation Management*, 11(5), 397–417.

Naman, JL and DP Slevin (1993). Entrepreneurship and the concept of fit: A model and empirical tests. *Strategic Management Journal*, 14(2), 137–153.

Narver, JC, SF Slater and DL MacLachlan (2004). Responsive and proactive market orientation and new-product success. *Journal of Product Innovation Management*, 21(5), 334–347.

Nohria, N and R Gulati (1996). Is slack good or bad for innovation? *Academy of Management Journal*, 39(5), 1245–1264.

Olson, EM, SF Slater and GTM Hult (2005). The performance implications of fit among business strategy, marketing, organization structure, and strategic behavior. *Journal of Marketing*, 69(3), 49–65.

Oppenheim, AN (1966). *Questionnaire Design, Interviewing and Attitude Measurement.* New York: Free Press.

Pattikawa, LH, E Verwaal and HR Commandeur (2006). Understanding new product project performance. *European Journal of Marketing*, 40(11/12), 1178–1193.

Pelled, LH, KM Eisenhardt and KR Xin (1999). Exploring the black box: An analysis of work group diversity, conflict, and performance. *Administrative Science Quarterly*, 44(1), 1–28.

Podsakoff, PM and DW Organ (1986). Self-reports in organizational research: Problems and prospects. *Journal of Management*, 12(4), 531–544.

Preacher, K and A Hayes (2004). SPSS and SAS procedures for estimating indirect effects in simple mediation models. *Behavior Research Methods*, 36(4), 717–731.

Rosenbusch, N, J Brinckmann and A Bausch (2011). Is innovation always beneficial? A meta-analysis of the relationship between innovation and performance in SMEs. *Journal of Business Venturing*, 26(4), 441–457.

Scott, SG and RA Bruce (1994). Determinants of innovative behavior: A path model of individual innovation in the workplace. *Academy of Management Journal*, 37(3), 580–607.

Seidler, J (1974). On using informants: A technique for collecting quantitative data and controlling measurement error in organization analysis. *American Sociological Review*, 39(6), 816–831.

Siegel, PA and DC Hambrick (2005). Pay disparities within top management groups: Evidence of harmful effects on performance of high-technology firms. *Organization Science*, 16(3), 259–274.

Simons, TL and RS Peterson (2000). Task conflict and relationship conflict in top management teams: The pivotal role of intragroup trust. *Journal of Applied Psychology*, 85(1), 102–111.

Simsek, Z (2007). CEO tenure and organizational performance: An intervening model. *Strategic Management Journal*, 28(6), 653–662.

Simsek, Z, JF Veiga, MH Lubatkin and RN Dino (2005). Modeling the multilvel determinants of top management team behavioral integration. *Academy of Management Journal*, 48(1), 69–84.

Smith, KG, KA Smith, JD Olian, HP Sims, DP O'Bannon and JA Scully (1994). Top management team demography and process: The role of social integration and communication. *Administrative Science Quarterly*, 39(3), 412–438.

Sorescu, AB, RK Chandy and JC Prabhu (2003). Sources and financial consequences of radical innovation: Insights from pharmaceuticals. *Journal of Marketing*, 67(4), 82–102.

Sorescu, AB and J Spanjol (2008). Innovation's effect on firm value and risk: Insights from consumer packaged goods. *Journal of Marketing*, 72(2), 114–132.

Spector, PE and MT Brannick (1995). The nature and effects of method variance in organizational research. In *International Review of Industrial and Organizational Psychology*, Cooper, CL and IT Robertson (Eds.), pp. 249–274. West Sussex: Wiley.

Srinivasan, S and DM Hanssens (2009). Marketing and firm value: Metrics, methods, findings, and future directions. *Journal of Marketing Research*, 46(3), 293–312.

Stinchcombe, AL (1965). Social structure and organizations. In *Handbook of Organizations*. March JG (Eds.), pp. 142–193. Chicago: Rand McNally & Company.

Talke, K, S Salomo and K Rost (2010). How top management team diversity affects innovativeness and performance via the strategic choice to focus on innovation fields. *Research Policy*, 39(7), 907–918.

Talke, K, S Salomo and A Kock (2011). Top management team diversity and strategic innovation orientation: The relationship and consequences for innovativeness and performance. *Journal of Product Innovation Management*, 28(6), 819–832.

Tellis, GJ, JC Prabhu and RK Chandy (2009). Radical innovation across nations: The preeminence of corporate culture. *Journal of Marketing*, 73(1), 3–23.

Tsai, K-H (2009). Collaborative networks and product innovation performance: Toward a contingency perspective. *Research Policy*, 38(5), 765–778.

Upton, N, EJ Teal and JT Felan (2001). Strategic and business planning practices of fast growth family firms. *Journal of Small Business Management*, 39(1), 60–72.

Wallace, JC, LM Little, AD Hill and JW Ridge (2010). CEO regulatory foci, environmental dynamism, and small firm performance. *Journal of Small Business Management*, 48(4), 580–604.

Wiklund, J and DA Shepherd (2003). Knowledge-based resources, entrepreneurial orientation, and the performance of small and medium-sized businesses. *Strategic Management Journal*, 24(13), 1307–1314.

Wood, MS and MD Michalisin (2010). Entrepreneurial drive in the top management team: Effects on strategic choice and firm performance. *Journal of Leadership & Organizational Studies*, 17(3), 222–239.

Worren, N, K Moore and P Cardona (2002). Modularity, strategic flexibility and firm performance: A study of the home appliance industry. *Strategic Management Journal*, 23(12), 1123–1140.

Wright, M and FW Kellermanns (2011). Family firms: A research agenda and publication guide. *Journal of Family Business Strategy*, 2(4), 187–198.

Zahra, SA and M Fescina (1991). Will leveraged buyouts kill U.S. corporate research & development? *The Executive*, 5(4), 7–21.

Zahra, SA, JC Hayton and C Salvato (2004). Entrepreneurship in family vs. non-family firms: A resource-based analysis of the effect of organizational culture. *Entrepreneurship Theory and Practice*, 28(4), 363–381.

Zhang, Y and N Rajagopalan (2010). Once an outsider, always an outsider? CEO origin, strategic change, and firm performance. *Strategic Management Journal*, 31(3), 334–346.

Chapter 12

ORGANISATIONAL LEARNING, NON-TECHNICAL INNOVATION AND CUSTOMER SATISFACTION OF SMEs

Dimitrios Kafetzopoulos and Evangelos Psomas

Department of Business Administration of Food and Agricultural Enterprises
University of Patras

The purpose of the present study is to analyse the role of organisational learning as an antecedent to non-technical innovation comprising organisational and marketing innovation. Examining the effect of these dimensions of non-technical innovation on customer satisfaction is also an aim of the present study. The paper formulates research hypotheses based on a comprehensive literature review. These hypotheses are tested using structural equation modelling (SEM) based on data collected from 226 Greek small and medium-sized enterprises (SMEs). The formulated conceptual model is strongly supported based on empirical evidence. The results confirm that organisational learning capability favours the development of organisational innovation and marketing innovation. Furthermore, the study shows that both dimensions of non-technical innovation promote the achievement of customer satisfaction.

Keywords: Organisational innovation; marketing innovation; customer satisfaction.

Introduction

The vast majority of enterprises in most European countries consist of small and medium-sized enterprises (SMEs) (Salavou *et al.*, 2004). SMEs play a significant role in promoting the fast and healthy economic and social development of every country (Chen, 2006). The survival and competitiveness of SMEs depends on the

This paper was first published in *International Journal of Innovation Management*, Vol. 20, No. 3 (April 2016), 1650041 (28 pages). DOI: 10.1142/S1363919616500419 satisfaction of their customers, who are sophisticated, segmented and demanding (Simon and Yaya, 2012). Customer satisfaction refers to the level of satisfaction with products and after sales services and to the level of customer complaints. To adapt to ever changing customer needs and satisfy them, organisations need to provide quality and innovative products. It is assumed that an SME can increase customer satisfaction by improving its innovativeness (Feng *et al.*, 2006; Hung, 2007). Innovation has long been considered a fundamental component of entrepreneurship, a key factor for the survival (Pallas *et al.*, 2013), growth and development of SMEs (Bigliardi and Dormio, 2009), and an important source of competitive advantage of any firm (Tan *et al.*, 1999; Jimenez-Jimenez *et al.*, 2014). Humphreys *et al.* (2005) also highlight the need for SMEs to increasingly innovate to compete in global and niche markets, while similarly, Akman and Yilmaz (2008), state that innovation is a key success factor in an increasingly competitive and global economy.

However, both technical and non-technical innovation should be taken into account to bring about benefits to organisations (Amalia and Nugroho, 2011). Technical innovation includes the development of products, services and processes that are new to the organisation, or changes in the way products are made, while non-technical innovation includes the application of new ideas to improve the organisational structure, system, and processes pertaining to the social structure of an organisation (Damanpour, 1991; Weerawardena, 2003). In this context, the management of innovation and customer satisfaction are considered essential for organisations to be able to compete in the markets.

Firms should also possess superior organisational learning capability based on which innovative products that meet customer requirements can be provided. More than ever, organisational learning has become a need rather than a choice (Garcia-Morales *et al.*, 2012). It is defined as the capability of an organisation to create, acquire, transfer, and integrate knowledge and to modify organisational behaviour to reflect new cognitive situations with a view to improving business performance (Jerez-Glomez *et al.*, 2005). New knowledge should also be acquired so that new and innovative ways of working more closely with customers are identified (Chaston *et al.*, 2001; Prieto and Revilla, 2006). So, innovation depends on the firm's capability to learn through which new knowledge is developed, distributed and used (Alegre and Chiva, 2008). The literature on management emphasises the key role that organisational learning plays in innovation (Darroch and McNaughton, 2002; Jimenez-Jimenez and Sanz-Valle, 2011). In addition, the literature emphasises the great importance of both organisational learning and innovation for a company's survival. More specifically, a firm which is capable from the knowledge management perspective, uses its resources efficiently and consequently it is more innovative and performs better (Darroch, 2005). So, how to turn the available organisational knowledge into profitable innovations and bring them to the market in a continuous manner has become a major concern both of industries and academics (Xu *et al.*, 2010).

Innovation as an organisational phenomenon has a long tradition of research and has been studied in many different fields (Quintane *et al.*, 2011). However, there is a general agreement among authors that there is considerable scope for further research in this area (Salavou *et al.*, 2004; Oke *et al.*, 2007; Lee and Ging, 2007). Abou-Zeid and Cheng (2004) and Quintane *et al.* (2011) state that few empirical studies operationalise innovation as a knowledge-based outcome, while, Jimenez-Jimenez *et al.* (2014), note that despite the acknowledged importance of knowledge management for innovation, further studies are still needed. In particular, recent studies have encouraged research on the development of models and theories of non-technical innovation to extend the existing models and theories, which are mainly drawn from research on technical innovation (Damanpour and Aravind, 2012). The impact of a firm's learning capacity on non-technical innovation has not yet been studied in detail either (Camison and Villar-Lopez, 2011). This means that, theoretical and empirical research on organisational learning that leads to non-technical innovation is limited (Camison and Villar-Lopez, 2011) and this relationship needs to be studied in greater detail (Armbruster *et al.*, 2008; Mol and Birkinshaw, 2009; Fiedler and Welpe, 2010). Furthermore, few studies have linked marketing and organisational innovation and customer satisfaction directly (Simon and Yaya, 2012), giving directions for future research.

The research topics that the previously published papers have not addressed as well as the respective research suggestions made by experts in the field, stimulated the authors to carry out the present study. The purpose of the study is to address the existing literature gap and add to the literature in two important ways. First, it provides empirical evidence for the role organisational learning plays in non-technical innovation by using the dimensions of marketing and organisational innovation. Second, it also provides evidence for the relationship between organisational and marketing innovation and customer satisfaction; such evidence has been limited to date, especially as far as SMEs are concerned. From the above mentioned, it is apparent that the present study simultaneously considers antecedents and consequences of non-technical innovation. This is in line with Camison and Villar-Lopez (2011) who suggest that the previously developed models should be extended to more complete models capturing the complexity of the relationships between all the variables involved in the non-technical innovation process. Knowing more about the dynamics of learning values, market information-processing behaviours and organisational action is relevant to understanding how organisations perform and improve (Sinkula *et al.*, 1997). Based on a survey of 226 managers from SMEs in Greece, this study applies the structural equation modelling (SEM) to investigate the research hypotheses.

It is worth noting that the present study is substantially different from those carried out so far, given that it is based on a large sample of SMEs exclusively, provides up to date empirical evidence to examine the relationships described in the literature (which are not all empirically supported through previous research studies), examines the innovativeness of SMEs from the non-technical point of view (organisational and marketing innovation), and finally it applies multiple regression analyses using SEM to test the goodness of fit of the theoretical model to the data. So, it is apparent that the present study contributes to the existing literature.

The remainder of this paper is organised as follows. The next section presents the literature review introducing the key constructs of the study and developing a research model that depicts the hypothesised relationships. This is followed by the research methodology describing the construct operationalisation and data collection. Data analysis and the findings of the study are reported in the next section. The findings related to testing the structural model and the managerial implications are then discussed. Finally, the conclusions, limitations, and future research proposals are presented.

Theoretical Background and Conceptual Foundation

Organisational learning

Organisational learning involves the acquisition of explicit and tacit knowledge and knowledge sharing and use (Dibella *et al.*, 1996). More specifically it includes obtaining and sharing information with regard to customer needs, market changes and competitors' actions, as well as developing new technologies to create new products that are superior to those of competitors (Calantone *et al.*, 2002). Organisational learning is the basis for gaining sustainable competitive advantage (Sinkula *et al.*, 1997) and a key factor for the improvement of organisational performance (Brockmand and Morgan, 2003). The role of learning in relation to the innovation and the performance of SMEs has become a major research focus (Garcia-Morales *et al.*, 2006). The basic assumption is that learning plays a key role in enabling SMEs to implement fast and flexible innovation processes (Weerd-Nederhof *et al.*, 2002). This paper focuses on organisational learning as an antecedent factor to non-technical innovation. So, what is already known with regard to the dynamic and enabling role of organisational learning is enhanced through the present study.

Non-technical innovation of SMEs

Literature distinguishes between different types of innovation and researchers have explored its classification in different ways (Jimenez-Jimenez and Sanz-Valle, 2011). Damanpour (1991) distinguishes between technical and non-technical innovation (administrative innovation). Nowadays, non-technical innovation plays an increasingly important role in better understanding innovation and its impact on the competitiveness of firms. However, the literature on non-technical innovation is diverse and scattered (Armbruster *et al.*, 2008). A recent review of the innovation literature of the past 27 years reveals that only 3% of the reviewed papers deal with innovation in administrative processes and management practices (Crossan and Apaydin, 2010). Thus, recent studies have encouraged research on the development of models and theories of non-technical innovation in organisations (Camison and Villar-Lopez, 2011). Non-technical innovation includes the dimensions of organisational and marketing innovation as they deal with changes in the organisational structure (Jimenez-Jimenez and Sanz-Valle, 2011). Many studies describe innovation based on marketing and organisational innovation (Evangelista and Vezzani, 2010; Wonglimpiyarat, 2010; Gunday *et al.*, 2011; Chang *et al.*, 2012). SMEs should organise the innovation process efficiently by combining organisational innovation with skills in marketing innovation. So, on the basis of the above, the present study adopts the concept of non-technical innovation that includes organisational and marketing innovation. By focusing on these dimensions, non-technical innovation is precisely conceptualised and its effects on performance dimensions are more clearly identified. More specifically, the impact of these innovation dimensions on customer satisfaction is a subject examined through the present study. The present study is in line with (Armbruster *et al.*, 2008) who recommend going further in the comprehension of distinct dimensions of innovation including organisational and marketing innovation.

Organisational innovation

Organisational innovation refers to the organisational coordination and cooperation of the R&D department (which deals with creating innovation) with the other departments of the firm, the organisational integration and simplification of the main functions, and the identification, adoption and support of innovation initiatives from the organisational capability perspective (Yam *et al.*, 2011; Gunday *et al.*, 2011 and Forsman, 2011). Organisational innovation deals with new organisational methods for business management not only in the workplace but also in the external business environment, for example, in the relationship between the firm and external agents (OECD, 2005). It involves changes to administrative processes and/or organisational structures relating to the basic activities of an organisation and its management. Changes in the organisational structure and procedures can facilitate the creation of new products and processes (Chang *et al.*, 2012). In addition, the introduction of innovative methods in the management of a firm's external relationships involves the implementation of new ways of organising the relationships with other firms or public institutions, such as close collaboration with research organisations or customers, methods for integration with suppliers, or outsourcing (Camison and Villar-Lopez, 2011). By conceptualising organisational innovation, its mediating role in the relationship between organisational learning and customer satisfaction can be more clearly determined.

Marketing innovation

Marketing innovation refers to the successful launch of new products in different market segments, new methods of product placement and promotion that increase sales-force and the diversification of the product distribution to the supply chain (Yam *et al.*, 2004, 2011). Based on a better understanding of consumer needs and competition (Yam *et al.*, 2011), new marketing methods can be established involving significant changes in product design, packaging, placement, promotion and pricing. By conceptualising marketing innovation, its mediating role in the relationship between organisational learning and customer satisfaction can be more clearly determined.

Customer satisfaction

Customer satisfaction is business terminology used to evaluate the degree to which a product or service provided by a firm satisfies its customers (Nemati *et al.*, 2010). It is considered an important factor in determining a firm's competitiveness and a reliable indicator of a firm's future profit as it is positively associated with return on investment (ROI) and market value (Ryding, 2010). In an economically challenging environment where survival is increasingly difficult across a whole range of businesses, understanding customers is crucial (Ryding, 2010). Thus, in order to achieve customer satisfaction, firms need to create an environment to better meet and even exceed customer needs, requirements, desires and expectations. For these reasons, innovation that aims at satisfying customer needs is becoming popular (Simon and Yaya, 2012).

Nowadays, firms, more than ever, implement quality management systems, which are process and customer oriented. This means that customer needs and requirements are taken into consideration as process inputs while the evaluation of customer satisfaction is taken into consideration as a process output. As a consequence, firms are in close contact with their customers, who, in turn, perceive a firm's organisational learning capability and the non-technical innovation practices it adopts. So, the effects of organisational learning and non-technical innovation are reflected in customer satisfaction.

Development of hypotheses

Organisational learning and innovation

Small businesses need to be innovative in order to survive in a dynamic and competitive environment (Georgellis *et al.*, 2000). Innovation initiatives depend heavily on employees' knowledge, skill and experience in the value creation process. Researchers conclude that organisational learning is associated with the development of new knowledge, which in turn, is crucial for firm innovativeness (Liu *et al.*, 2002). According to Cohen and Levinthal (1990), knowledge can be seen as a valuable input for innovation. An organisation committed to learning can enhance its innovation capability in three ways. First, it is more likely to be committed to innovation, have state-of-the-art technology, and use that technology in innovation. It is also more likely to build, promote, and take advantage of a technological breakthrough (Calantone *et al.*, 2002). Second, it has the knowledge and ability to understand and anticipate customer needs that leads to innovation initiatives (Cahill, 1996). Third, an organisation committed to learning is likely to have greater innovation capability than its competitors (Damanpour, 1991). From the above mentioned, it is apparent that the learning capability of a firm positively influences innovation as a whole. However, it is also a determinant of a dimension of innovation, namely non-technical innovation, as well as its sub-dimensions, namely organisational and marketing innovation.

Organisational learning and organisational innovation

Firms can benefit from cross-industry innovation if they are open to external knowledge sources and understand how to explore, transform, and exploit cross-industry knowledge. So, firms should establish certain organisational structures and processes to facilitate and operationalise organisational learning across industry boundaries (Bader, 2013). Organisational innovation depends on the knowledge base that the organisation possesses, generated by organisational learning (Cohen and Levinthal, 1990; Nonaka and Takeuchi, 1995). Goswami and Mathew (2011) and Liao and Wu (2010) also support the view that organisational learning competency contributes to organisational innovation potential. Camison and Villar-Lopez (2011) state that learning can favour the development of organisational innovation in business practices. Although the empirical findings which shape the dominant view that learning acts as a significant contributor to organisational innovation come from large firms, the industry experts state that learning is even more important for SMEs (Salavou *et al.*, 2004). Laforet (2011) points out that organisational learning has an effect on SME innovative performance, while a study by Zhang *et al.* (2006) links learning to innovation. Furthermore, several studies report the lack of skilled or trained personnel as being a major business obstacle to innovation in the manufacturing SMEs (Freel, 2005; Laforet and Tann, 2006). Many other authors (Weerawardena, 2003; Weerawardena *et al.*, 2006; Camison and Villar-Lopez, 2011) support the view that a firm's learning capability favours organisational innovation. These arguments are reflected in the following hypothesis:

H1: *There is a positive and strong association between organisational learning and organisational innovation.*

The literature gap identified, the research proposals suggested by experts in the field and the fact that the present study focuses on SMEs, enhance the value of testing the significance of the relationship between organisational learning and organisational innovation.

Organisational learning and marketing innovation

Learning capabilities can also favour marketing innovation. Based on organisational learning, creativity is enhanced and new knowledge and ideas are inspired (Aragon-Correa *et al.*, 2007). Organisational learning can lead to the development of innovative marketing instruments through which a firm's products can be better distributed, promoted, and invoiced (Camison and Villar-Lopez, 2011). Organisational learning involves a commitment to learning in itself, an open mind and exchange of knowledge. These characteristics promote a set of knowledge-questioning and knowledge-enhancing values that lead to the development of innovative products, services, and technologies, as well as to methods of exploration of new markets (Keskin, 2006). The framework for market-based organisational learning studied by Sinkula *et al.* (1997) suggests that learning orientation (determined by commitment to learning, shared vision, and open-mindedness) positively influences the generation and dissemination of marketing information which in turn enhances the dynamism of the marketing program. According to Sinkula *et al.* (1997), learning-efficient organisations are likely to be more nimble, changing their marketing strategies in a rapid and fluid manner to anticipate, neutralise, or possibly flourish from shocks incurred in unstable environments. Many other authors (Weerawardena, 2003; Weerawardena *et al.*, 2006; Camison and Villar-Lopez, 2011) support the view that a firm's learning capability favours marketing innovation. These arguments are reflected in the following hypothesis:

H2: *There is a positive and strong association between organisational learning and marketing innovation.*

The literature gap identified, the research proposals suggested by experts in the field and the fact that the present study focuses on SMEs, enhance the value of testing the significance of the relationship between organisational learning and marketing innovation.

Innovation and customer satisfaction

Reviewing the literature in the knowledge management field, knowledge and innovation are often seen as moderating or mediating variables of company performance (Quintane *et al.*, 2011). For example, innovation capability and customer satisfaction are closely associated. Customer satisfaction can be achieved and exceeded by providing customers with innovative and quality products (Tan *et al.*, 1999). Companies that bring about excellent and successful organisational changes and innovations attract and satisfy many customers (Nemati *et al.*, 2010). So, those firms that place an emphasis on innovation are more likely to improve the fit between their innovative offers and customer needs (Ngo and O'Cass, 2011). Innovation increases customer satisfaction as it maximises the value of their purchases. White (1996) defined a set of variables that influence customer satisfaction including innovation. In order to satisfy their customers, firms need to be innovative by managing the different systems in their organisation effectively (Simon and Yaya, 2012). The capacity to innovate can assist firms in the process of developing superior products that meet customers' changing needs and demands (Li and Mitchell, 2009; Rosenbusch *et al.*, 2011). Simpson *et al.* (2006) state that innovative firms focus on successfully creating and promoting novel ideas, products, and services in the market place; as well as on anticipating consumer needs and responding to them better than competitors (Laforet, 2011). Managers are willing to invest in innovation once they consider that customers' needs are not fully satisfied by the current products (Simon and Yaya, 2012). According to Knox (2002), it is important for company managers to adopt innovative thinking in their decision making in order to achieve superior customer value. Furthermore, Calantone *et al.* (2002) point out that firms with greater innovativeness are more successful in responding to customers' needs.

From the above mentioned, it is apparent that the adoption of innovation, in general, is a parameter that satisfies the customers of a firm. However, the literature suggests that it is not only innovation as a whole that improves customer satisfaction, but its specific dimension, namely non-technical innovation and more specifically its sub-dimensions, namely organisational and marketing innovation.

Organisational innovation and customer satisfaction

The underlying goal of managing organisational innovation is customer satisfaction (Krause, 2004). Organisational innovativeness has a positive impact on customer value and business performance (Han *et al.*, 1998; Hult *et al.*, 2004). Totterdell *et al.* (2002) support the view that organisational innovation positively affects a firm's financial indexes, the relations among employees and customer satisfaction. Empirical evidence suggests that an organisation's innovativeness positively affects market orientation in generating superior value to customers (Atuahene-Gima, 1996). Implementing organisational innovation practices results in meeting customer requirements, gaining new markets, reducing cost, and increasing production flexibility. Therefore, it is vital for a firm to evaluate the effectiveness of organisational innovation management in terms of customer satisfaction. These arguments are reflected in the following hypothesis:

H3: *There is a positive and strong association between organisational innovation and customer satisfaction.*

The value of testing the significance of such a relationship is enhanced given the literature gap identified, the research proposals suggested by experts in the field and the SMEs which the present study focuses on.

Marketing innovation and customer satisfaction

The response to customer needs through market driven innovation is compatible with organisational innovativeness (Slater and Narver, 1994). Marketing innovation is a continual process whereby managers establish a new marketing system which is expected to be more competitive and successful than those of the competitors (Ren *et al.*, 2010). Given that, it is difficult for competitors to imitate the marketing innovation practices of a firm, marketing innovation can be the primary source of customer satisfaction (Ren *et al.*, 2010). The existence of a link between marketing innovation and customer satisfaction is supported by several authors who state that marketing innovation is positively perceived and evaluated by customers. This, in turn, makes firms more willing to invest in marketing innovation (Kafetzopoulos and Psomas, 2015). Simon and Yaya (2012) find a positive effect between marketing innovation and customer satisfaction, meaning that customers are more satisfied if they perceive new and innovative ways of presenting and promoting the products. This finding is coherent with the innovation literature, according to which innovation has a positive impact on customers since firms are more likely to prioritize and satisfy customer demands through marketing innovation than through product innovation (Simon and Yaya, 2012). Thus, we postulate the following hypothesis:

H4: *There is a positive and strong association between marketing innovation and customer satisfaction.*

The literature gap identified and the research proposals suggested by experts in the field support the view that no recent empirical evidence on SMEs exists

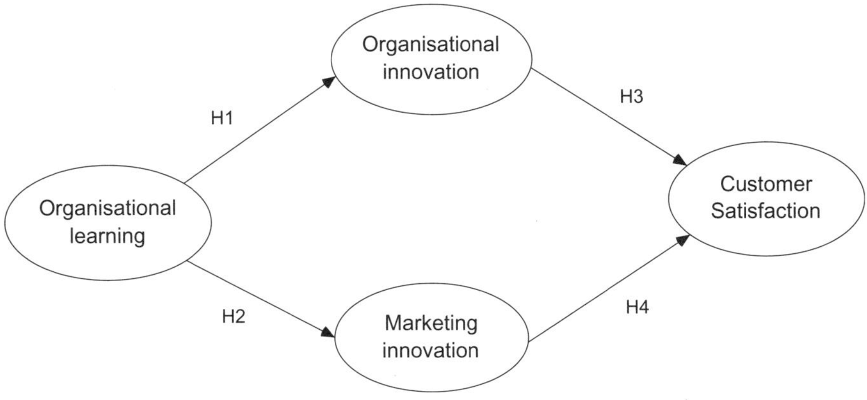

Fig. 1. Hypotheses and conceptual model to be tested.

confirming or refuting the relationship between marketing innovation and customer satisfaction. Thus, the value of testing the significance of such a relationship is enhanced.

Figure 1 presents a conceptual model that links organisational learning, organisational innovation, marketing innovation and customer satisfaction together.

Research Methodology

Sample and data collection

The population of firms chosen for this study consists of SMEs belonging to all economic sectors in Greece. For the purpose of the present study, an SME is defined as an enterprise with more than 20 and less than 250 employees, corresponding to the definition used by the European Union (Kalantaridis, 2004). A sample of 800 SMEs was randomly selected from the data base of ICAP (the largest business information and consulting firm in Greece). They are heterogeneous in terms of sub-sectors and product/process complexity.

To test the above formulated hypotheses, a structured questionnaire comprising 18 questions was developed through a multi-step process. All the questionnaire items are adopted based on a comprehensive literature review. More specifically, the questionnaire items describing organisational learning are drawn from the studies of Yam *et al.* (2011), Jimenez-Jimenez and Sanz-Valle (2011) and Camison and Villar-Lopez (2011). The items of organisational innovation used in the present study are drawn from the studies of Yam *et al.* (2004, 2011), Gunday *et al.* (2011) and Forsman (2011), while the items of marketing innovation are drawn from the studies of Yam *et al.* (2004) and Yam *et al.* (2011). Finally, the questionnaire items of customer satisfaction are drawn from the studies of Narver and Slater (1990), Zhang *et al.* (2003) and Yu *et al.* (2013).

The questionnaire was judged through interviews with experts in the field and finally it was pilot tested on 38 SMEs in Greece proving its appropriateness and achieving the content validity of the constructs used. The questionnaire items were measured on a seven point modified Likert scale (1 = very low to 7 = very high). The questionnaire was sent by e-mail to the sample SMEs. It was requested that the questionnaire be answered by the general manager or another manager designated by him/her because of the latter's familiarity with the issues dealt with in the questionnaire. Similarly, in order to study firm innovation practices and performance, Martinez-Costa and Martinez-Lorente (2008) rely on personal interviews with general managers, Inauen and Schenker-Wicki (2012) on responses of R&D managers, Verdun *et al.* (2012) on answers from CEOs and

Hung *et al.* (2010) on the perceptions of managing directors or CEOs. The identification of innovation practices by company managers/owners was also chosen by the Organisation for Economic Co-operation and Development (OECD) in the Oslo Manual as the basis for setting guidelines when collecting data on innovation (Massa and Testa, 2008).

Two waves of responses were received including 136 and 90 questionnaires, respectively. Finally, 226 valid questionnaires were collected satisfying the criterion for SEM analysis (Wu and Liu, 2010), and yielding a response rate of 28.25%. The responding firms represented the following sectors: 82 food and beverages (36.3%); 52 agricultural products (23%); 26 machinery and equipment (11.5%); 20 metal products (8.8%); 10 plastic, chemical, and associated products (4.4%); 9 pharmaceutical/cosmetics (4%), and 27 other industries (12%).

Non-response bias and common method bias

The Mann–Whitney test is used (Field, 2005) to determine if there are differences between the early and late respondents in terms of variables relevant to the research hypotheses (Fotopoulos and Psomas, 2009). In addition, the responding and non-responding companies are compared in terms of firm characteristics. These comparisons do not reveal any significant differences, suggesting no response bias. The possibility of common method variance is also a major issue in ensuring reliable results in empirical studies. For this reason, we assured respondents of the anonymity and confidentiality of the study (Verdun *et al.*, 2012). Moreover, in order to check for common method variance, the single-factor test is applied (Lau *et al.*, 2010). This test is run by loading all the questionnaire items into a principal component analysis and forcing them into one latent factor. This method produces poor results indicating 28% of the variance extracted, while many items suffer from poor factor loadings, which fall below 0.4. From the above, it is obvious that common method variance is not a problem in this study.

Measurement analysis

Exploratory factor analysis (EFA) (principal component extraction method with varimax orthogonal rotation) is applied to uncover the underlying structure (latent factors) of the measured variables. Confirmatory factor analysis (CFA) is also applied to refine the resulting scales in EFA and to determine if the latent factors extracted and the loadings of the measured variables on them conform to what is expected on the basis of pre-established theory (Narayan *et al.*, 2008). Factor analytic methods are useful for observing the underlying patterns or relationships of a large number of variables and they determine whether the information can be condensed or summarised in a smaller set of factors or components (Gunday *et al.*, 2011). So, based on EFA and CFA, the construct validity of the latent factors is assessed (Hair *et al.*, 2006). Moreover, it is ensured that the measurement items are reliable and unidimensional (Hair *et al.*, 2006). The research hypotheses are tested using SEM via path analysis, as it is a multivariate analytic methodology that gives insights into the causal ordering of variables in a framework of relationships (Fynes and Voss, 2001). The statistical package SPSS AMOS 20 is used for data processing.

Data Analysis and Results

Validation of the instrument

Applying EFA, we observe that all the factor loadings of the measured variables are well above 0.50. Items with factor cross-loadings of 0.40 or above are not reported. Four latent factors are extracted explaining 67.42% of the total variance. These four factors are labelled based on the respective items as follows: Organisational learning, organisational innovation, marketing innovation and customer satisfaction. Table 1 shows the results of EFA for all scale components (Kaiser–Meyer–Olkin = 0.912, Bartlett's test of Sphericity = 2231.459, $p = 0.00$, eigenvalue > 1, MSA > 0.80). All factors are practically significant regarding the sample size employed (Hair *et al.*, 2006). Cronbach's Alpha coefficient is calculated to assess the scale reliability of each latent factor in the research model. All Cronbach's alpha coefficients are greater than the suggested threshold value of 0.7 for an acceptable level of reliability (Kline, 2005; Hair *et al.*, 2006).

Based on the above, it is obvious that the unidimensionality of the measurement scales is well supported. It is also worth noting that multicollinearity problems do not appear in the present study (all correlation coefficients are lower than 0.80). Furthermore, the variance inflation factors (VIF) for all the latent factors range from 1.93 to 2.76, well below the threshold level of 10. This suggests that there is no harmful collinearity; hence all correlations among the four latent factors will not confound the results of the subsequent statistical tests (Newbert, 2008). Finally, the latent factors are checked for normality, randomness, and independency assumptions and thus, the data is validated for the

Table 1. The results of EFA.

	Latent factors			
Items	Marketing innovation	Organisational learning	Organisational innovation	Customer satisfaction
	Factor loading			
V010	0.799			
V012	0.749			
V013	0.700			
V011	0.687			
V014	0.675			
V002		0.779		
V004		0.743		
V005		0.708		
V003		0.702		
V001		0.657		
V006			0.854	
V009			0.738	
V008			0.729	
V007			0.729	
V017				0.807
V018				0.740
V016				0.730
V015				0.622
Eigenvalue	7.825	1.693	1.573	1.043
Cumulative variance %	43.474	52.880	61.621	67.415
Cronbach alpha	0.883	0.852	0.856	0.751

statistical tests. Table 2 provides an overview of the mean values, standard deviations, and the correlations between the latent factors.

In order to further validate the extracted latent factors, the CFA (maximum likelihood estimation technique) is applied. According to the results, the goodness of fit of the hypothesised CFA model to the empirical data is satisfactory (Table 3) (Hair *et al.*, 2006; Sadikoglu and Zehir, 2010). So, the four latent factors revealed using EFA are strongly confirmed through CFA.

Moreover, a series of tests are performed to determine the construct validity of the latent factors. Construct validation includes tests for face, convergent, and discriminant validity. Given that the survey instrument was formulated based on the existing literature, reviews by experts, and a pilot test, it is apparent that the

Table 2. Latent factors correlation matrix.

Latent factors	Mean value	Standard deviation	Correlation matrix			
			1	2	3	4
1. Organisational learning	4.82	1.20	1			
2. Organisational innovation	4.97	1.19	0.72	1		
3. Marketing innovation	5.06	1.07	0.60	0.71	1	
4. Customer satisfaction	5.33	1.09	0.51	0.50	0.52	1

Notes: All correlations are significant at 0.001.

Table 3. The fit indices of the measurement and structural model.

Fit indices	Measurement model (CFA)	Structural model	Acceptable fit indices[a]
Absolute fit indices			
Chi-square (χ^2)	185.126	212.680	$0 \leq \chi^2 \leq 2df$
Degrees of freedom (*df*)	125	127	
Root Mean Square Residual (RMR)	0.042	0.055	< 0.08
Root Mean Square of Approximation (RMSEA)	0.046	0.055	< 0.08
Incremental fit indices			
Incremental Fit Index (IFI)	0.972	0.961	> 0.90
Tucker-Lewis coefficient (TLI)	0.966	0.952	> 0.90
Comparative Fit Index (CFI)	0.972	0.960	> 0.90
Parsimonious fit indices			
Chi-square/ degrees of freedom (χ^2/df)	1.481	1.675	<3.0
Normed Fit Index (NFI)	0.920	0.908	> 0.50
Goodness of Fit Index (GFI)	0.919	0.906	> 0.50
Adjusted Goodness of Fit Index (AGFI)	0.890	0.874	> 0.50

[a]Hair *et al.* (2006); Sadikoglu and Zehir (2010).

latent factors revealed and their associated items possess a sufficient level of face validity (Hair *et al.*, 2006). As far as convergent validity is concerned, it is confirmed by evaluating the factor loadings (> 0.622), the average variance extracted (AVE > 0.503) and the construct reliability (CR > 0.757) in all cases (Table 4) (Hair *et al.*, 2006). Discriminant validity is evidenced by the fact that the Corr^2 (the highest squared correlation between the factor of interest and the

Table 4. The validity of the latent factors.

Latent factors	Average variance extracted*	Construct Reliability**	(Corr)^{2***}
Organisational learning	0.524	0.829	0.518
Organisational innovation	0.596	0.862	0.518
Marketing innovation	0.595	0.881	0.504
Customer satisfaction	0.503	0.757	0.270

Notes: *: AVE $= \Sigma\lambda_i^2 / n$, (number of items $i = 1, \ldots, n$, $\lambda i =$ standardised factor loading);
**: CR $= (\Sigma\lambda_i)^2/[(\Sigma\lambda_i)^2 + (\Sigma\delta_i)]$, (number of items $i = 1, \ldots, n$, $\lambda i =$ standardised factor loading, $\delta_i =$ error term);
***: The highest squared correlation between the factor of interest and the remaining factors.

remaining factors) is less than the AVE for each latent factor (Table 4) (Hair *et al.*, 2006). The nomological (significant correlations among the extracted latent factors) and criterion-related validity of the extracted latent factors are also supported.

Hypothesis testing

Finally, the structural model is tested. The structural model investigates the impact of organisational learning on organisational and marketing innovation and the impact of these innovation dimensions on customer satisfaction. The goodness of the model fit to the measured data is presented in Table 3. The basics of goodness of fit, the absolute fit indices, the incremental fit indices, and the parsimony fit indices indicate an acceptable fit of the proposed model (Hair *et al.*, 2006; Sadikoglu and Zehir, 2010). Moreover, the SEM procedure is applied (maximum likelihood method) to estimate the causal relations between the latent factors, and confirm or refute the hypotheses presented earlier (H_1–$H4$). Figure 2 presents the estimated standardised parameters for the causal paths, their level of significance, and the results of the squared multiple correlations for the endogenous factors.

Based on Fig. 2, it is apparent that organisational learning has a significant and positive effect, primarily, on marketing innovation (H_2: $b = 0.814$, $p \leq 0.001$) and secondarily on organisational innovation (H_1: $b = 0.683$, $p \leq 0.001$). Thus, the first (H_1) and second research hypotheses (H_2) of the present study are

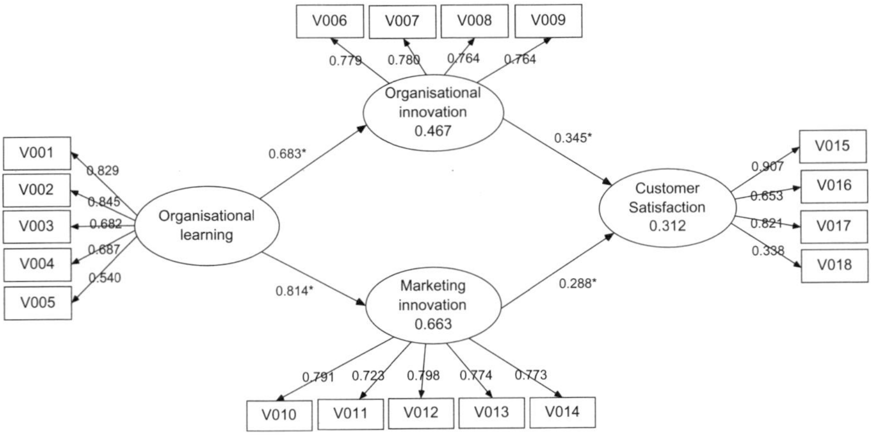

Source: *p-value ≤ 0.001.

Fig. 2. Theoretical model, showing estimated standardised parameters for the causal paths and the results of the squared multiple correlations.

accepted. Overall, 46.7% of the variation in organisational innovation and 66.3% of the variation in marketing innovation is explained by organisational learning.

The findings also indicate that organisational innovation has a positive and significant direct effect on customer satisfaction (H_3: $b = 0.345$, $p \leq 0.001$). So, hypothesis H_3 is confirmed. The standardised regression weight for the direct relationship between marketing innovation and customer satisfaction (H_4: $b = 0.288$, $p \leq 0.001$) is positive and statistically significant. So, hypothesis H_4 is also confirmed. Overall, 31.2% of the variation in customer satisfaction is explained by organisational innovation and marketing innovation. Thus, these findings support the view that organisational learning is a key driving force for the dimensions of non-technical innovation. In addition, organisational and marketing innovation plays a key role in enhancing customer satisfaction.

Discussion

The present study reveals several useful findings based on the hypothesised model and its empirical validation. It extends the existing research by exploring issues that have been missing in previous studies. More specifically, previous research suggests a positive relationship between the organisational learning, innovation, and competitive advantages of a firm, by focusing mainly on the case of technical innovation (Camison and Villar-Lopez, 2011). The model suggested in the present study presents simultaneously determinants and consequences of non-technical innovation by considering the effect of organisational learning on non-technical innovation and its effect on the customer satisfaction of SMEs. Two dimensions of non-technical innovation, according to OECD (2005), are taken into consideration in the present study, namely organisational and marketing innovation. Hence, the present research study gives insights into the management strategies of SMEs.

The present study findings suggest that organisational learning has a positive effect on organisational and marketing innovation. It is worth noting that, according to industry experts, the Greek SMEs do not pay due attention to the development of learning competencies. They adhere to the traditional authoritative management style, often displayed by family owned firms (Salavou *et al.*, 2004). So, based on the present study findings, managers of SMEs should be aware that the knowledge-based capability of a firm is crucial for the development of non-technical innovation. In other words, in order for the organisational and marketing innovation to be developed, the diffusion of new knowledge throughout the organisation, supported by managers and resources, is essential. By supporting organisational learning, a firm can renew and reinvent its processes with regard to firm organisation and marketing management, with the aim of avoiding falling into stagnation. So, from the above mentioned, it is apparent that promoting and advancing the "soft" aspect of firm innovativeness, namely non-technical innovation, depends on the development of a "soft" critical element, namely organisational learning. However, managers should bear in mind that the knowledge-based capability should not be characterised by precisely adopting what has been learned but by carefully adapting the new ideas and knowledge to the conditions of their internal business environment. It is the specific organisational structure and procedures of a firm as well as its existing marketing procedures that can be innovatively adapted based on the knowledge acquired from the literature, conferences, similar business practices, etc. The significant consequences of organisational learning do not concern only the non-technical innovation dimensions, but customer satisfaction also. So, the findings of the present study expand the literature on learning capability and reveal its specific effect not only on the internal business environment (firm innovation practices) but on the external business environment (customers), too.

The present study confirms the empirical study of Bolıvar-Ramos *et al.* (2012) and Garcia-Morales *et al.* (2012), according to which there is a positive relation between organisational learning and organisational innovation. Similarly,

Calantone *et al.* (2002), Hult *et al.* (2004), Keskin (2006), Aragon-Correa *et al.* (2007), Rhee *et al.* (2010) and Jimenez-Jimenez and Sanz-Valle (2011) support the view that innovation is related to organisational learning and is dependent upon the organisation's knowledge base. From the above mentioned, the significant difference of the present study from those carried out so far is evident. This is justified given that the present study focuses on non-technical innovation dimensions of SMEs.

The present study findings also demonstrate that the greater the non-technical innovation, the better the customer satisfaction of SMEs. Both dimensions of non-technical innovation, namely organisational innovation and marketing innovation, have a strong causal effect on customer satisfaction. This means that firms are more likely to prioritize and meet customer requirements based on their innovativeness from the organisational and marketing perspective. By adopting organisational innovation, a firm can establish new operations and change the existing organisational structure which in turn helps it penetrate new territorial or foreign market segments (Cheng *et al.*, 2010). This also helps a firm be more flexible and establish new marketing strategies based on which customers will be satisfied on a sustainable basis. From the above, it is apparent that SMEs that bring about many changes and innovations attract more customers and satisfy them better (Nemati *et al.*, 2010). However, it should be noted that it is the "soft" aspect of firm innovativeness that makes customers satisfied. It seems that the organisational and marketing innovative practices firms implement are detected by customers, who positively react to them. Their positive attitude is reflected in their satisfaction level. Thus, non-technical innovation can be considered to be a new source of innovation and competitive advantage for firms (Qingyu *et al.*, 2004). To the best of our knowledge, this is the first empirical study that determines and confirms the casual linkages between organisational and marketing innovation and customer satisfaction of SMEs. The findings of the present study are partially supported by Simon and Yaya (2012). More specifically, they support the positive relationship between marketing innovation and customer satisfaction, while the relationship between organisational innovation and customer satisfaction is not supported by them.

The findings of the present study have several practical implications for managers and owners of SMEs. Managers should focus on building a sound learning environment, given that it is a determinant of both organisational and marketing innovation. Thus, knowledge capability needs to be taken into consideration when non-technical innovation is established. Therefore, firms should promote the acquisition of new knowledge, for example, by making employees attend exhibitions and conferences regularly, encouraging work teams to identify opportunities for improvement, fostering the development of new ideas and approaches, and designing experiments within the firm. Non-technical innovation is also a significant determinant of customer satisfaction. Therefore, managers should strive to adopt innovation not only from the technical but also from the non-technical point of view. In other words, managers should adopt innovative organisational and marketing practices to increase customer satisfaction. They should consider strategies especially with regard to optimising the perceived product quality and after-sales services that meet and exceed customers' expectations. The above mentioned appear to constitute important strategic directions in the new competitive landscape.

Conclusion

The present study tests a model incorporating all the theoretical linkages between organisational learning, organisational and marketing innovation and customer satisfaction. In this context, scarce empirical evidence has been provided so far, based on SMEs. The present study, using a sample of SMEs, empirically supports the statistical significance of all theoretical relationships examined. So, the present study contributes to the theoretical development of innovation and fills the literature gap by demonstrating the determinant role of organisational learning in both organisational and marketing innovation and their determinant role in enhancing customer satisfaction.

The present study suffers from some limitations that give rise to suggestions for further research. First, the study is limited by its cross-sectional design. Further research study is proposed to explore the applicability and representativeness of the findings to a larger sample of firms including SMEs and non-SMEs, as well as to specific business sectors. Second, the effects of the internal business environment and endogenous business factors have not been taken into consideration in the present study. Thus, it is worth examining the effects of SME characteristics such as the leadership style, process management, the type of industry in which a firm operates and human resources on the relationships between learning, non-technical innovation and customer satisfaction. Third, the subjective data collected reflect the opinions of only one manager per firm studied, excluding, for example, the auditors, employees or the customers themselves. Further study should examine whether any variability exists in the responses describing one firm's attitude. Taking into consideration the opinions of a firm's managers, employees, and auditors, the conclusions would be more representative of the conditions in which a firm operates. Finally, although the measured variables used in the proposed model explain an acceptable amount of variance in customer satisfaction, future research studies should test models including additional variables with regard to quality management, transformational leadership or organisational memory.

Appendix

Table A.1. Constructs and associated items.

Constructs	Items	Code
Organisational learning		
	The company adopts new knowledge into its daily activities.	V001
	The company encourages work teams to identify opportunities for improvement.	V002
	Experiences and ideas provided by external sources (consultants, customers, training firms, etc.) are taken into consideration for learning purposes.	V003
	New ideas and approaches to work performance are experimented continuously.	V004
	The managers/employees attend conferences and exhibitions regularly.	V005
Organisational innovation		
	The departments of R&D, sales, marketing, and production frequently coordinate their activities from the organisational perspective.	V006
	The firm integrates and simplifies the main functions from the organisational perspective.	V007
	The firm successfully gains access to and adopts new technologies based on organisational and structural changes.	V008
	The firm identifies and responds to the organisational innovative strategies of competitors based on organisational and structural changes.	V009

Table A.1. (*Continued*)

Constructs	Items	Code
Marketing innovation		
	The firm has close relationships with the majority of customers.	V010
	The firm has good knowledge of how to penetrate different market segments (in which new products can be successfully launched).	V011
	The firm has a highly efficient sales-force based on new methods of product placement and promotion.	V012
	The firm's product distribution to the supply chain is efficient and differs from that of competitors.	V013
	The company has good knowledge of market conditions.	V014
Customer satisfaction		
	Overall customer satisfaction has been increased (during the last three years).	V015
	Customer satisfaction from after-sales services has been increased.	V016
	The customers are loyal to the firm's products.	V017
	During the last three years, customer complaints have been decreased.	V018

References

Abou-Zeid, E and Q Cheng (2004). The effectiveness of innovation: A knowledge management approach. *International Journal of Innovation Management*, 8(3), 261–274.

Akman, G and C Yilmaz (2008). Innovation capability, innovation strategy and market orientation: An empirical analysis in Turkish software industry. *International Journal of Innovation Management*, 12(1), 69–111.

Alegre, J and R Chiva (2008). Assessing the impact of organisational learning capability on product innovation performance: An empirical test. *Technovation*, 28(6), 315–326.

Amalia, M and Y Nugroho (2011). An innovation perspective of knowledge management in a multinational subsidiary. *Journal of Knowledge Manegement*, 15(1), 71–87.

Aragon-Correa, A, V Garcıa-Morales and E Cordon-Pozo (2007). Leadership and organisational learning's role on innovation and performance: Lessons from Spain. *Industrial Marketing Management*, 36(3), 349–359.

Armbruster, H, A Bikfalvib, S Kinkela and G Lay (2008). Organisational innovation: The challenge of measuring non-technical innovation in large-scale surveys. *Technovation*, 28(10), 644–657.

Atuahene-Gima, K (1996). Market orientation and innovation. *Journal of Business Research*, 35(1), 93–103.

Bader, K (2013). How to benefit from cross-industry innovation? A best practice case. *International Journal of Innovation Management*, 17(6), 65–90.

Bigliardi, B and AI Dormio (2009). An empirical investigation of innovation determinants in food machinery enterprises. *European Journal of Innovation Management*, 12(2), 223–242.

Bolıvar-Ramos, MT, V Garcia-Morales and E Garcia-Sanchez (2012). Technological distinctive competencies and organisational learning: Effects on organisational innovation to improve firm performance. *Journal of Engineering and Technology Management*, 29(3), 331–357.

Brockmand, B and F Morgan (2003). The role of existing knowledge in new product innovativeness and performance. *Decision Science*, 32(2), 385–419.

Cahill, DJ (1996). Entrepreneurial orientation or pioneer advantage. *Academy of Management Review*, 21(3), 603–605.

Calantone RJ, ST Cavusgil and YS Zhao (2002). Learning orientation, firm innovation capability, and firm performance. *Industrial Marketing Management*, 31(6), 515–524.

Camison, A and A Villar-Lopez (2011). Non-technical innovation: Organisational memory and learning capabilities as antecedent factors with effects on sustained competitive advantage. *Industrial Marketing Management*, 40(8), 1294–1304.

Chang Y, J Linton and M Chen (2012). Service regime: An empirical analysis of innovation patterns in service firms. *Technological Forecasting & Social Change*, 79(9), 1569–1582.

Chaston, I, B Badger and E Sadler-Smith (2001). Organisational learning: An empirical assessment of process in small UK manufacturing firms. *Journal of Small Business Management*, 39(1), 139–151.

Chen, J (2006). Development of Chinese small and medium-sized enterprises. *Journal of Small Business and Enterprise Development*, 13(2), 140–147.

Cheng, C, M Lai and W Wu (2010). Exploring the impact of innovation strategy on R&D employees' job satisfaction: A mathematical model and empirical research. *Technovation*, 30(7–8), 459–470.

Cohen, WM and DA Levinthal (1990). Absorptive capacity: A new perspective on learning and innovation. *Administrative Science Quarterly*, 35(1), 128–152.

Crossan, M and Apaydin M (2010). A multi-dimensional framework of organisational innovation: A systematic review of the literature. *Journal of Management Studies*, 47(6), 1154–1191.

Damanpour, F (1991). Organisational Innovation: A Meta-Analysis of Effects of Determinants and Moderators. *Academy of Management Journal*, 34(3), 555–590.

Damanpour, F and D Aravind (2012). Managerial innovation: Conceptions, processes, and antecedents. *Management and organisation Review*, 8(2), 423–454.

Darroch, J (2005). Knowledge management, innovation and firm performance. *Journal of Knowledge Management*, 9(3), 101–115.

Darroch, J and R McNaugton (2002). Examining the link between knowledge management practices and types of innovation. *Journal Intellectual Capital*, 3(3), 210–222.

Dibella, AJ, EC Nevis and JM Gould (1996). Understanding organisational learning capability. *Journal of Management Studies*, 33(3), 361–379.

Evangelista, R and A Vezzani (2010). The economic impact of technological and organisational innovations. A firm-level analysis. *Research Policy*, 39(10) 1253–1263.

Feng, J, DI Prajogo, KC Tan and AS Sohal (2006). The impact of TQM practices on performance: A comparative study between Australian and Singaporean organisations. *European Journal of Innovation Management*, 9(3), 269–278.

Fiedler, M and I Welpe (2010). How do organisations remember? The influence of organisational structure on organisational memory. *organisation Studies*, 31(3), 381–407.

Field, A (2005). *Discovering Statistics Using SPSS*, 2nd Edition. London: Sage Publications.

Forsman, H (2011). Innovation capacity and innovation development in small enterprises. A comparison between the manufacturing and service sectors. *Research Policy*, 40(10), 739–750.

Fotopoulos, C and E Psomas (2009). The impact of soft and hard TQM elements on quality management results. *International Journal of Quality and Reliability Management*, 26(2), 150–163.

Freel, MS (2005). The characteristics of innovation-intensive small firms: Evidence from Northern Britain. *International Journal of Innovation Management*, 9(4), 401–429.

Fynes, B and C Voss (2001). A path analytic model of quality practices, quality performance and business performance. *Production and Operation Management*, 10(4), 494–513.

Garcia-Morales, VJ, FJ LLorens-Montes and AJ Verdu-Jover (2006). Organisational learning categories: Their influence on organisational performance. *International Journal of Innovation and Learning*, 3(5), 518–536.

Garcia-Morales, VJ, MM Jimenez-Barrionuevo and L Gutierrez-Gutierrez (2012). Transformational leadership influence on organisational performance through organisational learning and innovation. *Journal of Business Research*, 65(7), 1040–1050.

Georgellis, Y, P Joyce and A Woods (2000). Entrepreneurial action, innovation and business performance: The small independent business. *Journal of Small Business and Enterprise Development*, 7(1), 7–17.

Gordon, L (2006). Innovation diffusion and relationship marketing: An empirical study of factors influencing the intention to adopt the innovation of modular facility technology. dissertation, University International, Touro.

Goswami, S and M Mathew (2011). Competencies for organisational innovation potential: An empirical analysis on Indian information technology organisations. *International Journal of Innovation Management*, 15(4), 667–685.

Gunday, G, G Ulusoy, K Kilic and Alpkan L (2011). Effects of innovation types on firm performance. *International Journal of Production Economics*, 133(2), 662–676.

Hair, JF, WC Black, BJ Babin, RE Anderson and RL Tatham (2006). *Multivariate Data Analysis*, 6th Edition. New Jersey: Pearson/Prentice Hall.

Han, JK, N Kim and RK Srivastava (1998). Market orientation and organisational performance: Is innovation a missing link? *Journal of Marketing*, 62(4), 30–45.

Hult, GT, RF Hurley and GA Knight (2004). Innovativeness: Its antecedents and impact on business performance. *Industrial Marketing Management*, 33(5), 429–438.

Humphreys, P, R McAdam and J Leckey (2005). Longitudinal evaluation of innovation implementation in SMEs. *European Journal of Innovation Management*, 8(3), 283–304.

Hung, HM (2007). Influence of the environment on innovation performance of TQM. *Total Quality Management*, 18(7), 715–730.

Hung, RYY, BYH Lien, SC Fang and GN McLean (2010). Knowledge as a facilitator for enhancing innovation performance through total quality management. *Total Quality Management*, 21(4), 425–438.

Inauen, M and A Schenker-Wicki (2012). Fostering radical innovations with open innovation. *European Journal of Innovation Management*, 15(2), 212–231.

Jerez-Gomez, P, J Cespedes-Lorente and R Valle-Cabrera (2005). organisational learning and compensation strategies: Evidence from the Spanish chemical industry. *Human Resource Management*, 44(3), 279–299.

Jimenez-Jimenez, D, M Martinez-Costa and R Sanz-Valle (2014). Knowledge management practices for innovation: A multinational corporation's perspective. *Journal of Knowledge Management*, 18(5), 905–918.

Jimenez-Jimenez, D and R Sanz-Valle (2011). Innovation, organisational learning, and performance. *Journal of Business Research*, 64(4), 408–417.

Kafetzopoulos, D and E Psomas (2015). The impact of innovation capability on the performance of manufacturing companies. *Journal of Manufacturing Technology Management*, 26(1), 104–130.

Kalantaridis, C (2004). Internationalization, strategic behaviour, and the small firm: A comparative investigation. *Journal of Small Business Management*, 42(3), 245–262.

Keskin, H (2006). Market orientation, learning orientation, and innovation capabilities in SMEs: An extended model. *European Journal of Innovation Management*, 9(4), 396–417.

Kline, RB (2005). *Principles and Practice of Structural Equation Modeling*, 2nd Edition. The Guilford Press, New York.

Knox, S (2002). The broadroom agenda: Developing the innovative organisation. *Corporate Governance*, 2(1), 27–36.

Krause, DE (2004). Influence-based leadership as a determinant of the inclination to innovate and of innovation-related behaviours: An empirical investigation. *The Leadership Quarterly*, 15(1), 79–102.

Laforet, S (2011). A framework of organisational innovation and outcomes in SMEs. *International Journal of Entrepreneurial Behaviour & Research*, 17(4), 380–408.

Laforet, S and J Tann (2006). Innovative characteristics of small manufacturing firms. *Journal of Small Business and Enterprise Development*, 13(3), 363–380.

Lau, AKW, RCM Yam and EPY Tang (2010). Supply chain integration and product modularity: An empirical study of product performance for selected Hong Kong manufacturing industries. *International Journal of Operations & Production Management*, 30(1), 20–56.

Lee, C and LC Ging (2007). SME innovation in the Malaysian manufacturing sector. *Economics Bulletin*, 12(30), 1–12.

Li, X and RK Mitchell (2009). The pace and stability of small enterprise innovation in highly dynamic economies: A China-based template. *Journal of Small Business Management*, 47(3), 370–397.

Liao, SH and CC Wu (2010). System perspective of knowledge management, organisational learning, and organisational innovation. *Expert Systems with Applications*, 37(2), 1096–1103.

Liu, SS, X Luo and Y Shi (2002). Integrating Customer Orientation in organisations–in-Transition: An Emprical Study. *International Journal of Research in Marketing*, 19(4), 367–382.

Martinez-Costa, M and AR Martinez-Lorente (2008). Does quality management foster or hinder innovation? An empirical study of Spanish Companies. *Total Quality Management & Business Excellence*, 21(2), 159–170.

Massa, S and S Testa (2008). Innovation and SMEs: Misaligned perspectives and goals among entrepreneurs, academics, and policy makers. *Technovation*, 28(7), 393–407.

Mol, MJ and J Birkinshaw (2009). The sources of management innovation: When firms introduce new management practices. *Journal of Business Research*, 62(12), 1269–1280.

Narayan, B, C Rajendran and LP Sai (2008). Scales to measure and benchmark service quality in tourism industry. *Benchmarking: An International Journal*, 15(4), 469–493.

Narver, JC and SF Slater (1990). The effect of a market orientation on business profitability. *Journal of Marketing*, 54(1), 20–35.

Nemati, A, K Khan and M Iftikhar (2010). Impact of innovation on customer satisfaction and brand loyalty: A study of mobile phones users in Pakistan. *European Journal of Social Sciences*, 16(2), 299–306.

Newbert, SL (2008). Value, rareness, competitive advantage, and performance: A conceptual-level empirical investigation of the resource-based view of the firm. *Strategic Management Journal*, 29, 745–768.

Ngo, LV and A O'Cass (2011). Winning through innovation and marketing: Lessons from Australia and Vietnam. *Industrial Marketing Management*, 40(8), 1319–1329.

Nonaka, I and H Takeuchi (1995). *The Knowledge-Creating Company: How Japanese Companies Create the Dynamics of Innovation*. New York: Oxford University Press.

OECD (2005). Oslo manual: Proposed guidelines for collecting and interpreting technological innovation data. Paris.

Oke, A, G Burke and A Myers (2007). Innovation types and performance in growing UK SMEs. *International Journal of Operations & Production Management*, 27(7), 735–753.

Pallas, F, F Bockermann, O. Goetz and K Tecklenburg (2013). Investigation organisational innovativeness: Developing a multidimensional formative measure. *International Journal of Innovation Management*, 17(4), 1–49.

Prieto, I and E Revilla (2006). Assessing the impact of learning capability on business performance: empirical evidence from Spain. *Management Learning*, 37(4), 499–522.

Qingyu, Z, L Jeen-Sun and G Mel (2004). Innovation-driven learning in new product development: A conceptual model. *Industrial Management & Data Systems*, 104(3), 252–262.

Quintane, E, RM Casselman, BS Reiche and PA Nylund (2011). Innovation as a knowledge-based outcome. *Journal of Knowledge Management*, 15(6), 928–947.

Ren, L, G Xie and K Krabbendam (2010). Sustainable competitive advantage and marketing innovation within firms. A pragmatic approach for Chinese firms. *Management Research Review*, 33(1), 79–89.

Rhee, J, T Park and DH Lee (2010). Drivers of innovativeness and performance for innovative SMEs in South Korea: Mediation of learning orientation. *Technovation*, 30(1), 65–75.

Rosenbusch, N, J Brinckmann and A Bausch (2011). Is innovation always beneficial? A meta-analysis of the relationship between innovation and performance in SMEs. *Journal of Business Venturing*, 26(4), 441–457.

Ryding, D (2010). The impact of new technologies on customer satisfaction and business to business customer relationships: Evidence from the soft drinks industry. *Journal of Retailing and Consumer Services*, 17(2), 224–228.

Sadikoglu, E and C Zehir (2010). Investigating the effects of innovation and employee performance on the relationship between total quality management practices and firm performance: An empirical study of Turkish firms. *International Journal of Production Economics*, 127(1), 13–26.

Salavou, H, G Baltas and S Lioukas (2004). Organisational innovation in SMEs: The importance of strategic orientation and competitive structure. *European Journal of Marketing*, 38(9–10), 1091–1112.

Simon, A and LH Yaya (2012). Improving innovation and customer satisfaction through systems integration. *Industrial Management & Data Systems*, 112(7), 1026–1043.

Simpson, PM, SA Judy and A Enz-Cathy (2006). Innovation orientation outcomes: The good and the bad. *Journal of Business Research*, 59(10–11), 1133–1141.

Sinkula, JM, WE Baker and T Noordeweir (1997). A framework for market-based organisational learning: Linking values, knowledge, and behaviour. *Journal of the Academy of Marketing Science* 25(4), 305–318.

Slater, S and JC Narver (1994). Does competitive environment moderate the market orientation performance relationship? *Journal of Marketing*, 58(1), 46–55.

Tan, KC, M Xie and XX Shen (1999). Development of innovative products using Kano's model and quality function deployment. *International Journal of Innovation Management*, 3(3), 271–286.

Totterdell, P, D Leach, K Birdi, C Clegg and T Wall (2002). An investigation of the contents and consequences of major organisational innovation. *International Journal of Innovation Management*, 6(4), 343–368.

Verdun, AJ, I Tamayo and A Ruiz-Moreno (2012). The moderating effect of environmental uncertainty on the relationship between real options and technological innovation in high-tech firms. *Technovation*, 32(9–10), 579–590.

Weerawardena, J (2003). Exploring the role of market learning capability in competitive strategy. *European Journal of Marketing*, 37(3–4), 407–429.

Weerawardena, J, A O'Cass and C Julian (2006). Does industry matter? Examining the role of industry structure and organisational learning in innovation and brand performance. *Journal of Business Research*, 59(1), 37–45.

Weerd-Nederhof, P, B Pacitti, J Da Silva Gomes and A Pearson (2002). Tools for the improvement of organisation learning processes in innovation. *Journal of Workplace and Learn*, 14(8), 320–331.

White GP (1996). A meta-analysis model of manufacturing capabilities. *Journal of Operations Management*, 14(2), 315–331.

Wonglimpiyarat, J (2010). Innovation index and the innovative capacity of nations. *Futures*, 42(3), 247–253.

Wu, SI and SY Liu (2010). The performance measurement perspectives and causal relationship for ISO-certified companies: A case of optoelectronic industry. *International Journal of Quality & Reliability Management*, 27(1), 27–47.

Xu, J, R Houssin, E Caillaud and M Gardoni (2010). Macro process of knowledge management for continuous innovation. *Journal of Knowledge Management*, 14(4), 573–591.

Yam, CM, JC Guan, KF Pun and PY Tang (2004). An audit of technological innovation capabilities in Chinese firms: Some empirical findings in Beijing, China. *Research Policy*, 33(8), 1123–1250.

Yam, R, W Lo, E Tang and A Lau (2011). Analysis of sources of innovation, technological innovation capabilities, and performance: An empirical study of Hong Kong manufacturing industries. *Research Policy*, 40(3), 391–402.

Yu, W, MA Jacobs, WD Salisbury and H Enns (2013). The Effects of Supply Chain Integration on Customer Satisfaction and Financial Performance: An organisational Learning Perspective. *International Journal of Production Economics*, 146(1), 346–358.

Zhang, M, A MacPherson and O Jones (2006). Conceptualizing the learning process in SMEs. *International Small Business Journal*, 24(3), 299–323.

Zhang, Q, MA Vonderembse and JS Lim (2003). Manufacturing flexibility: Defining and analyzing relationships among competence, capability, and customer satisfaction. *Journal of Operations Management*, 21(2), 173–191.

Chapter 13

WHAT'S SMALL SIZE GOT TO DO WITH IT? PROTECTION OF INTELLECTUAL ASSETS IN SMEs

Heidi Olander

School of Business
Lappeenranta University of Technology

Pia Hurmelinna-Laukkanen

School of Business
Lappeenranta University of Technology, Finland
and
Oulu Business School, University of Oulu, Finland

University of Oulu, Oulu Business School
PO Box 4600, F1-90014
University of Oulu, Finland
pia.hurmelinna@oulu.fi

Jukka Mähönen

Faculty of Law, University of Turku

The knowledge protection/sharing dilemma related to innovation activities is becoming known to all firms, even though it is generally more notable for SMEs: the small size of the firms inherently creates a need for inter-organizational collaboration, but it also makes dealing with the related contradictories more challenging. One factor behind this is that the needed tools — such as the protection mechanisms of intellectual assets — may be more limited. In line with this notion, we examine the protection available for and used by small firms in their innovation activities. We approach the issue by conducting a review on the

This paper was first published in *International Journal of Innovation Management*, Vol. 13, No. 3 (Sept., 2009), pp. 349–370.

relevant literature, and use a qualitative multiple case study conducted in eight small companies to empirically study the issue. Departing from prior research, we consider protection of innovations by distinguishing between the intangibles needed in innovation activities and the actual innovation outputs, and combine these considerations to the knowledge protection/sharing dilemma.

Keywords: Intellectual property; SME; innovation; protection strategy.

Introduction

Managing innovation activities is an area where — despite all the work already done — unanswered questions reside. There are various premises for successful innovation (see, e.g., Nickerson *et al.*, 2007), and an equally wide range of possibilities to organize for innovation activities. In addition to this, there are different types of innovations, which increase the managerial complexity from idea generation and value creation to value capture. In such settings, it is relatively easy to comprehend that equally positive — or unsuccessful — outcomes can result from following different paths. Some of these can even seem quite contradictory. For instance, it may not be easy to combine exploration and exploitation even if both elements would be required in innovation activities (March, 1991). The dilemmas become even more pronounced as the innovation activities become more open. Let us consider, for example, coopetition, the simultaneous competition and cooperation between firms; combining autonomy and control in such situations is needed, yet challenging (Tidd, 1993). Indeed, it can be said that innovation activities hold many paradoxes. The problem is that it can be very difficult to predict direction to be taken, and whether some characteristics of the industry or a firm affect the possibilities to succeed.

Both the likelihood of having to deal with contradictory elements, and the ability of a firm to do so may indeed depend on the characteristics of the firm itself. Among the innovation related paradoxes, knowledge sharing vs. knowledge protection is closely linked to a central feature of any firm — the size. This dilemma may be noticeably intensified especially by small firm size (see, e.g., Dicksona *et al.*, 2006). Also, managing this dilemma may be challenging due to size-related factors: using certain protection (and appropriability) mechanisms as a remedy may be difficult for small and medium-sized enterprises (SMEs) because of the limited resources. This, again, may influence the incentives to invest in innovation (Van Dijk, 2000).

In this paper, we describe the reasons why choosing between knowledge sharing and protection, or combining them, is particularly challenging for SMEs.

For this, we bring together prior discussion on the protection of intellectual assets and innovativeness of SMEs, and augment it. In fact, prior work calls for expansion. Kitching and Blackburn (1999, p. 328) note that although "many SMEs undoubtedly create intellectual property, research has focused mainly on large enterprises and, in particular, patenting activity (e.g., Hayward and Greenhalgh, 1994). Yet, such research has limitations in providing an understanding of how SME owners manage their intellectual property." Indeed, quite a few studies that have touched SMEs have compared the ways in which innovation protection is done in large and small firms (instead of taking the features of SMEs as the starting point) (see, e.g., Arundel and Steinmueller, 1998; Arundel, 2001), or have put emphasis on evaluating patent protection in particular (see, e.g., Hall *et al.*, 1999; Iversen, 2004; MacDonald, 2004; Reitzig *et al.*, 2007; Bessen, 2008; see also, e.g., Byma and Leiponen, 2007 on the gap in prior research regarding innovation protection in SMEs). Research is controversial: Arundel (2001) claims that a higher percentage of R&D performing firms in all size classes find secrecy to be a more effective means of appropriation than patents. On the other hand, March (2005) claims that neither copyrights nor trade secrets provide a useful mechanism that allows SMEs to appropriate the values of their inventions. Only patents give significant positive benefits for SMEs. In the attempt to approach SMEs with a fresh approach, Kitching and Blackburn (1999) consider formal and informal protection practices of SMEs in their study (see also Byma and Leiponen, 2007).

We continue their work in the sense that we consider varying protection mechanisms that SMEs can utilize. However, we consider the protection mechanisms from a different angle, and in addition to examining a range of protection mechanisms, we look at the protection of intellectual assets — limited in this study to intellectual property, such as human capital and innovations and especially the knowledge embodied within intellectual assets (Bismuth and Tojo, 2008; Leiponen, 2008) — at two distinct phases of innovation activities and process of SMEs. We examine their role in utilizing intellectual assets as an input (i.e., building a basis for value creation and capture) and an output (the phase of the realization of value capturing). Furthermore, we combine the use of protection mechanisms to the knowledge protection/sharing dilemma. Also, while Kitching and Blackburn (1999) — like many other researchers — have relied on the quantitative approach, we consider these issues through a literature review and a case study of eight SMEs, which has potential to reveal more in-depth information on the elements and features of the protection mechanisms. In our attempt to fill a gap from prior research, we try to find answers to the question: *How does the small size of the firm relate to innovation protection possibilities and the subsequent incentives to innovate*?

Protecting Intellectual Assets for Innovativeness

Knowledge management paradoxes in innovation activities

Innovation activities can simultaneously incorporate dimensions such as stability and dynamism, autonomy and interdependence, and separateness and collaboration. Such paradoxes are present in all the stages of innovation processes, and there may be notable feedback loops. For instance, considering the incentives to engage in the innovation process, the shadow of the future can have strong effects: the possibility of firms to profit from their innovations has an effect on their willingness to invest in carrying out innovation activities (Van Dijk, 2000). Appropriation and profiting possibilities, for their part, may depend on how well knowledge can be protected or shared. On the one hand, innovating firms need to *create value* through combining existing knowledge in innovative ways (Crossan and Inkpen, 1995; Kogut and Zander, 1996). Thus, *knowledge sharing* is needed. For example, firms can cooperate through standardization systems, based on voluntary disclosure and assurance (Hemphill, 2007). On the other hand, firms need to be able to make profit and *capture value*. Capturing returns on investments in innovation is naturally a multi-dimensional issue. Nevertheless, one of the central dimensions is the possibility to *protect innovations* against imitation, i.e., prevent unwanted knowledge flows. If competitors were able to extract ideas and knowledge related to commercialized products and services without too much costs or effort and produce their offerings based on that, they would gain excessive advantage compared to the innovating firm that has put time and money in creating the underlying knowledge. In other words, the outputs of innovation activities need to be protected.

However, *knowledge protection* also plays a role in relation to *value creation*, and *knowledge sharing* is relevant for *capturing value*. For instance, value creation requires producing, introducing, and spreading new offerings to the markets. Knowledge protection relates to value creation, for example, through a fact that a company that engages in collaborative forms of innovation may be relatively vulnerable if the knowledge that it contributes to the joint effort is expropriated by other parties (Heiman and Nickerson, 2004). Thus, companies need to consider the protection of existing innovative assets — assets that will be used as an input for further innovation.

	Value/Innovation creation	*Value capturing/ Profiting from innovation*
Knowledge sharing	Needed for creating new innovative combinations and ideas; may cause losing core knowledge assets to rivals	Needed for getting access to markets; may cause losing competitive edge
Knowledge protection	Needed for safe disclosure of existing knowledge; may prevent adequate knowledge flows	Needed for protecting the investments in innovation, needed for safe disclosure; may cause slower diffusion

Fig. 1. Knowledge sharing vs. protection dilemma in innovation.

In sum, value creation and capture require that companies choose between knowledge sharing and protection, or try to find some way of incorporating these two alternatives. Succeeding in managing the knowledge sharing vs. protection dilemma may have a huge effect on the company performance (Mazzarol and Reboud, 2008). Figure 1 illustrates the need for finding the right balance.

Controllability over strict protection

Choosing the strategy that the firm adopts in relation to knowledge protection and sharing is followed by the need to make the choice about the mechanisms that are used to allow protection and/or exchange of intellectual assets. This applies both in the input and output phases in innovation: to the existing background knowledge of the firm including intellectual property and know-how, and the generated foreground knowledge.

The means that enable the appropriation of the returns from an innovation include a wide range of mechanisms, many of which are industry-specific (Burk and Lemley, 2003). Intellectual property rights (IPRs such as patents, copyright, and trademarks) and contracts enable the company to hold on to its original creations and delay their imitation. Similarly, human resource management practices (e.g., minimizing personnel turnover, monitoring communication), labor legislation,[1] relying on mechanisms such as access restriction and passwords (i.e., technical means of

[1]Labor legislation incorporates, for example, institutionally secured possibilities to use non-competition contracts, or to refer to the loyalty requirement set to employees and the transfer of rights in the employment relationship.

concealment), or on tacitness of knowledge, or lead time may facilitate innovation appropriability in terms of making the boundary between the firm and outside organizations less permeable (Hurmelinna-Laukkanen and Puumalainen, 2007).

The discussion above suggests that firms need to have room to maneuver according to emerging opportunities and threats. They need to be able to enhance the flow of knowledge when value is created through the expansion of the markets or by generating new innovations, and restrict imitation when value needs to be captured. Thus, when selecting knowledge protection mechanisms, the key is to find such mechanisms that do not dwell solely around notable protective power and inherent difficulty of knowledge transfer, but that enable the company to *control* the knowledge flows (see Hurmelinna-Laukkanen, 2009).

The ways to protect ideas and innovations differ depending on the chosen model of innovating, and the economic and legal environment the firm operates in. If a firm chooses separateness, for example, in the form of internal R&D preventing imitation often becomes the central goal, and a variety of protection mechanisms is used to serve this purpose.[2] On the other hand, when open innovation models are utilized, the role of protection is diminished. In such forms of innovation activities the idea is to make use of external ideas in a firm's business and to let ideas generated within the firm to be used by others (Chesbrough, 2006). However, even with regard to open innovation activities, protection issues are relevant: the ways and goals of using protection mechanisms just change. It is not so much a question of preventing others from copying knowledge assets, as it is about making them more transferable in the sense that transaction costs connected to the exchange are lowered and that the innovations are turned into more liquid, yet controllable, assets (see, e.g., Cohen *et al.*, 2000). It is easier and safer to trade or license (or even litigate) when the innovation has certain boundaries and definable features. In this way the existing IPRs, for instance, can support the start of collaboration — the basic logic being that the companies need not be overly concerned about the misappropriation of existing knowledge assets and future innovations that require using them. However, the use of formal protection mechanisms (IPRs most notably), if enforced, has been criticized for being bureaucratic and for creating distrust between collaborating partners, among other things (Macaulay, 1963). Also, while property rights consist of information that decreases transaction costs of contracting, the negative side of them is that they

[2]Benefits can also be achieved by choosing the most cost-efficient legal playing field (e.g., by choosing a jurisdiction with a more cost-efficient patent system). On cross-border markets, markets of law (i.e., forum shopping) are an important factor that can bring advantages to an individual firm.

are monopolies, creating holdup problems and, through them, other transaction costs (Lemley and Weiser, 2007). Other appropriability mechanisms can similarly have both advantageous[3] and strategy impeding effects. Therefore, it is important to align their usage with the company's goals. However, in certain cases the alternatives are limited, such as when resources are scarce. Therefore, most SMEs may face relatively big challenges in value creation and capture.

Value Creation and Capture in SMEs

Characteristics of SMEs defining the presence of knowledge protection/sharing dilemma

SMEs are seen as the engines of dynamic efficiency, and as such, they are an important source not only of technological progress and employment, but also of innovation (Pavitt *et al.*, 1987; Audretsch, 2002; Blomqvist, 2002; Davis, 2004; De Jong and Marsili, 2006). Indeed, SMEs can enjoy a number of behavioral advantages over larger firms in the innovation process. These include a rapid response to external threats and opportunities, efficient internal communication, and interactive management style.

However, SMEs can also suffer from a number of disadvantages, most of which are immaterial. Such impediments include, for example, the inability to spread risk over a portfolio of new products, difficulties in entering markets or in market start-up abroad, and problems in funding longer-term R&D and innovation activities (Rothwell and Dodgson, 1991). These disadvantages frequently lead to situations where SMEs that are trying to limit risks and costs and to get access to lacking complementary resources[4] are pushed towards collaboration (Muller and Zenker, 2001; Simon *et al.*, 2002; Blomqvist *et al.*, 2005; Street and Cameron, 2007; Mazzarol and Reboud, 2008). Due to the limited and scarce resources, SMEs are likely to need external networks to complement them in their activities — including innovation creation and commercialization. Often in such cases SMEs can only gain access to the needed resources if they are able to convince other organizations of their capabilities and to attract potential partners

[3]For example, contracts have been argued to create trusting attitudes between collaboration partners as they form and clarify the necessary ground rules (Burchell and Wilkinson, 1997). In that sense, contracts are also the primary tool to create incentives against value destructing opportunism: the most important thing in collaboration is the mutually advantageous reciprocity created by partnerships (Burnham *et al.*, 2000).

[4]For instance, delivery channels, legitimacy, and financing are often needed.

by offering their expertise and innovations to others (Blomqvist *et al.*, 2008). This also means that the very nature of SMEs makes them face the knowledge sharing/protection dilemma. While sharing their knowledge assets for the above mentioned purposes, SMEs would benefit from using protection mechanisms that allow them to disclose their ideas and innovations safely, without having to worry about losing their core knowledge assets. Similarly, they need to be able to make sure that they will get their share of the generated benefits and innovation outputs.

Characteristics of SMEs defining the means to deal with knowledge protection/sharing dilemma

The small size not only has an effect on the likelihood that SMEs become influenced by the knowledge sharing/protection dilemma, but it also affects the possibilities to find remedies for dealing with the related challenges. Recognizing SMEs as innovative implies that they should have intellectual assets that are relatively easily protectable (for example, novelty and an innovative step that are the basic premises of patent protection are more easily reached when fundamentally new ideas are involved; see Levin *et al.*, 1987; Hurmelinna-Laukkanen *et al.*, 2008). SMEs should also have the incentives to protect their creations: unique innovations and keeping them from the competitors' reach can be essential for their survival (Simon *et al.*, 2002). It may even be that the whole company is based on a single ground breaking innovation.[5] However, achieving protection that allows both knowledge protection and sharing may be quite challenging — again because of the limited resources.

Obtaining patent protection is inherently costly and requires capabilities that enable dealing with complex rules and regulations (Thumm, 2001; Kitching and Blackburn, 1999; Lanjouw and Schankerman, 2001). Patent law is uniform and technology-neutral only in theory. In practice, such as in relation to biotechnology, pharmaceutics and software, it is highly industry-specific (Burk and Lemley, 2003; Thomas, 2003; MacDonald, 2004). Opportunity costs from patenting are especially high for SMEs as patenting diverts their focus from the core business, and the benefits of excluding competitors are limited as an SME cannot necessarily exploit the relevant technology due to its scarce resources (Mann, 2005).

[5]See also Siquieira and Cosh (2008, p. 113), who note while studying SMEs that "the association between capabilities and competitive advantages is moderated by innovations."

Larger firms can rely on their internal expertise, whereas such protection can typically be accessed by SMEs only by using costly external experts.

Similarly, with regard to using contracts as an appropriability mechanism, there is a real problem in the weak bargaining power of SMEs compared to larger players. SMEs often face standard form contracts on a take-it-or-leave-it basis. Furthermore, it is not only the acquisition of IP rights and contracts that creates relatively greater costs for an SME, but also maintaining them. For instance, the legal costs of monitoring and enforcing their own IPRs and defending against accusations of violating others' IPRs may turn out to be too much for SMEs (Hall *et al.*, 1999, 2000; Kingston, 2006; Byma and Leiponen, 2007). On a larger scale, SMEs have a weaker lobbying and bargaining power. Intellectual property related rules are often created in account of bigger firms, and changes in them do not necessarily benefit SMEs (Hanel, 2006). Again, patent law can be taken as an example: new disclosure rules benefiting bigger firms may be disadvantageous to SMEs as they force them to expose the details of their inventions before the patents have been granted (Gallini, 2002). Furthermore, there may emerge competition law related problems that the usage of IPRs sometimes creates (Lea and Hall, 2004).

Questions asked, propositions presented

Considering all these aspects, a question arises whether the innovativeness of SMEs stems from the fact that the lack of protection pushes them to generate innovations continuously, or whether they actually have found cost-effective forms of protection that are available despite the limited resources and that provide adequate incentives to continue innovating (see Van Dijk, 2000). For instance, instead of using intellectual property rights, they may rely on lead time, tacitness of knowledge, and human resource management (HRM). Some previous studies support this, showing that SMEs are more likely to rely on informal than formal protection (see, e.g., Kitching and Blackburn, 1999; Hall *et al.*, 1999; Byma and Leiponen, 2007). However, not all SMEs follow the same patterns in protecting their knowledge. The empirical literature on the relationship between firm size and innovation activity is inconclusive, like it is also on the intensity to use IP rights and firm size (Jensen and Webster, 2006). The differences between industries, technologies, and legal systems as well as the intensity of cross-border activities make it difficult to draw generalized conclusions. Yet a deeper understanding of the logic behind the SMEs' innovation activities and knowledge protection possibilities is needed for managerial and policy purposes. The managers of both small and large firms could benefit from seeing in what kind of a playground they play and by which rules considering collaboration, acquisitions, and mergers. In similar vein, a broader view could be beneficial for legislators, because it is important to understand where the small firms are coming from.

Summarizing the above discussions, it could be expected that, first, SMEs often face a knowledge sharing/protection dilemma because of the limited resources. Second, SMEs should rely more on informal protection mechanisms than formal ones when dealing with the dilemma (except for a few potential examples, such as biotech firms that inherently need to patent their creations). Finally, although not much evidence exists, it can be assumed that the protection is targeted more on protecting existing knowledge assets — and thus premises for future innovation — than on protecting innovation output. In the following, we will address these issues through a case study.

Methodology

Data collection and case companies

As the research problem of this study is to learn about the possibilities and problems faced by SMEs in relation to innovation protection, a qualitative research approach was chosen which aims, in its context relatedness, data collection and its analysis, at a more holistic understanding of the studied issue (Eriksson and Kovalainen, 2008). Qualitative research has been defined in social science and business research as the first phase of research which is followed by a quantitative phase (Silverman, 2000). On the other hand, qualitative research complements quantitative research by providing a better understanding of issues that have remained unclear in quantitative studies (Silverman, 2000; see also Hall *et al.*, 2000, applying a qualitative approach for studying SMEs). With this point in mind, this research aims at increasing understanding on the issues behind SMEs' selection and use of protection mechanisms.

A multiple case study method was adopted for the empirical analysis of the innovation and knowledge protection related issues. Case studies emphasize the real life and rich context of the phenomenon (Eisenhardt and Graebner, 2007). Multiple-case studies retain only those relationships that are replicated across most or all of the cases, which makes the resulting theory often more robust and widely applicable (Eisenhardt and Graebner, 2007; Yin, 2003).

As the purpose of qualitative study is not to make statistical generalization, systematic sampling methods are usually not necessary (Eriksson and Kovalainen,

2008). Hence, the firms were chosen by using a combination of theoretical sampling (information rich cases including theoretical issues; Patton, 2002) and convenience sampling (access to the cases as the defining factor; Silverman, 2000). Firms from different industries were acquired in the sample in order to enable comparisons. As a result, eight small Finnish firms were selected from the fields of information technology (IT), engineering, content and advertising, and recreation. Choosing the case firms from different industries enabled taking into account the industry effects on the type of created innovations and innovativeness in more general.

Semi-structured interviews characterized by open questions[6] and Likert-scaled item sets[7] for triangulation (Eriksson and Kovalainen, 2008) were carried out in the selected firms during fall 2007. The Likert-scaled items were used for testing the coherence of the answers in comparison to interview answers, and to make sure that the interviewees understood which mechanisms were counted as protection mechanisms. The interviews lasted up to three hours and they were recorded and later transcribed. There were two interviewers present in each interview in order to decrease the risk of single observer bias, and either one or two interviewees. The interviewees held positions as CEOs or senior managers in their firms, and when two interviewees were involved, the other interviewee typically was a long-time employee with either technical or marketing executive background. The interviewed persons had excellent knowledge on the areas discussed. The seniority and long expertise of the interviewees indicate that they were well informed not only on the issues in their own firms, but also more widely in their industries. Company websites and other documents that were received from the case companies were used in order to analyze the coherence of the data and thus improve the reliability of the data (see e.g., Patton, 2002, for triangulation of sources). Table 1 illustrates some features of the case firms.

The content of the cases was analyzed using thematic analysis (e.g., Hardy and Bryman, 2006). As is usual in case study research, we started the analysis of each individual case separately (Eriksson and Kovalainen, 2008; Patton, 2002). In the first round of analysis, we went through the data, coding it with themes derived from the theory, such as the need for knowledge sharing and protection, core

[6]The questions concerned the types of knowledge that might need protection in the firm, the ways the firm has to prevent leakages of such knowledge in different kinds of relationships, the use of contracts internally and in collaborative relationships, and the role of employees in knowledge protection, for example.

[7]The questions aimed at finding out to what extent the mechanisms were available to the firms, and how often they were used. Also the strength in protecting the type of innovations concerned in each company, and usefulness of the mechanisms in the competitive sense were evaluated this way.

Table 1. Descriptions of case firms.

Industry	Firms (# of personnel)	Business	Innovativeness
IT	F(12), G(6), H(2)	Product/service	Improvements
Engineering	D(50), B(8)	Product/service	Improvements/new offerings
Content	A(5), E(12)	Service	New offerings
Recreational	C(2)	Service	New offerings

Note: "New offerings" here refer to innovations that are related to introducing new solutions (new products, services) to new markets (new customers; sometimes in addition to existing ones), and "Improvements" refer to improvements in existing products and services offered mainly to the existing clientele (sometimes also new customers).

know-how, innovativeness of the firm, and the use of different mechanisms. After that we moved forward to make the analysis stronger, looking for interconnections between the themes in the separate cases (Pope *et al.*, 2006). This was done by first comparing cases within each industry, and then between industries. In this cross-case analysis phase, we compared the cases in search for similarities or differences across cases and in contrast to theory (see Eriksson and Kovalainen, 2008; Patton, 2002).

Innovativeness and knowledge protection in SMEs

The size of firms, together with their central business and offerings (products/services), may have an effect on the innovativeness of the creations that they produce. This, in turn, affects the knowledge protection possibilities and the use of different mechanisms. In the following, the analyses of the companies are shortly presented in this respect.

Creating innovations in SMEs

Firms in different industries may produce more or less radical innovations, and this also shows in our case companies. In the IT industry, firms face a rapidly developing environment where new entrants have low entry costs because they do not have to make large investments in machinery. In the attempt to stay among the leading firms, the IT firms were committed to making incremental and continuous improvements to their products, mostly based on the needs of the clients. Because in principle it would be relatively easy for a large sized entrant to copy the products, the IT firms felt that they needed to be innovative all the time and try to keep ahead of competing firms. The ideas often came up as feedback from the customers, and sometimes through independent discovery by the employees. One interviewee from Firm F noted that "*We have all the time some 500 pieces of client feedback that we go through as fast as we can.*" Responding to client feedback by improving the product was considered to play a role in tying the clients to the created product.

In engineering firms the innovativeness was not as explicit and radical as in some other industries. The two engineering firms designed machinery, processes, and services solutions based on the guidelines provided by their clients, and they considered subcontracting and process deliveries for customers to involve incremental improvements. Overall, the firms concluded that almost everything had already been invented in their core businesses. The more radical innovations that the engineering firms generated were related to totally new and separate businesses. The knowledge involved was not as technically advanced as in the IT sector, but the firms considered their knowledge to involve a tacit dimension that would be difficult for an entrant to imitate. An interviewee from Firm D noted that "*the know-how of your clients' processes is far more important than knowing how to build up machinery. Anyone can build up machinery — but it is the firms with know-how that make the machinery actually work in the client's process.*" In other words, innovativeness was illustrated not so much in technology itself, but in matching technologies with customer needs. Both firms had a long history — and with that wide client bases — and they noted that client- and offer-related knowledge would be difficult to copy. The competition for customers, however, was considered fierce among established rivals, and the role as a small player in the field was sometimes found to be an obstacle for getting a deal. Thus expertise and innovativeness got a bigger role.

Content providers and advertisers, for their part, need to be highly innovative in order to keep their clients and attract new ones. The service begins from an idea, the quality of which is strongly related to the personnel's skills. The interviewees were especially concerned about an event that a key employer would leave and start a firm of his/her own: the knowledge and innovativeness of the firm is so extensively employee-involved that in such a case the firm could be seriously injured for some time. This was stated by an interviewee from Firm A: "*If a sales person would leave us for another firm, she would take all our contract and offerings information with her, which is really critical.*" The interviewees also feared the loss of their client base in relation to losing a key employer, and the loss of other employees with the first leaving person.

The firm providing recreational services in the Finnish wild (Firm C) has its innovative challenges in continuously developing something new that the clients have not yet seen as well as in bringing in new customers. The firm was active in taking part in international seminars and doing internet searches in its attempt to capture new ideas that are already in use somewhere else in the world and that it could apply, modify, and implement as the first firm in Finland. The nature of the industry involves copying ideas from others and transforming them into one's own offering: "*We are trying to explore new ideas, and when we travel somewhere we look around and study new things.*" Knowledge extraction is therefore very active from outside the firm. As the other side of this, there is also a need to protect the core elements of the innovation against the imitation of (closely located) rivals.

Knowledge protection mechanisms used

The level of innovativeness can have an effect on the need and possibilities of companies to utilize certain ways of knowledge protection. Regarding the availability of different mechanisms, the interviewees indicated that obtaining IPRs, with the exception of copyright, is often difficult, whereas informal protection mechanisms (e.g., secrecy) are more readily at the SMEs' disposal. The actual use of different mechanisms depended to a large extent on the innovations created and the industry characteristics. Figure 2 illustrates the use of different mechanisms.[8]

Among the analyzed firms, the IT firms used a wide variety of mechanisms[9] and they were also well aware of the effects of protection issues on their customer relationships. They even used protection as a marketing tool that they advertise on their websites. Technical data, source code, human capital, and client information are well covered. On the other hand, as one software firm (Firm F) noted in the interview, they tend to communicate almost everything about their product on their websites and in discussions with clients: "*We believe it is best to be open since 99.9 percent of the people that come to the website are our clients that come to buy our product.*" As the firms have the important things protected, they can use more open sharing as an advantage.

[8]The interviewees were asked to evaluate the use of different mechanisms on a Likert-scale from 1 (= used very rarely) to 5 (= used very often). Leaving an item unmarked indicated that such a mechanism was not used at all.

[9]The IT firms seem to have contracts, HRM, practical concealment, lead time and tacitness of knowledge among the most used mechanisms. Especially employee related contracts and HRM were emphasized in the discussions. Perhaps somewhat surprising is that IPRs were not among the top mechanisms for IT firms. Only trade secrets and copyright were used.

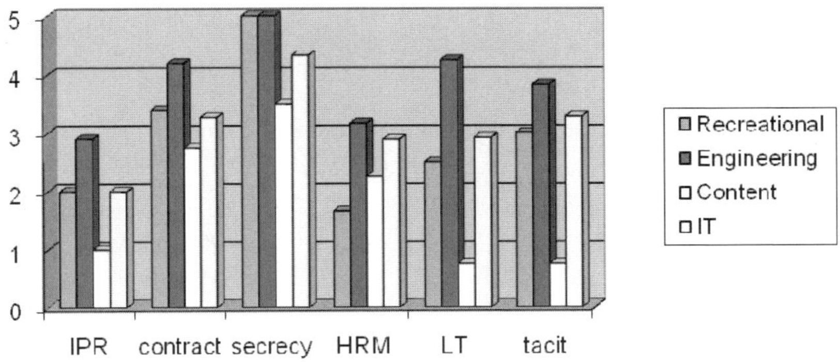

Fig. 2. The use of protection mechanisms.

The core competence of engineering firms lays in the knowledge of how to put together planning, machinery, and engineering works to benefit the customers. The technology used is not that complicated. The knowledge is thus highly bound to the skillful personnel that the firms want to keep. The protective means that were considered to be available include contracts, HRM, the tacitness of knowledge, and lead time. The mechanisms related to keeping the employees and their knowledge within the firm were considered to be the most useful ones for the firms, although there were some challenges in using them. Tacitness of knowledge is considered a valuable protection mechanism. On the other hand, IPRs, although available for radical innovations in particular, were rarely used in this industry, as they were not considered very useful for protective purposes and were believed to be even unavailable in the mature machinery building business.

The content providers valued protection mechanisms highly. One of the central reasons behind this was that they sometimes have access to quite sensitive client information that should not be leaked out by the supplier. Because of the central role of client information and preferences, it is particularly important to take care of practical concealment. Besides, tying the employees into the firm is important. Although copyright of the created content usually belongs to the content provider and the client buys the right to use the innovative outcome, the advertisements and other content are often one-time products that cannot be used again as such. Consequently, the content providing firms considered practical concealment, contracts, and HRM the most valuable protection mechanisms for their knowledge, and those mechanisms were also actively used. Since the firms feared that employees would leave them, they had figured that they had to hold on to them informally. This was expressed by the interviewee from Firm A: "*I guess we just need to be the best workplace that we can be.*" In addition, trade secrets and copyright were important mechanisms, whereas the other IPRs were not found usable in that particular industry.

In the recreational industry, concealment regarding the products and services is not always the best option, as recreational firms often get most of their customers by references. Word-of-mouth is one of the best marketing mechanisms. However, secrecy is very much needed in relation to information from customers — whether it be about the customers themselves, or their feedback related to the offering of the firm. Tacitness, for its part, makes differentiation easier, because others have a hard time copying things from the company. Lead time is also understandably an important mechanism for SMEs providing recreational services. Employee related protection mechanisms (HRM and employment contracts) were considered important as well, since much of the service is closely related to the personal characteristics of the service providers.

The interplay of innovativeness and knowledge protection

Summarizing the above discussion, certain common features connecting knowledge protection potential and innovativeness can be seen. *All the SMEs seem to be balancing between knowledge sharing and protection*, which shows, for example, in the use of both secrecy and contracting; collaboration is needed, but certain issues need to be kept in the firms. With things that need to be disclosed, contracts are relied upon. In more general, *informal means of protection are actively used*, especially compared to IPRs that do not seem to be so easily at the SMEs' disposal. Regarding the incentives to innovate, many firms brought up the need to keep creating something new. These companies *did not necessarily worry about protecting their latest offerings against imitation, but they were worried about keeping core knowledge, trade secrets, and skills from flowing out with their employees. HRM related issues* thus achieved a central role when the usefulness of different mechanisms was considered. Naturally, some industry related differences exist: *Engineering firms* worked around established technologies and products, which created obstacles for using IPR protection (especially IPR with novelty requirements) efficiently. For this reason the firms had needs to rely on mechanisms such as HRM and tacitness. Lead time is important for these SMEs, but since it is sometimes difficult to achieve, it is not used to protect the firms' competitive edge as actively. Because the engineering firms were considered to be small players, they had to work harder in order to prove to the clients repeatedly that they were trustworthy suppliers in large deals. Thus, those protection mechanisms that promoted their skills were needed. *IT firms*, for their part, work in a very competitive field. Innovativeness is their only way to keep ahead of competitors. This is reflected in their use of knowledge protection. All forms of it are familiar to IT firms and they use them flexibly. Innovativeness is supported by keeping personnel turnover low, and by using protection mechanisms for embedding valuable knowledge in firm practices and technology instead of individuals solely. The innovativeness of IT firms is also promoted, when customers that can rely on the firm to keep central intellectual assets concealed are more willing to reveal knowledge to the IT firms. For *content providers*, innovativeness is the core of their business. Holding on to skillful employees who know the firm's business secrets was noted to be very important for the innovativeness of the companies. If a key employee were lost, there would definitely be difficulties in keeping the clients. Innovativeness was highly personified. The case of employees leaving a content providing firm was feared to seriously hurt its innovativeness: an interviewee from Firm A noted that "*this kind of situation could damage us for a whole year.*" Finally, in *recreational services* copying from rivals is common. It is so notable, that for firms working in this field the protection of new ideas before their public introduction is essential. Secrecy, in particular, is important in safeguarding innovative ideas long enough to turn them into offerings, and to make profit from them. Also, in this field, personal characteristics are of importance, and thus the means of contracting, labor legislation and HRM were of relevance.

Discussion and Conclusions

Innovation activities are inherently characterized by the need to extract and combine existing knowledge and new ideas, and the need to protect such knowledge and the outcomes of the process. However, the small size of firms intensifies the knowledge protection/sharing dilemma. SMEs inherently need to disclose some of their knowledge in exchange for lacking resources, and they should cover their most valuable intellectual assets in order to succeed commercially or even survive. Their ability to handle these issues is also affected by size-related factors, because limited resources may make it more difficult for them to adquately control knowledge transfer. Based on this reasoning we propose that:

Proposition 1: *The limited resources of SMEs simultaneously create* (1) *the need to disclose their knowledge to varying partners, and* (2) *impediments to using expensive and complex protection mechanisms.*

Thus the paradox is escalated compared to larger firms with more room to decide on the strategy that they want to follow (e.g., they do not necessarily need partners in order to access markets) and with more sophisticated systems for knowledge protection (based on a larger resource base).

Since SMEs are, in principle, disadvantaged against large firms in appropriating returns on their innovation activities (see also MacDonald, 2004), their incentives for innovation could be diminished. However, since SMEs are considered to be innovative, there may be other factors that make innovation lucrative for them. Although it is possible that the lack of protection actually is what pushes firms to improve their innovative performance[10] — not the potential rewards secured with effective barriers to imitation — continuous innovation and achieving lead time is not easy.

Indeed, our results indicate that even though lead time is among the most useful protection mechanisms for small firms (see also Byma and Leiponen, 2007), SMEs might not achieve it. Lead time may not be necessary, however, since SMEs also have other mechanisms to rely on. Especially practical secrecy (see also Hanel, 2006) and HRM (like labor legislation) can be very central in keeping the innovation related knowledge within the firm. Contracts are used actively as well, as the SMEs need to work with other organizations. However, the discussions revealed that many contracts were drafted by the entrepreneurs themselves, and that so far the strength or validity of them had not really been put to the test. It may be that emerging conflict situations would push the SMEs towards using external experts — if they had resources for that. In any case, traditional, formal protection mechanisms are often replaced by informal mechanisms. For instance, patenting is mostly done to achieve legitimacy (see also Gick, 2008, on patent filing as a disclosure strategy for SMEs), and the case studies suggest that HRM, in particular, is of high relevance among all the SMEs. Knowledge in SMEs is often very much embedded in people, and the loss of a key employee could shutter the firms of small size more than it would in large firms. Thus, it can be expected that:

Proposition 2: *Human resource management as a protection and appropriability mechanism is essential for SMEs.*

The reasons not to use formal protection mechanisms vary, however, and they are not solely dependent on the size. In our case study, the decisive element for

[10]Keeping ahead of other firms is considered to be among the most important and efficient mechanisms to enhance SMEs' competitiveness (Blind *et al.*, 2006).

some firms was the limited resource base, whereas for others the lack of competitive usefulness and efficiency of certain protection mechanisms in their line of business was what led to the same result. Even if certain mechanisms were available — or even used — they did not really generate notable benefits. Indeed, SMEs (like their larger counterparts) have different strategies for protection issues, some companies taking more notice of the issue than others. Moreover, we believe that the industry has an effect on how the innovations are protected, and thus we propose:

Proposition 3: *The field of industry is relevant in the protection efforts of SMEs: the more competitive and information-specific the business, the more emphasis on protection.*

According to our findings, the key in managing knowledge protection/sharing is that SMEs are not copying the protection models used by bigger players, but they are innovative also in choosing their own toolboxes. Balancing with the elements decreasing and increasing the transaction costs of a protection mechanism is relevant for SMEs. A rational entrepreneur seeks the marginal optimum which might mean abandoning a conventional patent and seeking a more efficient mechanism with a better protection/cost ratio. In particular, the protection mechanisms need to be selected in line with the innovative assets that are of concern. For the SMEs in our study, the most valuable innovative assets were located in the front-end of innovation activities. Thus, we propose that:

Proposition 4: *SMEs protect innovations by concentrating on safeguarding innovative input, rather than innovations as an output.*

This research has added to the existing literature by building four propositions based on the combination of existing literature and a multiple case study. We have demonstrated here that it is not so much the protection of innovation output where small firms can win out, but it is the background knowledge that they need to protect — for knowledge sharing, among other things. This creates differences in the use of protection mechanisms; even though a variety of them is used by SMEs, such mechanisms as HRM get a highlighted role. A comparison of SMEs from several industries provides further support for this.

Managerially this research benefits managers both in SMEs and in large firms. As firms need to collaborate, and as the success of collaboration requires understanding the motives and fears of the partner(s), it is beneficial for firms of varying sizes to know the factors influencing the operations of small firms. However, further qualitative and quantitative studies are needed to fully reveal the interplay between firm size, innovativeness and the knowledge protection/sharing dilemma.

References

Arundel, A (2001). The relative effectiveness of patents and secrecy for appropriation. *Research Policy*, 30, 611–624.

Arundel, A and E Steinmueller (1998). The use of patent databases by European small and medium-sized enterprises. *Technology Analysis and Strategic Management*, 10(2), 157–173.

Audretsch, DB (2002). The dynamic role of small firms: Evidence from the US. *Small Business Economics*, 18, 13–40.

Bessen, J (2008). The value of US patents by owner and patent characteristics. *Research Policy*, 37, 932–945.

Bismuth, A and Y Tojo (2008). Creating value from intellectual assets. *Journal of Intellectual Capital*, 9(2), 228–245.

Blind, K, J Edler, R Frietsch and U Schmoch (2006). Motives to Patent: Empirical evidence from Germany. *Research Policy*, 35, 655–672.

Blomqvist, K (2002). Partnering in the dynamic environment: The role of trust in asymmetric technology partnership formation. A doctoral thesis for Lappeenranta University of Technology.

Blomqvist, K, P Hurmelinna-Laukkanen, N Nummela and S Saarenketo (2008). The role of trust and contracts in the internationalization of technology-intensive born globals. *Journal of Engineering and Technology Management*, 25(1/2), 123–135.

Blomqvist, K, P Hurmelinna and R Seppänen (2005). Playing the collaboration game right — Balancing trust and contracting. *Technovation*, 25, 497–504.

Burchell, B and F Wilkinson (1997). Trust, business relationships and the contractual environment. *Cambridge Journal of Economics*, 21, 217–237.

Burk, DL and MA Lemley (2003). Policy levers in patent law. *Virginia Law Review*, 89(7), 1575–1696.

Burnham, T, KA McCabe and VL Smith (2000). Friend-or-foe intentionality priming in an extensive form trust game. *Journal of Economic Behavior and Organization*, 43, 57–73.

Byma, J and A Leiponen (2007). Can't block, must run: Small firms and appropriability. Mario Einaudi Center for International Studies. Working paper series, WP 1-07.

Chesbrough, H (2006). *Thrive in the New Innovation Landscape*. Boston, Mass: Harvard Business School Press.

Cohen, WM, RR Nelson and JP Walsh (2000). Protecting their intellectual assets: Appropriability conditions and why US manufacturing firms patent (or not). Working Paper 7552, National Bureau of Economic Research, Inc.

Crossan, MM and AC Inkpen (1995). The subtle art of learning through alliances. *Business Quarterly*, 60(2), 68–78.

Davis, L (2004). Intellectual property rights, strategy and policy. *Economics of Innovation and New Technology*, 13(5), 399–415.

De Jong, JPJ and O Marsili (2006). The fruit flies of innovations: A taxonomy of innovative small firms. *Research Policy*, 35(2), 213–229.

Dicksona, PH, KM Weaverb and F Hoyc (2006). Opportunism in the R&D alliances of SMEs: The roles of the institutional environment and SME size. *Journal of Business Venturing*, 21(4), 487–513.

Eisenhardt, K and M Graebner (2007). Theory building from cases: Opportunities and challenges. *Academy of Management Journal*, 50(1), 25–32.

Eriksson, P and A Kovalainen (2008). *Qualitative Methods in Business Research*. London: Sage Publications.

Gallini, NT (2002). The economics of patents: Lessons from recent US patent reform. *Journal of Economic Perspectives*, 16(2), 131–154.

Gick, W (2008). Little firms and big patents: A model of small-firm patent signaling. *Journal of Economics and Management Strategy*, 17(4), 913–935.

Hall, M, C Oppenheim and M Sheen (1999). Barriers to the use of patent information in UK small and medium-sized enterprises, Part I: Questionnaire survey. *Journal of Information Science*, 25(5), 335–350.

Hall, M, C Oppenheim and M Sheen (2000). Barriers to the use of patent information in UK small and medium-sized enterprises, Part 2 (1): Results of in-depth interviews. *Journal of Information Science*, 26(2), 87–99.

Hardy, M and A Bryman (2006). *Handbook of Data Analysis*. London: Sage Publications.

Hanel, P (2006). Intellectual property rights business management practices: A survey of the literature. *Technovation*, 26(8), 895–931.

Hayward, PA and CA Greenhalgh (1994). *Intellectual Property Research Relevant to Science and Technology Policy*, ESRC, Swindon.

Heiman, BA and JA Nickerson (2004). Empirical evidence regarding the tension between knowledge sharing and knowledge expropriation in collaborations. *Managerial and Decision Economics*, 25, 401–420.

Hemphill, TA (2007). Firm patent strategies in US technology standards development. *International Journal of Innovation Management*, 11(4), 469–496.

Hurmelinna-Laukkanen, P (2009). The availability, strength and efficiency of appropriability mechanisms — Protecting investments in knowledge creation. *International Journal of Technology Management*, 45(3/4), 282–290.

Hurmelinna-Laukkanen, P, L-M Sainio and T Jauhiainen (2008). Appropriability regime for radical and incremental innovations. *R&D Management*, 38(3), 278–289.

Hurmelinna-Laukkanen, P and K Puumalainen (2007). The nature and dynamics of appropriability — Strategies for appropriating returns on innovation. *R&D Management*, 37(2), 95–112.

Iversen, EJ (2004). Norwegian SMEs and the IPR-system or a small country's small companies in an increasingly big and IPR intensive world. 13th Nordic Conference on Small Business Research.

Jensen, PH and E Webster (2006). Firm size and the use of intellectual property rights. *Economic Record*, 82(256), 44–55.

Kingston, W (2006). Limited incontestability for small-firm patents. *European Intellectual Property Review*, 28(9), 463–468.

Kitching, J and R Blackburn (1999). Intellectual property management in the small and medium enterprise (SME). *Journal of Small Business and Enterprise Development*, 5(4), 327–335.

Kogut, B and U Zander (1996). What firms do? *Coordination, Identity, and Learning Organization Science*, 7(5), 502–518.

Lanjouw, JO and M Schankerman (2001). Characteristics of patent litigation: A window on competition. *RAND Journal of Economics*, 32(1), 129–151.

Lea, G and P Hall (2004). Standards and intellectual property rights: An economic and legal perspective. *Information Economics and Policy*, 16, 67–89.

Leiponen, A (2008). Control of intellectual assets in client relationships: Implications for innovation. *Strategic Management Journal*, 29(13), 1371–1394.

Lemley, MA and PJ Weiser (2007). Should property or liability rules govern information? *Texas Law Review*, 85, 783–841.

Levin, RC, AK Klevorick, RR Nelson and SG Winter (1987). Appropriating the returns from industrial research and development. *Brookings Papers on Economic Activity*, 3, 783–831.

Macaulay, S (1963). Non-contractual relations in a business: A preliminary study. *American Sociological Review*, 28(1), 55–67.

MacDonald, S (2004). When means become ends: Considering the impact of patent strategy on innovation. *Information Economics and Policy*, 16, 135–158.

Mann, RJ (2005). Do patents facilitate financing in the software industry? *Texas Law Review*, 83, 961–1030.

March, J (1991). Exploration and exploitation in organizational learning. *Organization Science*, 2(1), 71–87.

Mazzarol, T and S Reboud (2008). The role of complementary actors in the development of innovation in small firms. *International Journal of Innovation Management*, 2(12), 223–253.

Muller, E and A Zenker (2001). Business services as actors of knowledge transformation: The role of KIBS in regional and national innovation systems. *Research Policy*, 30(9), 1501–1516.

Nickerson, JA, BS Silverman and T Zenger (2007). The 'problem' of creating and capturing value. *Strategic Organization*, 5(3), 211–225.

Patton, MQ (2002). *Qualitative Research and Evaluation Methods*, 3rd edition. London: Sage Publications.

Pavitt, K, M Robson and J Townsend (1987). The size distribution of innovating firms in the UK. *Journal of Industrial Economics*, 35(3), 297–316.

Pope, C, S Ziebland and N Mays (2006). Analysing qualitative data. In *Qualitative Research in Health Care*, 3rd edition, Catherine, P and N Mays (eds.), pp. 63–81. Blackwell Publishing.

Rothwell, R and M Dodgson (1991). External linkages and innovation in small and medium-sized enterprises. *R&D Management*, 21(2), 125–138.

Silverman, D (2000). *Doing Qualitative Research. A Practical Handbook*. London: Sage Publications.

Simon, M, B Elango, SM Houghton and S Savelli (2002). The successful product pioneer: Maintaining commitment while adapting to change. *Journal of Small Business Management*, 40(3), 187–203.

Siquieira, AC and AD Cosh (2008). Effects of product innovation and organizational capabilities on competitive advantage: Evidence from UK small and medium manufacturing enterprises. *International Journal of Innovation Management*, 2(12), 113–137.

Street, CT and A-F Cameron (2007). External relationships and the small business: A review of small business alliance and network research. *Journal of Small Business Management*, 45(2), 239–266.

Thomas, S (2003). Intellectual property in biotechnology firms. In *Intellectual Property and Innovation Management in Small Firms*, Blackburn, RA (ed.), pp. 75–91. Routledge.

Thumm, N (2001). Management of intellectual property rights in European biotechnology firms. *Technological Forecasting and Social Change*, 67, 259–272.

Tidd, J (1993). Technological innovation, organizational linkages and strategic degrees of freedom. *Technology Analysis and Strategic Management*, 5(3), 273–284.

Van Dijk, M (2000). Technological regimes and industrial dynamics: The evidence from Dutch manufacturing. *Industrial and Corporate Change*, 9(2), 173–194.

Yin, RK (2003). *Case Study Research, Design and Methods*, 3rd edition. London: Sage Publications.

SME Networks for Open Innovation

Chapter 14

GOVERNANCE OF INNOVATION AND GROWTH IN SME NETWORKS

Tove Brink

University of Southern Denmark

The aim of this article is to reveal what organising of innovation and growth can be enabled in small and medium-sized enterprise (SME) networks. This is especially interesting for the governance of loosely coupled networks. The research is conducted in the Danish food industry with three food-producing firm networks. The posed hypotheses are tested in a structural equation modelling (SEM) approach with the available data of 60 SMEs.

The findings show that the governance of preferred behaviours of SMEs has a considerable significant positive impact on product and systematic innovation, which have a significant impact on growth. Governance of network connections is also found to have a significant positive impact on growth. The findings contribute to an understanding of how both governance of preferred SME behaviour and network connections enable innovation. This study provides SMEs, SME network organisations and policy bodies with an enhanced understanding of governance approaches.

Keywords: Innovation; SMEs; network; preferred behaviour; governance.

Introduction

The move towards innovation-based sustainable growth is at the heart of the EU's response to the economic challenges caused by the financial crisis in Europe, as noted at the EU Innovation Convention (2014). According to this aim, small and medium-sized enterprises (SMEs) in Europe represent approximately 99% of

This paper was first published in *International Journal of Innovation Management*, Vol. 20, No. 3 (April 2016), 1650052 (23 pages). © World Scientific Publishing Europe Ltd. DOI: 10.1142/S1363919616500523 the enterprises, employ 67% of workers and contribute 58% of value added[1] (European Commision, 2012). SMEs are therefore of essential interest for innovation and growth in the European area. Moreover, literature streams perceive SMEs' innovative activities as the missing link to enhancing existing growth theories, as highlighted by Braunerhjelm *et al.* (2010). The anticipation of the importance of SMEs as growth agents is stated within both entrepreneurship theory (Davidsson, 2004; Sarasvathy, 2008; Trettin and Welter, 2011) and network theory (Freel, 2003; Powell and Grodal, 2005; Huggins and Johnston, 2010). In these theories, SMEs are acknowledged for their innovative activities, which can lead to growth. However, in other literature streams, SMEs are noted for having limited resources and, thus, hindering innovation and growth in SMEs, as highlighted by (Edwards *et al.*, 2005; Murphy, 1996). SMEs can overcome their resource limitations through participation in networks, where they can have access to enhanced resources and knowledge (Powell and Grodal, 2005). There is a call for research on the understanding of SME networks and their aim for innovation to contribute to the understanding of SME growth. Therefore, the following research question is addressed in this paper: *What organizing of innovation and growth can be enabled in SME networks?* Organizing represents a process approach (Weick, 1995), which typically has an underlying antecedent of governance to support the organising process. The governance literature highlights the complementary nature of three perspectives towards governance (Müller, 2012), namely, economic through transaction cost economics (TCE) (Coase, 1937; Williamson, 1996), human through agency theory (Eisenhardt, 1989b; Sundaramurthy and Lewis, 2003) and organisational through trust and control (Puranam and Vanneste, 2009). The governance perspectives are complementary, which means that they can be present in coordination but not necessarily all of them need to be in coordination. However, governance takes its point of departure in at least one of them. In this paper, the focus will be on the economic perspective of governance in relation to growth in turnover in SMEs.

[1]The concept of added value is based on official country statistics on the share of the operating surplus of SMEs. Certain qualifications, however, are in order. A large share of the labour input (on average 20% in SMEs) consists of self-employed and unpaid family workers (EIM. Audretsch *et al.*, 2009, p. 16). These workers are not included in the wage bill of enterprises; thus, their labour input is not included in labour costs. Therefore, a correction has to be made through the calculation of the number of self-employed and unpaid family workers times the corresponding sectorial labour cost per employee. In addition, the net operating surplus should preferably be used; unfortunately, only data on the gross operating surplus adjusted for the imputed wage of the self-employed, as a percentage of the gross value added, were statistically available for measurement of the profitability differences between SMEs and large-scale enterprises (LSEs), (EIM. Audretsch *et al.*, 2009, pp. 16–18).

A contribution is hereby made to the understanding of governance to support the organising of innovation in SME networks for growth.

In Schumpeter's seminal papers on innovation, he stressed the importance of both the behaviour of the entrepreneur and the SMEs connections to society in general. Schumpeter's (1934) mark I highlighted the importance of preferred behaviour through the intuition and skills of SME managers to enable innovation. Later, Schumpeter's (1942) mark II emphasised the significance of innovation as embedded in the institutional framework through connections and governance structures related to SMEs. These marks showed the early acknowledgement of the importance of antecedents of SME behavioural preferences and network connections for innovation and growth. As highlighted earlier, such thinking has developed within literature streams on entrepreneurship theory, network theory and growth theory. However, the combined impact of these issues has remained underdeveloped.

The SME firm network context provides an opportunity for SMEs to pursue innovation. The SME participants in this research aim for innovation. The Danish SME networks are organised as loosely coupled systems providing a context for drawing on project governance frameworks with multi-project structures and multi-firm settings (Söderlund, 2004) and focusing on the economic perspective on growth. This context provides interesting opportunities for an enhanced contribution to innovation theory through a further understanding of governance to support the organising of innovation aims in SME networks to enable innovation and growth.

This paper starts with a review of the literature streams of growth, innovation, preferred behaviour and connections in SMEs to enable innovation. The hypotheses are developed concurrently. Then, the method and data collection are described. The statistical model and findings are analysed and discussed. The conclusion based on the main findings finalises the article.

Literature Review and Hypotheses

Literature on growth is the first issue to review in increasing the understanding of the relationship between innovation and growth. Furthermore, a literature review on innovation is conducted to reveal different innovation content. Thus, a literature review on aims of innovation is conducted. Then, a literature review on the antecedents of innovation on preferred organisational behaviours are conducted. Hereby, the internal organisational sources for innovation are revealed. Then, a literature review on external network connections that enable innovation is conducted. Both internal and external sources are important and often highlighted for innovation and growth (Tidd and Bessant, 2013; Dodgson *et al.*, 2014; Brink,

2014). Finally, a review on governance finalises the review and a summary provides an overview of the posed hypotheses.

Growth

Authors within different literature streams have addressed growth. Penrose (1959) notes, in her early notion of growth within the entrepreneurial field, the importance of a continuous focus on new productive opportunities, which drive the growth of the firm. Davidsson *et al.* (2009) indicate the importance and necessity of having a resource-based advantage and behaviour that allows firms to be profitable and form a sound basis for their growth. Ramezani *et al.* (2002) argue that SMEs should secure a sound level of profitability before they utilise their growth opportunities for further growth. The literature thus highlights an interrelation between innovation and growth.

In this research, growth is measured as the increase in firm turnover. This is the most frequently employed indicator of growth (Delmar, 1997). When research examines very different operating firms, there is a growing consensus that the increase in turnover should be the preferred choice (Weinzimmer *et al.*, 1998; Wiklund, 1998) because it is the most general of the alternatives to measuring growth. All commercial firms need to have turnover to survive. The networking SMEs in the present research operate very differently, even though they are all situated in the food industry. Therefore, in the absence of a better measure of growth, the increase in turnover in the SME is used as the measure of growth in this paper.

The literature review reveals that *innovation has a positive impact on growth* in turnover. The content of innovation is now examined more closely.

Innovation and different innovation aims

In this paper, innovation is used in accordance with the definition by Amabile *et al.* (1996), which highlights both the development of the new creative idea and the ability to successfully implement the new idea. This definition refers to first getting the new idea, integrating the new idea in an organisational learning approach, and finally executing the new idea in a collective, successful implementation. Innovation, viewed as an isolated new creative idea, is therefore not sufficient. It also requires organisational coordination, learning and implementation. This refers to an organisational approach to innovation with a multidimensional framework, as elaborated by Lam (2005) and by Crossan and Apaydin (2010) in their literature review on organisational innovation.

In the seminal paper by Cohen and Levinthal (1990), the authors highlight that absorption capacity is important for innovation to prosper through recognition of the value of new information to turn into innovation capabilities in the organisation. The notion of absorptive capacity overlaps with notions in other fields and has since been cited by researchers in various fields. As highlighted in Volberda *et al.*'s (2010) literature review on absorptive capacity, the notion has been used in the fields of learning, innovation, managerial cognition, knowledge-based view of the firm, dynamic capabilities and co-evolution. Among others, Volberda *et al.* (2010) reveal the important research gap between how inter-organisational knowledge interacts with intra-organisational knowledge according to Cohen and Levinthal's (1990) distinction between diverse and overlapping knowledge. This highlights the notion of innovation as reaching from the individual to the organisational level and overlapping across the levels within the organisation.

Schumpeter (1934) provided an early consideration of different innovation aims. Schumpeter (1934) understands innovation aims as doing "new and difficult things" and obtaining economic development and growth through this effort. This definition is in line with the definition highlighted by Amabile *et al.* (1996) but focuses on different aims for innovation.

Schumpeter (1934, p. 66) notes five typical innovation aims:

- New products
- New processes of production
- New markets
- New sources of supply
- New organisational approaches.

The listed innovation aims are very different in content, ranging from the development of the product itself to how the product is made, how the market for the product is identified/employed and how the supply of sources for production is identified to how the whole organisation surrounding the product is constructed for innovation. The five innovation aims can overlap or stand alone for any innovation, as interpreted by Schumpeter (1934). It must be noted that Schumpeter's original categories of innovation aims are widely found and employed today, e.g., the European Innovation Union Scoreboard Index (2014).

The hypothesis constructed based on the literature review on innovation and growth therefore emphasises the positive impact of innovation on growth and the presence of different innovation aims that enable innovation:

H1: *Innovation aims have a positive impact on growth.*

Now, focus is set on the effect of SME behaviour on innovation and growth.

Behaviour of SMEs

Individual preferences shape behaviour within SMEs and have an impact on creativity and innovation (Penrose, 1959; Davidsson, 2004; Sarasvathy, 2009; Trettin and Welter, 2011; Brink, 2014). This notion is supported by research conducted by the psychologist Csikszentmihaly (1996), who emphasises the preferences that distinguishes the behaviour of creative people from others. Creative people have a complex personality and show a tendency towards integration of thought and action, which are segregated in most people. This means that they typically can contain contradictory extremes and embrace the structural law of the psyche, as elaborated by Jung (1968) and Jacobi (1973), where e.g., the two attitude types (extrovert and introvert) are often viewed as mutually exclusive.

The findings can be extended from individual creativity to teams at the organisational level. The balancing of opposite and complementary preferences shaping organisational behaviour and the performance of the organisation, as noted by McWilliam and Dawson (2008), can enable innovation (Amabile *et al.*, 1996). The zone of 'sense-making and joint problem-solving' is essential at the team and organisational levels (Weick, 1995; Crossan and Apaydin, 2010). It refers to the notion that highly complementary opposed preferences result in higher collaborative innovation capabilities, which is shown as antecedents of innovation and growth. The complementary preferred behaviour of the SME organisation is important for governance to support its focus on the economic perspective.

The hypothesis constructed based on the literature review on SME behaviour emphasises the need for complementary behaviour in the SME to enable different innovation aims to enhance innovation and growth:

H2: *The higher the level of complementary behaviour in the SME organisation, the more positive the impact on innovation aims and the more positive the impact on growth.*

Finally, the review turns to external network connections.

SME networks and connections

Connections in networks have a positive impact on innovation and growth, as highlighted by Powell and Grodal (2005). Polenske (2004) enhances the understanding of the connection activities in firm networks as a triangle that consists of the three relations termed as competition, collaboration and cooperation. According to Polenske (2004), the triangle results in two cost notions to support economic governance, one from cooperation and competition and another from collaboration.

The cost notion from cooperation and competition is termed "transaction costs" (Coase, 1937; Williamson, 1996). Essentially, the term compares costs of competition and coordination in relation to their origin — either the market or the organisation. Sometimes it is cheaper to coordinate transactions through the external market (competition) and sometimes it is cheaper to have an organisation oversee and manage cooperation. Value is added through these transactions in the whole value system.

The cost notion from collaboration results, according to Polenske (2004), in another type of cost, which she terms "adaption costs". According to Polenske (2004), collaboration arrangements often lead to internal economies of scale shown on the firms long-run average cost curve. In other words, by collaborating in designing and producing a product, two or more firms can lower their adaption costs and innovate new products faster, have workers acquire new skills and obtain more capital investment.

The three network relations all have an impact on growth but in different ways. Competition and cooperation have a direct impact on growth through more short-term actions on products and services. Collaboration has an indirect, more long-term impact on growth through collaborative, innovative reciprocate activities, which fuel long-term growth.

Support for the positive impact from loosely coupled network connections is provided by Inauen and Schenker-Wick's (2011) study of SMEs. This study stressed the importance of connections to customers and competitors for innovation and growth in sales. Connections are thus important for the governance of SMEs. Connections are multifaceted and change over time (Hite and Hesterly, 2001); however, the connections valued and thereby mentioned by the SME participants play a role in growth. Hereby, a measure of connection activities can be provided through the number of connections acknowledged and mentioned by the SMEs.

The hypothesis constructed based on the network literature review is therefore as follows:

H3: *The higher the number of connections held in the organisation, the more positive the impact on innovation aims and on growth.*

Summary of literature review and hypotheses constructed

The previous literature review demonstrates that growth theories draw on cross-disciplinary approaches from very different fields, such as economics, innovation, organisation, psychology and network theories. From the economic perspective, the literature review reveals that governance is an integrated issue.

The previous literature review highlights the antecedents of SME behaviour and network connections as impacting aims for innovation activities and growth. This cross-disciplinary approach requires an integration of these issues for the construction of hypotheses.

In summary, the following hypotheses are constructed:

H1: *Innovation aims have a positive impact on growth*

H2: *The higher the level of complementary behaviour in the SME organisation, the more positive the impact on innovation aims and the more positive the impact on growth.*

H3: *The higher the number of connections held in the organisation, the more positive the impact on innovation aims and on growth.*

The hypotheses are listed in Fig. 1.

In Fig. 1, the three hypotheses are connected from antecedents to innovation aims and growth and from innovation aims to growth. This is the fundamental approach utilised in the latter test of the hypotheses.

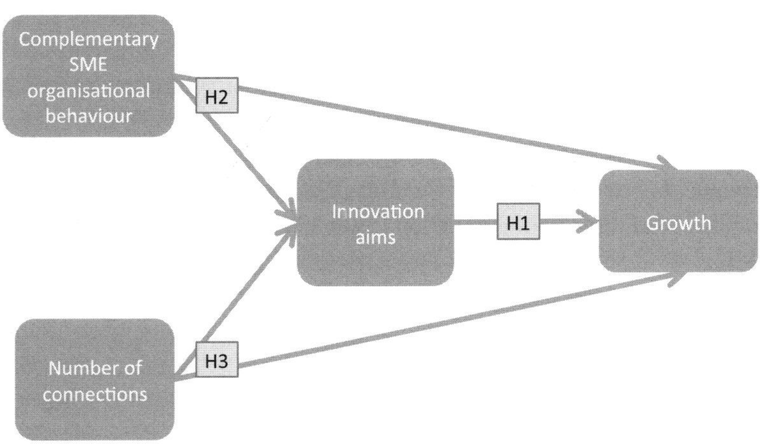

Fig. 1. Overview of hypothesis.

Method and Data Collection

In collaboration with a project team at CLF — Danish Centre for Rural Research at the University of Southern Denmark, a questionnaire was developed for three different food networks, with 93 SMEs within the food industry participating. The three networks are members of an umbrella organisation called 'Smagen af Danmark' (*www.smagenafdanmark.dk*) (in English, called 'The taste of Denmark'), which contains approximately 20 food networks that all aim for their members to engage in innovation.

The questionnaire intended to reveal information on basic economic information, growth, organisational culture and innovation initiatives in the participating companies in the network. This article only addresses data related to the research question. Data were collected through an online questionnaire with fixed quantifiable responses combined with open comments for each question. In other words, the data were primarily quantitative and supplemented by qualitative opportunities to provide comments. In the quantitative part, a seven-point Likert scale was used, making it easy for the respondents to take a stance on the statements. To ensure an adequate response rate, stances were made easy to relate to. The stances were worded such that they could be easily understood, and respondents were encouraged to respond impulsively.

With regard to validity and reliability, the questionnaire had to be accessible and understandable to all relevant respondents. This required pre-tests of the online system for answering the questions. Furthermore, a pre-test of the wording and structure of the questionnaire and a logical test of the questions were conducted with the research team to gather comments and improve the understanding of the questionnaire. After the questionnaire was revised, a pilot test of the questionnaire was conducted with three members of each of the three networks to gather comments and information regarding accessibility and understanding of the questions. The questions were primarily based on previous research that was conducted on one of the three participating networks from November 2007 to April 2008. This provided explorative, in-depth observations and interviews with seven members of the board of this network concerning their previous innovation and aims for future innovation. Innovation is context dependent (Tidd and Bessant, 2013); therefore, an explorative approach to the research constructs is needed, as elaborated by Churchill and Gilbert's (1979) notion of how to generate sample items.

During the pilot test, valuable comments and suggestions were received from the SMEs that tested the questionnaire, e.g., the test persons had problems understanding all of Schumpeter's (1934) innovation categories. The SME test persons were able to clearly identify and give information on product and process innovation. They understood the other innovation aims in Schumpeter's categories as a more systematised approach involving several innovation aims. Furthermore, they mentioned the innovation aim of 'the new idea coming out of the blue sky'. In this paper, this approach is called 'impulsive innovation', as it arises impulsively and typically can relate to several aims but is not reached in a systemised way. Here, the SMEs typically told stories about this approach to innovation. The comments from the pilot test persons were incorporated into the questionnaire and further discussed with the test persons prior to the final distribution of the questionnaire to all participants in the three networks. The questionnaire was distributed in late March 2009, and data gathering was finalised in the beginning of May 2009. The respondents were sent a reminder via email after 14 working days and again reminded via telephone seven working days after the email reminder. No further reminders were sent. The respondents were all busy working in their SMEs. When the research assistant called the participants via telephone for the last reminder, the participants primarily mentioned time constraints as reasons for not answering the questionnaire. The chairman in each network sent an email to the respondents with a call to answer the questionnaire close to the launch of the questionnaire. This initiative supported respondents' willingness to answer the questionnaire. Most responses came in during the first 14 working days. The later responses were not considerably different from the first responses; thus, no response bias is anticipated.

The analyses of the data were carried out through factor analysis in SPSS to reveal the underlying structure of the data and to 'purify the measures', as noted by Churchil (1979). The aim is to create parsimony in the data construction of the structural equation modelling (SEM). Furthermore, maximum likelihood estimation is employed in the SEM tool called AMOS (added tool to SPSS data) to test and reveal the significance of the data gathered.

The response rate for the data collection was satisfactory. In total, 65% of the respondents who could answer the questionnaire replied. Five members overlapped different networks, which meant that they participated in two of the three networks. All three networks had relatively the same number of 'double members'; thus, no bias is introduced due to this issue. The five companies participated in each of the networks that they were members of and, thus, were double counted in the analysis. This method was applied to generally show the networks independently of other affiliations that the members might have. Therefore, a total of 60 SME members responded, but five where overlapping; thus, 55 individual responses were received.

Description of Variables

A short description of the network profile in economic terms provides an understanding of which financial resources are available in the networks. The three networks represent an estimated total turnover that corresponds to approximately 100 million EURO, calculated as average figures from the responses. A similar calculation can be carried out on the total workforce of full-time staff, which shows a total of approximately 700 employees in the networks. The networks are therefore an interesting resource in the Danish economy. Overall, approximately 20 equivalent SME food networks exist in Denmark.

Growth

The questionnaire asked about the SMEs' average growth rate for the past two years and what they expect the growth rate to be in the next two years. It was easy for the SMEs to provide information about growth rates in turnover. A statistical factor analysis was conducted to reduce the data and create parsimony in the model on the growth issue.

- Growth rate in turnover for the past two years — variable: **Growth last two years**
- Growth rate in turnover for the next two years — variable: **Growth next two years**

The computation revealed a significant correlation between the variables of 0.872. The two variables are thus coherent measures of growth and can be combined in an index of growth called **I-growth** for the SEM model for MLE analysis. In the analysis, growth is thereby defined as the average combination of growth rate in turnover for the past two years with the expected growth rate in turnover for the next two years in standardized form. The growth variables are skewed, with many observations around the mean growth rate and a few observations with very high growth rates. Therefore, a logarithm of LN is taken to reduce the weight of these outliers with very high observations of growth rates. Thus, the diminishing impact of very high growth rates is taken into account.

Innovation aims in the SME organisations

As noted earlier, context shapes the use of innovation aims, as explained in the literature review. In the exploratory research phase (Churchil, 1979), the most directly observed innovation parameters are product and process innovation. In this paper, these two innovation variables will be measured and tested, and the rest of the innovation aims (market, resources and organisation) will be merged and addressed in a more systemic innovation approach. The SME respondents typically refer to the rest of the innovation aims in a more integrated systemic approach according to their own understanding. The opposite of systemic innovation is the individual impulsive idea, which the SMEs often discuss in stories on innovation. The SMEs often perceive impulsive ideas as a specific aim of innovation. These four aims of innovation were found in the pilot test of the wording that SMEs typically use and understand regarding innovation. From the beginning, innovation aims were inspired by the literature review on Schumpeter's (1934) thinking. However, according to the SMEs' understanding of innovation, the innovation aims in the literature were reformulated and enhanced to include four types: product, process, impulsive and systematic innovation. They perceived product innovation as new products, process innovation as new ways to provide products and services, impulsive innovation as ideas that one person randomly has and systematic innovation as innovation integrating several issues (market, resources and organisation — as noted by Schumpeter (1934)) in a continuous manner, which sets innovation in a systemic approach. The questionnaire was thus formulated as follows:

How does your innovation happen?
By developing new products — variable: **New products**
By developing new processes of manufacturing and distribution — variable: **New processes**
Innovation happens impulsively — variable: **Impulsive**
Innovation is planned and systematic — variable: **Systematic**

The four types of innovation are kept as separate variables to analyse the impact of these innovation types on growth. The higher the value of the innovation types, the more positive the impact on growth in turnover. The innovation variables for the statistical analyses are thereby constructed to test the anticipated positive implication on growth.

Preferences for complementary behaviour in the SME organisations

Preferences for complementary behaviour are identified here by four variables, which reveal how SME managers behave in their organisation. The stances are developed from the measurement characteristics in the Insights tool based on Jung's (1968) notion:

> How do you see yourself in a normal work situation in your company?
> Accurate, systematic, analytical — variable: **accurate**
> Amenable, helpful, loyal — variable: **amenable**
> Positive, inventive, impulsive — variable: **positive**
> Efficient, resolute, exacting — variable: **efficient**

The variables show people's preferences and how they behave and act in daily life in the context of Jung's theory on complementary preferred behaviours. The notion is that the higher and complementarity of the preferred behaviours, the more innovation is anticipated in the networking SMEs. A statistical factor analysis is conducted on the four variables to reduce the data and create parsimony in the model on the issue of preferred behaviour. The variables range from 0.829 to 0.882, indicating that a clear positive correlation between the four variables of accurate, amenable, positive and efficient is found, reducing the data and creating parsimony in the model. Preferences are thus combined in an index of preferences called **I-preferences** that is entered into the statistical analyses. In the analyses, preferences are defined as the average combination of the four variables in standardised form.

Number of connections in the SME organisations

Information on connections is provided through the variability of functions in the organisations and the number of connections. The stances are as follows:

> *What activities can be found in the company?*
> — variable: **Number of different activities in the organisation**
> *Major connections on these activities?*
> — variable: **Number of connection partners**

Table 1. Bivariate correlation analysis on variables in the model.

Pearson correlation	I-preferences	I-connections	I-growth
I -preferences	1		
I-connections	−0.093	1	
I-growth	0.150	0.233*	1

*Correlation is significant at the 0.05 level (two-tailed).
**Correlation is significant at the 0.01 level (two-tailed).

The number of different activities from agriculture to sales and support activities was computed. The result shows the variety of activities performed by the organisations. The number of connecting partners was computed to show the connecting intensity of the different activities in the company. Some of the networking companies have many connecting partners; others seem to be much more isolated. A statistical factor analysis was done on the two variables to reduce the data and create parsimony in the model on the issue of preferred behaviour. **Number of different activities** and the **number of connecting partners** have a correlation of 0.878. Thus, the data can be reduced for parsimony in the model. The variables are constructed in an index called **I-connections**. To support the validity of the underlying variables in the model, a bivariate correlation analysis of the mentioned variables on growth, preferences and network connections is shown in Table 1.

Table 1 illustrates that preferences and connections, the two antecedents in the hypotheses, are not significantly correlated at p value < 0.05. Furthermore, the causal factors in the hypothesis are correlated. The model thereby has validity.

The Statistical Model and Results Revealed

The hypotheses are elaborated in a model containing all hypotheses in a path diagram construction. In other words, the model contains a composition of all hypotheses involving the direct impact of preferences and connections on innovation aims and on growth. This integrated model gives the opportunity of decomposition by tracing paths backwards in the path diagram. The tool that we employed is an addition to SPSS called AMOS, which is based on the notion of SEM utilising MLE for estimation. A prerequisite for running the SEM analyses is the existence of a time difference in the occurrence of the involved variables. The literature review shows that preferences for behaviour and connections are antecedents of both innovation aims and growth. Furthermore, innovation aims are anticipated as antecedents of growth, as elaborated in the literature review, highlighting the positive impact of innovation on growth. These anticipations create the statistical model revealed in Fig. 2.

The hypotheses in Fig. 2 are tested in a stepwise backward elimination of the least significant factor combinations. The tests are all one-tailed, which provides a significance level for causal acceptance of p value < 0.05. The analysis is conducted using the AMOS tool based on maximum likelihood estimation with stepwise backwards elimination. MLE is employed, and '*simulation studies suggest that under ideal conditions, MLE provides valid and stable results with sample sizes as small as 50*' (Hair *et al.*, 1992 (2010), p. 661). The sample size in our research is 55 individual SMEs, which can provide sufficient, valid and stable results.

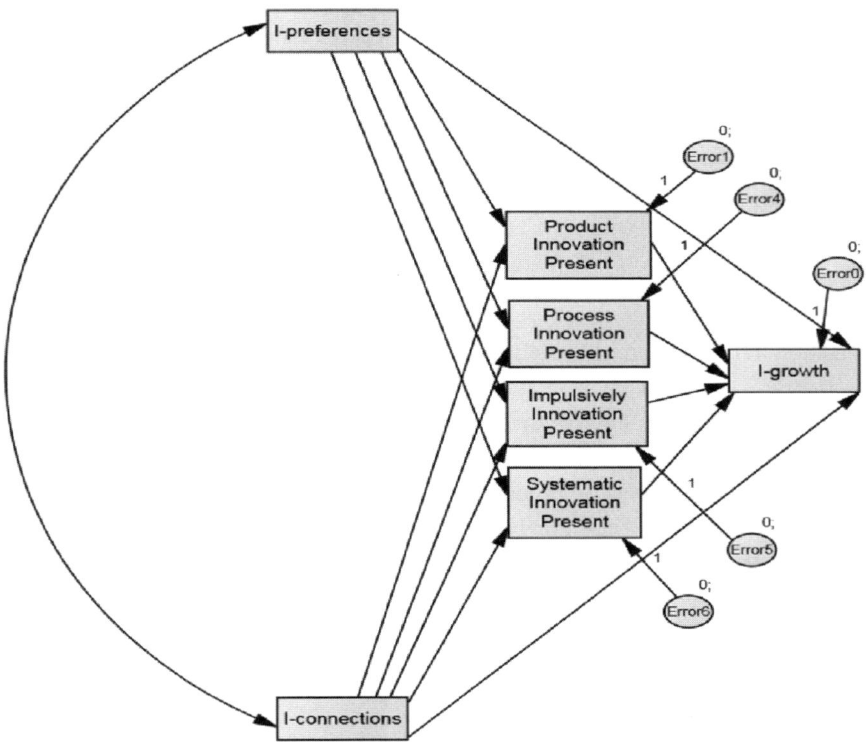

Fig. 2. Path diagram of preferences and connections relating.

Table 2. Significant level estimations in path diagram.

			Estimate	S.E.	C.R.	P
Product innovation	<—	I-preferences	0.750	0.119	6.278	***
Process innovation	<—	I-preferences	0.533	0.139	3.826	***
Systematic innovation	<—	I-preferences	0.507	0.141	3.599	***
I-growth	<—	I-connections	0.177	0.053	3.352	***
I-growth	<—	Product innovation	0.485	0.108	4.492	***
I-growth	<—	Process innovation	−0.395	0.106	−3.742	***
I-growth	<—	Systematic innovation	0.226	0.105	2.142	0.032

Note: Estimates not standardized in the table. Regression Weights: (Group number 1 - Default model).

The significance levels of the parameters in the model are revealed in Table 2, where p ranges from 0 to 0.032. Several p values thus show significance. The positive impact of preferences on innovation aims is significant. The impact of three of the four innovation aims on growth and the positive impact of connections on growth are also revealed. The significant model exhibits goodness of fit according to the comparative fit index (CFI) of 0.801, which is at the lower end for a good fit (Hair *et al.*, 1992 (2010), p. 669) but is an acceptable fit for the model.

The relations of significance shown in Table 2 are graphically revealed in Fig. 3. The significant variables in the model show considerable impact on innovation aims and considerable impact on growth.

Figure 3 shows a considerable positive impact of preferences for complementary behaviour on three of the four innovation aims, ranging from proportions of 0.75 to 0.51. It is thus revealed that preferences for complementary behaviour have a significant impact on product innovation, process innovation and systemic innovation — two of the original Schumpeterian aims of innovation and the other Schumpeterian aims redefined in one factor called systemic innovation. The impulsive innovation aim, which is often discussed in entrepreneurial stories and present in the network SMEs' stories, is not influenced by preferences for complementary behaviour. Furthermore, impulsive innovation does not have a significant impact on growth. The three innovation aims of product, process and systematic innovation have a significant impact on growth, ranging from a positive impact of 0.49 and 0.23 from product and systemic innovation to a negative impact of −0.40 from process innovation. As noted before, growth is measured as growth in turnover. It can thereby be revealed that product and systemic

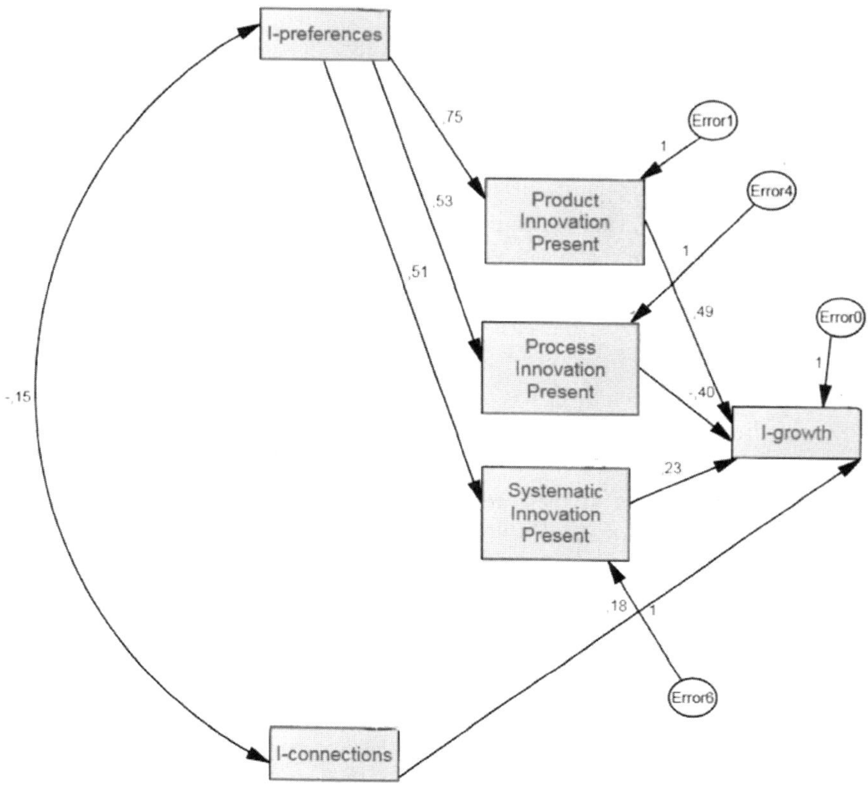

Fig. 3. Path diagram with significant relations from preferences and connections to innovation types and growth.

innovation increase turnover. The opposite is the case for process innovation. Process innovation is revealed to reduce turnover in SME networking companies.

There is no direct impact of preferences for complementary behaviour on growth. However, an indirect impact is revealed through innovation aims, which provide a positive impact through product innovation of $0.75*0.49 = 0.37$, systematic innovation of $0.51*0.23 = 0.12$ and a negative impact from process innovation of $0.53*(-0.40) = -0.21$ on growth in turnover. The innovation aims thus have very different impacts on growth in turnover. Awareness of these internal antecedents of preferences for complementary organisational behaviour is therefore important to enable innovation in this relatively informal and loosely coupled SME network context. This awareness can contribute to more deliberate aims for innovation on turnover and on profit. The governance structures regarding preferred behaviour and innovation aims are thus revealed as primarily grounded in human and organisational approaches.

Connections have a direct positive impact on growth, with a proportion of 0.18. Surprisingly, no significant positive relation is found between connections and innovation aims. This indicates that these networking SMEs do not collaborate as understood in the Polenskian sense, i.e., with direct interaction on collaborative activities between SMEs on different joint innovation initiatives. What they employ seems to be the easier economic approach of cooperation as understood within the transaction costs approach. This represents interaction in the value system where each SME has a role to play in the joint existing value system and growth can thereby be created. Within the network typology of Moller and Rajala (2007), the networking SME participants are primarily situated within Business renewal Nets in their employment of connections. The networking SMEs reap the value creation in the short term, where they can easily fit in and complement one another. The networking SMEs do not seem to reap the more long-term benefits of collaboration through joint new development of products and processes integrating innovation aims, which in the terminology of Moller and Rajala (2007) refers to employment of their connections as New Business Nets. The governance structure regarding network connections thus is revealed as primarily grounded in more short-term economic approaches. Awareness of these external antecedents of connections is therefore important for opportunities to enable innovation in this relatively informal and loosely coupled network context. An interesting difference is hereby found in the impact of preferences for complementary behaviour of SMEs and of connections on innovation aims and growth. Both antecedents of preferences and connections have a positive impact on growth but in very different ways. Therefore, it is beneficial to be able to employ both for governance in the SME networking context to support innovation and growth.

In short, the results of the testing of the hypotheses are as follows:

H1: *Innovation aims have a positive impact on growth*

H1 is only partly confirmed because impulsive innovation is not significant and process innovation is significant but with a negative impact and, thus, not confirmed.

Confirmation of the hypotheses has been achieved for product and systemic innovation.

H2: *The higher the level of complementary behaviour in the SME organisation, the more positive the impact on innovation aims and the more positive the impact on growth.*

H2 is only partly confirmed because preferred behaviour of SMEs does not have a direct significant effect on growth and impulsive innovation. Moreover, process innovation does not have a positive impact on growth.

Confirmation of the hypothesis has been achieved for product and systemic innovation.

H3: *The higher the number of connections held in the organisation, the more positive the impact on innovation aims and on growth.*

H3 is only partly confirmed because no impact is significant on innovation aims.

Confirmation of the hypothesis has been achieved directly for growth.

The implication of the confirmed hypotheses is a need for a dual approach to self-governance within and across SME food networks. Here, both internal preferred behaviours with focus on innovation aims and external connections represents opportunities for growth. This refers to a blend of governance structures employed with an economic approach to network connections and a human and organisational approach to innovation aims.

Discussion

The context in this research is open, loosely coupled food-producing SME networks. The ability to build on complementary SME behaviour and connections enable innovation aims and growth.

Continuously changing external pressures will naturally set governance structures under pressure to change and develop new business models, as highlighted in the EU Innovation Convention (2014). Thus, the findings of this paper must be anticipated to change over time. However, interesting antecedents are highlighted here for future research to explore innovation and growth in SME networks further.

Moreover, this paper shows that connections that are used for cooperation between organisations regarding efficient transactions have a positive impact on growth. This is an interesting contribution to governance of the SME network. An unused potential could be present in the Polenskian (2004) notion of collaboration as making a contribution to growth as well. In this paper, this notion could not be verified. In future research, it would be interesting to pursue this notion of an enhanced contribution to governance of SME networks on innovation and growth.

Innovation and growth are significantly enabled by the antecedent complementary preferred behaviour of SMEs for product and systemic innovation. Here, a contribution is made regarding the governance of SME networks with a focus on this antecedent and these two innovation aims. Moreover, the aim of process innovation seems to have a negative impact on growth. It could be anticipated that process innovation primarily has an impact on profit and thus a more indirect longer term derived impact on increased turnover. In future research, it would be interesting to enhance the research on process innovation and the impact on profit and turnover for a more thorough contribution in terms of governance, innovation and growth. Impulsive innovation does not have a significant impact on innovation, although the SMEs discuss occasions of impulsive innovation. This also provides a contribution to governance with a focus on more structured approaches to innovation aims that enable innovation and growth.

The context of open, loosely coupled SME networks is an interesting area for future research to pursue. A relatively small sample was used in this research. It could be interesting to conduct the same research with more participating SME networks and in other industrial contexts to enhance the understanding of the findings revealed in this paper.

Moreover, the research was conducted in the aftermath of the financial crises; further research could now reveal a possible impact of the financial crisis on the findings.

Conclusion

This paper shed light on what organizing of innovation and growth can be enabled in SME networks. Data were drawn from a survey that was conducted in three networks with 60 responding SMEs. Hypotheses were developed from a cross-disciplinary literature review modelling the antecedents of preferred behaviour of SMEs and SME connections on the intermediate innovation aims and on growth in turnover. The hypotheses were tested in a SEM employing AMOS and MLE to reveal significance.

The findings reveal the antecedents of preferred behaviours of SMEs and SME connections as important in enabling innovation and growth. This calls for governance to support both antecedents. In relation to SME connections, the findings reveal a direct positive impact of connections on growth. In relation to preferred behaviour of SMEs, the findings reveal a significant positive impact on product and systemic innovation, which again have a significant positive impact on growth. Preferred behaviour of SMEs thus has a much more indirect, but considerable, positive impact on growth through these two innovation aims. This is an important and underdeveloped issue for the enhanced contribution to governance of loosely coupled SME networks. More awareness of this underdeveloped area of governance and the impact on innovation and growth is called for in SME networks. Through such awareness, there is a potential for enhanced innovation and growth, which can be utilised by SME managers, SME network governing organisation and policy bodies.

Further research needs to be conducted to validate these findings in other contexts. Moreover, future research can pursue the potential underdeveloped issues of, e.g., enhanced understanding of connections to directly support innovation aims and the role of process innovation in growth, as measured by increased turnover.

References

Amabile, TM, R Conti, H Coon, J Lazenby and M Herron (1996). Assessing the work environment for creativity. *Academy of Management Journal*, 39(5), 1154–1184.

Braunerhjelm P, ZJ Acs, DB Audretsch and B Carlsson (2010). The missing link: Knowledge diffusion and entrepreneurship in endogenous growth. *Small Business Economics*, 34, 105–125.

Brink, T (2014). The Impact on growth of outside-in and inside-out innovation in SME network contexts. *International Journal of Innovation Management*, 18(4), 25.

Churchil, GA (1979). A paradigm for developing better measures of marketing constructs. *Journal of Marketing Research*, 16(1), 64–73.

Coad, A (2009). *The Growth of Firms: A Survey of Theories and Empirical Evidence*. Cheltenham, UK: Edward Elgar Publishing.

Coase, RH (1937). The nature of the firm. *Economica*, 4, 386–405.

Cohen, WM and DA Levinthal (1990). Absorptive capacity: A new perspective on learning and innovation. *Administrative Science Quarterly*, 35(1), 128–152.

Crossan MM and M Apaydin (2010). A multidimensional framework of organisational innovation: A systematic review of the literature. *Journal of Management Studies*, 47(6), 1154–1191.

Csikszentmihaly, M (1996). *Creativity — Flow and the Psychology of Discovery and Invention*. New York: Harper Perennial.

Davidsson, P (2004). *Researching Entrepreneurship*. Boston, MA: Springer.

Davidsson, P, P Steffens and J Fitzsimmons (2009). Growing profitable or growing from profits: Putting the horse in front of the cart. *Journal of Business Venturing*, 24(4), 388–405.

Delmar, F (1997). Measuring growth: Methodological considerations and empirical results. In *Frontiers of Entrepreneurship Research: On its Way to the Next Milennium*, PD Reynolds, WD Bygrave, S Manigart, C Mason, GD Meyer, HJ Sapienza and KG Shaver (eds.), pp. 190–216. Aldershot, UK and Brookfield, VA: Ashgate.

Dodgson M, DM Gann and N Philips (2014). Perspectives on innovation management. In *The Oxford Handbook of Innovation*. M Dodgson, DM Gann and N Philips (eds.), Oxford: Oxford University Press.

Dosi, G (1988). Sources, procedures and micro economic effect of innovations. *Journal of Economic Literature*, 26, 1120–1171.

Edwards, T, R Delbridge and M Munday (2005). Understanding innovation in small and medium-sized enterprises: A process manifest. *Technovation*, 25(10), 1119–1127.

Audretsch, EIM, D, RYD Horst, KT Kwaa and R Thurik (2009). *First Section of the Annual Report on EU Small- and Medium-Sized Enterprises*. EIM Business and Policy Research.

Eisenhardt, KM (1989b). Agency theory: An assessment review. *The Academy of Management Review*, 14(1), 57–74.

EU Innovation Convention (2014). Available at http://www.youtube.com/watch?v=zjJY-PRuJKMQ&feature=youtu.be.

European Commision (2012). EU SMEs in 2012: At the crossroads. Rotterdam: ECORYS.

Eisenhardt KM (1989). Agency theory: An assessment review. *The Academy of Management Review*, 14(1), 57–74.

Freel MS (2003). Sectoral patterns of small firm innovation, networking and proximity. *Research Policy*, 32, 751–770.

Hair JF Jr, WC Black, BJ Babin and RE Anderson (2010). *Multivariate Data Analysis. A Global Perspective*. New Jersey: Pearsson.

Helfat, CE, S Finkelstein, W Mitchell, MA Peteraf, H Singh, DJ Teece and SG Winter (2007). *Dynamic Capabilities. Understanding Strategic Change in Organisations*. USA: Blackwell Publishing.

Hite JM and WS Hesterly (2001). The evolution of firm networks: From emergence to early growth of the firm. *Strategic Management Journal*, 22, 275–286.

Huggins R and A Johnston (2010). Knowledge flow and inter-firm networks: The influence of network resources, spatial proximity and firm size. *Entrepreneurship and Regional Development*, 22(5), 457–484.

Inauen, M and A Schenker-Wick (2011). The impact of outside-in open innovation on innovation performance. *European Journal of Innovation Management*, 14(4), 496–520.

Jacobi, J (1973). *The Psychology of C.G. Jung*. New Haven and London: Yale University Press.

Jung, CG (1968). *Archetypes of the Collective Unconscious*, In Read H, Fordham M, Adler G.

Lichtenthaler, U and E Lichtentaler (2009). A capability-based framework for open innovation: Complementing absorptive capacity. *Journal of Management Studies*, 46, 1315–1338.

March, JG (1991). Exploration and exploitation in organisational learning. *Organisation Science*, 2(1), 71–87.

March, JG (2008). *Explorations in Organisations*. Stanford, USA: Stanford University Press.

McCorriston, S (2011). Competition, agricultural trade reform and developing countries-*Margin: The Journal of Applied Economic Research*, 5(1), 31–46.

McWilliam, E and S Dawson (2008). Teaching for creativity: Towards sustainable and replicable pedagogical practice. *Higher Education*, 56, 633–643.

Moller, KK and A Rajala (2007). Rise of strategic nets — new modes of value creation. *Industrial Marketing Management*, 36(7), 895–905.

Murphy, M (1996). *Small Business Management*. London: Financial Times.

Müller, R (2009). *Project Governance*. Aldershot: Gower Publishing.

Müller, R (2012). Project Governance. In *The Oxford Handbook of Project Management*, PWG Morris, JK Pinto and J Söderlund (eds.), pp. 297–320. Oxford: Oxford University Press.

OECD, (2010). *Ministerial Report on the OECD Innovation Strategy: Key Findings and Policy Messages*. OECD.

Penrose, E (1959), *The Theory of the Growth of the Firm*. Oxford: Oxford University Press.

Polenske, K (2004). Competition, collaboration and cooperation: An uneasy triangle in networks of firms and regions. *Regional Studies*, 38(9), 1029–1043.

Powell, WW and S Grodal (2005). Networks of innovators, In *The Oxford Handbook of Innovation*, J Fagerberg, D Mowery and R Nelson (eds.), pp. 56–86, Oxford: Oxford University Press.

Puranam P and BS Vanneste (2009). Trust and governance: Untangling a tangled web. *Academy of Management Review*, 34(1), 11–31.

Ramezani, CA, L Soenen and A Jung (2002). Growth, corporate profitability and value creation. *Financial Analysts Journal*, 62(6), 56–67.

Rampersad, GC, PG Quester and I Troshani (2010). Managing innovation networks: A cross-industry investigation of ICT and nano-bioscience networks. *Industrial Marketing Management*, 39, 793–805.

Rosenbusch, N, J Brinckmann and A Bausch (2011). Is innovation always beneficial? A meta-analysis of the relationship between innovation and performance in SMEs. *Journal of Business Venturing*, 26(4), 441–457.

Rowley, T, D Behrens and D Krackhardt (2000). Redundant governance structures: An analysis of structural and relational embeddedness in the steel and semiconductor industries. *Strategic Management Journal*, 21(3), 369–387.

Sarasvathy, SD (2008). *Effectuation: Elements of Entrepreneurial Expertise*. Cheltenham UK: Edwar Elgar.

Schumpeter, JA (1934). *The Theory of Economic Development*. New Brunswick, New Jersey: Transaction Publishers.

Schumpeter, JA (1942). *Capitalism, Socialism and Democracy*. London and New York: Harper Perennial.

Schumpeter, JA (1947). The creative response in economic history. *Journal of Economic History*, 7, 149–159.

Sundaramurthy C and M Lewis (2003). Control and collaboration: Paradoxes of governance. *Academy of Management Review*, 28(3), 397–415.

Söderlund, J (2004). On the broadening scope of the research on projects: A review and a model for analysis. *International Journal of Project Management*, 22(8), 655–667.

Teece, D, G Pisano and A Shuen (1997). Dynamic capabilities and strategic management. *Strategic Management Journal*, 18, 509–542.

Tidd J and J Bessant (2013). *Managing Innovation. Integrating Technological, Market and Organisational Change*. West Sussex, England: Wiley & Sons.

Tidstrom, A and S Ahman (2006). The process of ending inter-organisational cooperation. *Journal of Business and Industrial Marketing*, 21(5), 281–290.

Trettin, L and F Welter (2011). Challenges for spatially oriented entrepreneurship research. *Entrepreneurship and Regional Development*, 23(7–8), 575–602.

Weick, KE (1995). *Sensemaking in Organisations*. Thousand Oaks: Sage Publication Series.

Williamson, OE (1996). *The Mechanisms of Governance*. Oxford, New York: Oxford University Press.

Weinzimmer, LG, PC Nystrom and SJ Freemann (1998). Measuring organisational growth: Issues, consequences and guidelines. *Journal of Management*, 24(2), 235–262.

Volberda, HW, NJ Foss and MA Lyles (2010). Absorbing the concept of absorptive capacity: How to realize its potential in the organisation field? *Organisation Science*, 21(4), 931–951.

Chapter 15

KNOWLEDGE INFLOWS FROM MARKET- AND SCIENCE-BASED ACTORS, ABSORPTIVE CAPACITY, INNOVATION AND PERFORMANCE: A STUDY OF SMEs

Graciela Corral De Zubielqui

Entrepreneurship, Commercialisation and Innovation Centre (ECIC)
The University of Adelaide

Janice Jones

Flinders Business School, Flinders University

Laurence Lester

Centre for Housing, Urban and Regional Planning
The University of Adelaide

Small to medium-sized enterprises (SMEs) are increasingly opening up their innovation processes in order to improve innovation outcomes. Exactly how openness influences enterprises' ability to innovate and benefit from external sources of innovation is not well understood. Despite theoretical models explicating knowledge inflows as the antecedent and innovation as the consequence of absorptive capacity, to date, the relationships between external knowledge inflows from specific external actors, absorptive capacity, innovation and SME performance remain underexplored. We address this gap, focussing on market- and science-based knowledge flows, absorptive capacity, innovation and SME performance. Using structural equation modelling on a sample of 838 Australians SMEs, we find evidence indicating that the mechanisms by which external knowledge inflows from market- and science-based actors affect innovation differ: external knowledge inflows from market-based actors influence innovation directly or "serendipitously", while external knowledge inflows from science-based actors influence innovation indirectly via absorptive

This paper was first published in *International Journal of Innovation Management*, Vol. 20, No. 6 (August 2016) 1650055 (31 pages). © World Scientific Publishing Europe Ltd. DOI: 10.1142/S1363919616500559 capacity. We also find that their effect on innovation differ, with external knowledge inflows from market-based actors exerting a positive direct effect on innovation vis-à-vis external knowledge inflows from science-based actors where there is no significant direct effect. In contrast, external knowledge inflows from science-based actors influence innovation through absorptive capacity. In addition to this mediation effect, absorptive capacity has a positive direct effect on innovation, and an effect on firm performance through innovation. This research contributes to our understanding of how, and the extent to which, external knowledge inflows from specific external actors influence absorptive capacity, innovation and firm performance in SMEs.

Keywords: Open innovation; knowledge inflows; external actors; absorptive capacity; SME performance.

Introduction

Open innovation (OI) is defined as "the use of purposive inflows and outflows of knowledge to accelerate internal innovation, and to expand the markets for external use of innovation, respectively" (Chesbrough, 2006: 2). The basic premise of OI is to facilitate the innovation process by, firstly, sourcing external ideas, knowledge and technology from suppliers, customers, higher education institutions (HEIs) and other external actors and integrating these internally in the focal firm's production process(es), thereby increasing firm innovativeness. Secondly, the focal firm can interact with external actors that are better able to commercialise its technology, for example, through selling or licencing out technology. These two processes are referred to as inbound OI and outbound OI, respectively (Chesbrough and Crowther, 2006). Common to both processes is the opening up to, and the establishment of, relationships with external actors, although the aims differ between processes: the objective of inbound OI is to access external ideas, knowledge and technology to complement, strengthen or speed up in-house R&D activities (Spithoven *et al.*, 2013); outbound OI is concerned with establishing relationships with external actors in order to transfer proprietary technologies developed in-house for commercial exploitation (Bianchi *et al.*, 2011).

In the small- to medium-sized enterprise (SME) context, the central tenet that underlies much of the OI literature is that through access to, and utilisation of, external resources enterprises can overcome resource constraints (Teirlinck and Spithoven, 2013; Terziovski, 2010), or reduce some of their liability of "smallness" (Gassmann *et al.*, 2010). Yet while SMEs are increasingly opening up their innovation processes in order to improve innovation outcomes (Gassmann *et al.*, 2010; Spithoven *et al.*, 2013), exactly how openness influences enterprises' ability to innovate and benefit from external sources of innovation is not well understood (e.g., Chesbrough, 2003; Helfat, 2006; Laursen and Salter, 2006).

With few exceptions, there is a dearth of empirical studies on external knowledge flows into SMEs (Huggins and Johnston, 2009; Hughes *et al.*, 2009; O'Regan *et al.*, 2006), including SMEs in Australia (Huang and Rice, 2009). The use of HEIs and related providers such as research institutes, and the channels by which knowledge is acquired from these sources by SMEs for innovation remains underexplored, particularly from the SME perspective. This is surprising, given that a significant proportion of SMEs produce product, service and process innovations, both in Australia (Department of Industry, 2014) and elsewhere (European Commission, 2009), and that external actors can help enterprises, particularly SMEs, overcome size-related disadvantages, by providing them with access to a wider pool of knowledge and resources (Mackinnon *et al.*, 2004).

The extent to which an SME's absorptive capacity influences its ability to benefit from external knowledge flows has also not received much attention in the empirical literature, despite considerable theoretical models explicating knowledge inflow as the antecedent and innovation as the consequence of absorptive capacity (Kostopoulos *et al.*, 2011; Lane *et al.*, 2006). With few exceptions (Moilanen *et al.*, 2014), empirical work that does exist tends not to distinguish between different sources of external knowledge flows (e.g., customer, supplier, competitor and HEI linkages) (Escribano *et al.*, 2009; Kostopoulos *et al.*, 2011), nor the channels through which external knowledge is acquired by SMEs from these different sources. It is unhelpful to group external knowledge flows from different sources, as there may be important differences between them that contribute to our understanding of their relationships to absorptive capacity, innovation outcomes and firm performance.

These issues are particularly salient given that for some SMEs at least, absorptive capacity can be a source of competitive advantage, enabling them to increase returns from external knowledge inflows (Kostopoulos *et al.*, 2011), and all the more important in Australia, where both large and small businesses perform poorly on a range of innovation metrics relative to other developed countries (see Cutler, 2008). The purpose of this research is to begin to fill these gaps, through an empirical study that examines how, and the extent to which, external knowledge inflows from specific external actors influence absorptive capacity, innovation and firm performance in SMEs.

In so doing, this paper contributes to the innovation management literature in two aspects: firstly, we extend the extant literature by examining knowledge inflows from market-based actors *and* knowledge inflows from science-based actors, and their relationships to absorptive capacity, innovation and business performance. Most empirical studies investigate the impact of *aggregated* measures of external knowledge inflows (e.g., Kostopoulos *et al.*, 2011; Escribano *et al.*, 2009); in other words, knowledge inflows from market- and science-based actors. By delineating the effects of knowledge inflows from specific external actors (i.e., market- *or* science-based actors) on absorptive capacity, innovation and firm performance, this study responds to a call in the recent OI literature (Grimpe and Sofka, 2009; Kostopoulos *et al.*, 2011) for a more fine-grained investigation of knowledge inflows from specific external actors. This issue is of considerable practical relevance since a firm may use different sources to acquire knowledge, including interacting with actors within and outside the supply chain. Moreover, knowledge from different actors may affect innovation performance differently. By distinguishing between different sources of knowledge inflows we seek to determine if knowledge inflows from particular actors are more critical than others in facilitating an enterprise's innovation performance. This is important since the search for external knowledge is not costless (Mina *et al.*, 2014) — requiring significant investment in critical search capabilities and incurring potentially high transaction costs (Christensen *et al.*, 2005) — particularly in the resource constrained SME context, where the resources and capabilities which allow a firm to invest in external knowledge sourcing are more limited vis-a-vis large firms.

Secondly, by examining the effects of external knowledge inflows from specific actors, absorptive capacity and innovation on SME performance, we add to the very few studies that have attempted to answer the question of whether firms' relationships with external actors translate into superior firm performance, a question that Mina *et al.* (2014) notes is an important remaining question, including in the SME context. Prior research has mainly focused on the effects of OI practices on innovation outcomes only in organisations in general (e.g., Kang and Kang, 2009; Laursen and Salter, 2006), in SMEs (e.g., Theyel, 2012), or the relationship between external knowledge inflows, absorptive capacity, innovation and firm performance (Kostopoulos *et al.*, 2011), and the generalisability of these findings to SMEs has not yet been tested. By investigating these relationships in SMEs, we aim to fill this gap.

The paper proceeds as follows: the first section reviews research on external knowledge inflows from market- and science-based actors, absorptive capacity and innovation in the context of SMEs. The second section describes the research design. The results are presented in the third section and findings discussed thereafter. The paper concludes with policy implications, limitations and suggestions for future research.

Literature Review

External knowledge inflows from market- and science-based actors and innovation

Prior research distinguishes between external knowledge inflows from market-based actors (e.g., customers, suppliers, and competitors) and science-based actors (e.g., HEIs, research institutes and/or government research organisations) typically conceptualised as providing market- and science-based (which includes technical/scientific) knowledge respectively (Danneels, 2002; Du *et al.*, 2012). Empirical work indicates that market-based actors are important sources of knowledge for innovation outcomes (Faems *et al.*, 2005; Hughes *et al.*, 2009; Lasagni, 2012; West and Bogers, 2014), including for SMEs in the UK (Huggins and Johnston, 2009; Hughes *et al.*, 2009) and in Norway (Moilanen *et al.*, 2014), although there is no clear consensus about whether suppliers or customers are the most important sources, with some research finding that customers are most important (Davenport, 2005; Enkel *et al.*, 2009; Mina *et al.*, 2014; Moilanen *et al.*, 2014; Theyel, 2012), while other work suggests suppliers are more important (Laursen and Salter, 2006). Sourcing knowledge and ideas from customers and end users during new product development in particular is a generally accepted practice on the basis that these actors are typically well informed given that they are close to the market (Pittaway *et al.*, 2004) and in a position to provide insights with respect to market needs and future demands (Von Hippel, 2005). This also enables firms to develop customised and commercially viable products thereby reducing risks with innovation, and economising on resources. Suppliers on the other hand, can help frame the possibilities for innovation based on their knowledge of the materials, equipment and techniques that they supply (Kaufman *et al.*, 2000). Theyel (2012) makes the point that despite preferences, knowledge inflows from customers and suppliers exist although the level of involvement varies from one value chain activity to the next. Accordingly, we suggest:

> **H1a:** *External knowledge inflows from market-based actors are positively related to innovation.*

While consensus exists that universities are ranked low as a source of innovation in comparison with market-based actors (Fernández-Esquinas *et al.*, 2015; Mina *et al.*, 2014; Perkmann *et al.*, 2013), knowledge inflows from science-based actors may enhance innovation (Cohen *et al.*, 2002; Fernández-Esquinas *et al.*, 2015; Larsen, 2011; Mina *et al.*, 2014; Uyarra, 2010). For example, in an exploratory qualitative study of innovative SMEs in New Zealand, SMEs accessed knowledge in publicly funded research institutes and universities, particularly in the early stages of the innovation process (Davenport, 2005). Similarly, in a UK study Mina *et al.* (2014) found that scientific and technical knowledge sourced from universities and the public research base is more important for innovation in business services relative to manufacturers. Based on a survey of 737 innovative SMEs in peripheral Spain that investigated university–firm interactions, Fernández-Esquinas *et al.* (2015) conclude that universities are an important source of tacit knowledge transfer. According to Narula (2004), however, SMEs may have too few technological assets and expertise to be of interest to most science-based actors, and lack the requisite human and other resources to search systematically for technical knowledge. Hughes *et al.* (2009) ascribe the very limited role of HEIs as a source of knowledge for SMEs to the fact that they are not (organisationally, professionally or geographically) close to SMEs. The authors conclude that universities are more "remote" compared with other more proximate sources of information within the SME supply chain, and moreover, less effective. Thus, on the basis of conflicting results, questions pertaining to external knowledge inflows from science-based actors and innovation in SMEs persist.

Fernández-Esquinas *et al.* (2015) point to methodological limitations as a potential reason for inconclusive findings, with quantitative evidence based on secondary data sources (e.g., Huang and Rice, 2009) or on official innovation surveys such as the Community Innovation Survey (CIS) (e.g., Kostopoulos *et al.*, 2011; Escribano *et al.*, 2009). While the CIS has the advantage of a large sample size covering a full range of firms, the questionnaire is not specifically designed to gather detailed information on the forms of knowledge transfer for the firms involved, with analyses based on "proxy" variables that may not reflect the range and/or intensity of such links (Mohnen and Hoareau, 2003). Huang and Rice (2009) identify the need for more robust measures to address this issue of relying on "proxies". Another methodological issue that may account for the inconsistent findings is that empirical work specifically designed to study university–firm interactions tends to focus on industrial sectors that are closely linked to research (e.g., biotechnology, firms with R&D departments, high tech start-ups, etc.), the results of which are difficult to extrapolate to innovative SMEs that do not have formal in-house R&D but are commonplace in many peripheral regions (Fernández-Esquinas *et al.*, 2015). As Moilanen *et al.* (2014) point out, a substantial number of innovative SMEs in low-tech industries are not reporting doing their own R&D at all, yet rely heavily on external knowledge for innovation

(Ortega-Argile's *et al.*, 2009; Rammer *et al.*, 2009). In summary then, it is difficult to arrive at a definitive conclusion regarding the relationship between knowledge inflows from science-based actors and innovation, given the mixed empirical evidence. In line with the contention that knowledge transfers from science-based actors may enhance the capacity to innovate (Cohen *et al.*, 2002; Larsen, 2011; Uyarra, 2010) including in SMEs (Fernández-Esquinas *et al.*, 2015; Davenport, 2005; Mina *et al.*, 2014), we state following:

H1b: *External knowledge inflows from science-based actors are positively related to innovation.*

External knowledge inflows from market- and science-based actors, absorptive capacity, and innovation

While the foregoing implies a serendipitous relationship between external knowledge flows and innovation (Zahra and George, 2002, p. 190), theory suggests that in order to benefit from external knowledge inflows, firms must develop the ability to recognise the value of new external knowledge, and then assimilate and apply the knowledge for commercial ends. In other words, a firm must develop its absorptive capacity including in the OI context, since inbound OI is characterised by the acquisition and internal application of external knowledge (Flatten *et al.*, 2011). Cohen and Levinthal (1990, p. 128) define absorptive capacity as "the ability of a firm to recognise the value of new, external information, assimilate it, and apply it to commercial ends". Zahra and George (2002, p. 186) expand and further develop the concept to include: "... a set of organizational routines and processes by which firms acquire, assimilate, transform, and exploit knowledge to produce a dynamic organizational capability." But in order to develop absorptive capability, firms need exposure to external knowledge within their environment (Kostopoulos *et al.*, 2011; Moilanen *et al.*, 2014). This flow of external knowledge into the firm contributes to its knowledge base thereby developing the firm's absorptive capacity and as a result, the enterprise is better prepared to engage in acquiring, assimilating and exploiting knowledge (Kostopoulos *et al.*, 2011; Moilanen *et al.*, 2014).

In general, research indicates that SMEs are likely to more easily absorb market-based and applied knowledge acquired from market-based actors (Spithoven *et al.*, 2013). This is most likely because they have experience with, and knowledge about, something similar (Saviotti, 1998). The literature shows that SMEs tend to focus on a few key actors in the value chain with whom they have strong ties (Lowik *et al.*, 2012) and are cognitively close (Knoben and Oerlemans, 2006) or proximate (Boschma, 2005). Cognitive proximity (and distance) between firms engaged in collaborative innovation is an organisational level construct used to describe similarities (and differences) in organisational focus (Nooteboom, 2000), including similarities (and differences) in knowledge.

While absorptive capacity provides important insights into the processing of knowledge inflows close to a firm's existing knowledge base (Cohen and Levinthal, 1990) — in this study, market-based actors — there is a dearth of empirical research on whether the concept is directly applicable when firms reach for knowledge from distant contexts (Nooteboom *et al.*, 2007), for example, science-based actors. As Heil and Enkel observe (2015) observe, this is an uncertain process which typically requires the absorption of new knowledge inflows (Nooteboom, 2000), and there is high cognitive distance in terms of knowledge sourced from science-based actors and the SME applying the knowledge (Nooteboom *et al.*, 2007). In one of the few studies to utilise disaggregated measures of external knowledge inflows (i.e., separate knowledge flows from competitors, customers, trade organisations, personal networks and R&D institutions), Moilanen *et al.* (2014) find that collaboration with R&D institutions (together with trade organisations, competitors, and personal networks) is positively related to absorptive capacity, implying that interaction with R&D institutions (along with trade organisations, competitors, and personal networks) increases absorptive capacity in non-R&D SMEs. Such interaction broadens the SME's knowledge base, thus improving its absorptive capacity by reducing the cognitive distance between the enterprise and science-based actors, and facilitating the enterprise's capacity to interact and use external knowledge. Thus on the basis of this theory, we suggest the following hypotheses:

H2a: *External knowledge inflows from market-based actors are positively related to an SME's absorptive capacity.*

H2b: *External knowledge inflows from science-based actors are positively related to an SME's absorptive capacity.*

While the ability to recognise, acquire and assimilate external knowledge with the enterprise's existing knowledge enables the firm to create a new and enlarged knowledge base and *inter alias*, develop and increase its absorptive capacity, the new knowledge must still be exploited and translated into new products, services, or processes (Zahra and George, 2002) in order to generate important firm outcomes (Fosfuri and Tribó, 2008). However, there is a paucity of empirical evidence on whether absorptive capacity translates external knowledge inflows into innovation outcomes (Kostopoulos *et al.*, 2011; Todorova and Durisin, 2007), particularly in the SME context.

In a study that examined this relationship in organisations in general in Greece, Kostopoulos *et al.* (2011) provide empirical support for the mediating role of absorptive capacity in the relationship between external knowledge inflows (from suppliers, clients, competitors, universities, other research institutions, specialised journals, conferences and meetings) and innovation, underlining the key theoretical assumption of absorptive capacity theory that, in order for enterprises to derive innovation benefits from new external knowledge, they will need to recognise the value of this knowledge, internalise it and exploit it. However, Moilanen *et al.* (2014) find that knowledge inflows from trade organisations and personal networks only are partially mediated by absorptive capacity and therefore have a significant positive indirect effect on innovation. This suggests the need to investigate knowledge inflows from which actors absorptive capacity transforms into higher innovative performance (Moilanen *et al.*, 2014). In line with existing models of absorptive capacity (Zahra and George, 2002), and the limited empirical evidence (Kostopoulos *et al.*, 2011) we expect that:

H3a: *An SME's absorptive capacity mediates the relationship of external knowledge inflows from market-based actors with innovation.*

H3b: *An SME's absorptive capacity mediates the relationship of external knowledge inflows from science-based actors with innovation.*

The goal of absorptive capacity is to enable firms to realise commercial benefits from external knowledge, typically in the form of innovation outcomes (Cohen and Levinthal, 1989, 1990). Recent work (Kostopoulos *et al.*, 2011; Moilanen *et al.*, 2014) has noted that absorptive capacity can contribute to a firm's innovation performance by acting as a tool for processing new external knowledge, and as a conduit or pathway for transferring knowledge within the firm that can facilitate cross-organisational innovation activities (Hargadon and Sutton, 1997; Tsai, 2001).

However, the relationship between absorptive capacity and innovation outcomes has received comparatively little attention in the empirical literature (Kostopoulos *et al.*, 2011). Research that does exist provides conflicting empirical evidence about this relationship: according to Kostopoulos *et al.* (2011, p. 1341) "the outcomes of absorptive capacity materialize ... as a straightforward improvement in innovation". In contrast, in a study of Australian manufacturing

SMEs, Huang and Rice (2009) demonstrate that the relationship between absorptive capacity and current innovation performance is negative, and suggest further that investment in absorptive capacity over the short term does not lead to direct economic returns from innovation. Despite this finding and on the basis of theoretical arguments outlined above and the majority of the limited empirical evidence (Kostopoulos *et al.*, 2011), we suggest:

H3c: *An SME's absorptive capacity is positively related to innovation.*

Absorptive capacity, innovation and firm performance

The purpose of absorptive capacity is to apply externally sourced knowledge to commercial ends (Cohen and Levinthal, 1990). Thus absorptive capacity can help to promote firm performance and contribute to the achievement of competitive advantage (Lane *et al.*, 2006; Zahra and George, 2002). For example, the ability to distil, acquire and assimilate external knowledge with firm specific knowledge enables the focal enterprise to create, introduce and commercialise innovative products, services and processes (Morgan and Turnell, 2003; Murray and Peyrefitte, 2007) which are regarded as a primary source of value and competitive advantage (Barney, 1991). Drawing on the resource-based view of the firm (Barney, 1991), Zahra and George (2002) argue that, as a bundle of knowledge-based capabilities, absorptive capacity can be a source of competitive advantage (Grant, 1996), to the extent that a focal firm is better able to distil the critical elements of value out of public knowledge than other firms, giving the focal firm an edge over competitors (Torkkeli *et al.*, 2009). In other words, firms exposed to the same amount of external knowledge inflows may not derive equal returns, because they differ in their ability to identify and exploit external knowledge inflows (Escribano *et al.*, 2009). Thus, both the amount and effect of external knowledge flows are unequally distributed across firms, implying that absorptive capacity can be a source of competitive advantage (Escribano *et al.*, 2009). Hence:

H4a: *An SME's absorptive capacity is positively related to firm performance.*

In addition to the direct effect of absorptive capacity on firm performance, absorptive capacity is also able to contribute to firm performance indirectly through innovation (Barney, 1991). This is achieved through absorptive capacity processing, assimilating, and exploiting and translating new knowledge into new products, services, or processes (Zahra and George, 2002), leading to tangible financial outcomes (Kostopoulos *et al.*, 2011). In other words, absorptive capacity is able to achieve competitive advantage and influence firm performance indirectly through innovation (Barney, 1991). Thus:

H4b: *Innovation mediates the relationship between an SME's absorptive capacity and firm performance.*

Our final hypothesis relates to the effect of innovation on firm performance. Much of the scholarly and practitioner literature on OI highlights the benefits of openness, and inter alias, a potential positive effect on an enterprise's business success and financial performance (Laursen and Salter, 2006; Torkkeli *et al.*, 2009; West and Bogers, 2014). This focus on openness and interaction of enterprises with external actors within the innovation system is a reflection of a wider trend in studies of firm behaviour that suggest that the firm's relationships with external actors in its external environment plays a central role in determining performance (Laursen and Salter, 2006, p. 132). A firm's innovativeness can affect the enterprise's performance in terms of both cost and revenue. Few enterprises, particularly SMEs, have the financial and human resources to undertake all R&D projects by themselves, thus significant cost savings can accrue by leveraging economies of scale in R&D and accumulated production experience (i.e., learning effects) through OI (Torkkeli *et al.*, 2009). An ability to identify, acquire and assimilate knowledge from external sources, recombine it with the existing internal knowledge and apply this new knowledge to commercial ends enhances a firm's revenue potential and, ultimately, business performance (e.g., Tsai, 2001). Recent empirical evidence provides support for the positive effect of innovation on firms' financial performance. For example, Spithoven *et al.* (2013) report that SMEs experience a substantial effect on firm performance from the sale of new products/services generated through OI. Kostopoulos *et al.* (2011) suggest that while innovation may initially have no (or even a negative) effect on firm performance, over time, it enhances financial performance. Yet on the basis of a meta-analysis of empirical research on the innovation–performance relationship in SMEs, Rosenbusch *et al.* (2011, p. 453) "find that the innovation projects that focus on external collaboration do not increase the performance of SMEs". However, this study did not distinguish between different types of external collaborations, and whether external collaboration is beneficial to SME performance may to a large extent depend upon the types of external partners SMEs collaborate with (Belderbos *et al.*, 2004; Rosenbusch *et al.*, 2011).

Using longitudinal data from 1,435 SMEs in Australia, Gronum *et al.* (2012) observe a positive relationship between innovation and firm performance, but caution that SMEs should only focus on cultivating and maintaining external relationships if they lead to innovations. Ahn *et al.* (2015) examined the relationship between OI partners and firm performance and found that broad and intensive engagement with non-competing partners, such as customers, consultancy/intermediaries and public research institutes are positively associated with firm performance. A theme emerging from the recent literature then, is the need to distinguish between different types of external innovation collaborations, since, as noted above, the types of external partners SMEs collaborate with may influence performance (Belderbos *et al.*, 2004; Rosenbusch *et al.*, 2011).

Even though the above discussion points to a complex relationship between innovativeness and performance, we posit that in the SME context, a firm's innovativeness due to OI will improve a firm's performance. There exists substantial empirical evidence that innovativeness is an important correlate or determinant of firm performance (e.g., Crepon and Duguet, 1998; Klomp and van Leeuwen, 2001, with some evidence showing that innovators are persistently more profitable than non-innovators (e.g., Love *et al.*, 2009).

H5: *A positive relationship exists between innovation and firm performance.*

Figure 1 presents the conceptual framework that depicts the relationships between knowledge inflows from market- and science-based actors, absorptive capacity, innovation and firm performance.

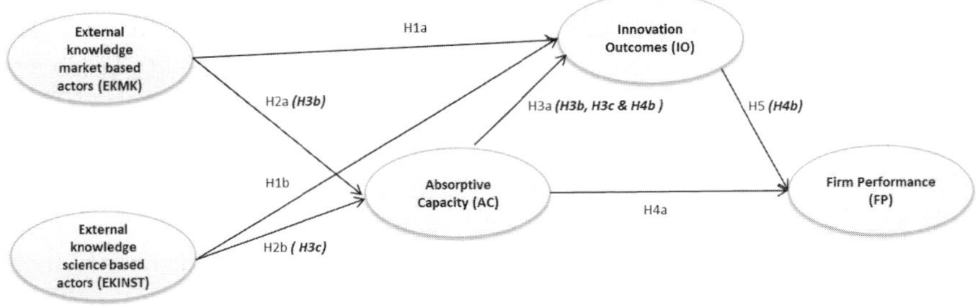

Note: H represents hypothesis and H in bold represents indirect (mediating) effect hypothesis.

Fig. 1. Conceptual framework.

Research Methodology

Sample

The study used a purpose-designed self-administered questionnaire conducted via the internet to survey SME owner/managers in the Adelaide Metropolitan area of South Australia (the capital of South Australia has a population of approximately 1.3 million people). The data collection was undertaken during 2007. SMEs are firms with less than 200 employees (ABS, 2007). The questionnaire was an adaptation of the Australian Innovation Survey (ABS, 2003) which was based on the Organisation for Economic Co-operation and Development's (OECD) Oslo Manual. Some minor modifications and additional questions were introduced in some areas to meet the study's research objectives. Questions relevant to this study included business demographics and other firm-level characteristics, sources and methods of external knowledge inflows, innovation outcomes and firm performance.

The population for the study consists of all SMEs in the Adelaide Metropolitan area.

The South Australian State Government supplied the dataset from which the sample was selected. The dataset was originally developed by the Australian Taxation Office (ATO) and includes all businesses registered in Australia. A subset of records was extracted from this database that met the following criteria: active businesses in the Adelaide Metropolitan area, entities with contact addresses with post (zip) codes applicable to the Adelaide metropolitan area, annual turnover of more than $50,000, Australian Business Number (ABN) registration for taxation purposes, and an email address. The sample was further restricted to commercial enterprises by excluding entities in agriculture, forestry and fishing, public administration and defence, education and training, health care and social assistance, and other services. This population subset comprises 14,206 Adelaide businesses. Of these, 33.7% (4788) of the email addresses proved to be inoperable. This resulted in a final Adelaide Metropolitan area population dataset comprising 9418 active firms.

A pilot study was first undertaken with a group of SMEs to determine SME reaction to the survey and to ensure that participants understood the instructions and survey questions. A personalised (i.e., with individual contact and company details) email invitation was sent to participants with a link to the survey. The email also included a unique ID and password to access the survey. This was accompanied by a cover letter from the local government municipalities and State Government that provided support for the project, encouraged participation, and explained the project's rationale. Clear instructions and explanations appeared in the cover letter and questionnaire (based upon feedback received from the pilot). Prospective respondents were also encouraged to view the website where an overview of the project could be found. A tailored, individual report of the research results for each respondent was also offered. These techniques were aimed at capturing the attention of the responders and encouraging questionnaire completion (Newby *et al.*, 2003).

A total of 1226 questionnaires were received, however only 838 were used in the analysis because some questionnaires contained missing data. Thus, the active response rate was 8.9% (Neuman, 2000). There is potential common method bias in single person per firm surveys of this type, but single-person or single unit surveys are a common tool for business research as it provides a quick and inexpensive way of obtaining information (Zikmund, 2000).

Variable definition and measurement

Measures used in this research are ordinal and categorical based questions. Some questions are transformed to summary measures (i.e., counts as explained below — a more detailed explanation is provided in Appendix A). A benefit of this approach is that there are fewer variables in the model a drawback is that some detail is lost.

Independent variables

External knowledge inflows from market- and science-based actors are two latent variables with three indicators each respectively. The three indicators for External Knowledge Inflows from Market-based actors (EKINST) and External Knowledge from Science-based actors are equivalent to summative indexes based on several questions as outlined below.

External knowledge inflows from market-based actors (*EKMK*): The first indicator is a count of the number of methods/channels of external knowledge inflows a firm had used from market-based actors (Moilanen *et al.*, 2014). It has a possible range of 0–15. The second indicator is a count of whether the firms had used sources of ideas from market-based actors for innovation activities (Veugelers and Cassiman, 1999). It has a possible range of 0–4. The third indicator is also a count of which market-based actors the firms collaborated with (Escribano *et al.*, 2009). It has a possible range of 0–6.

External knowledge from science-based actors (*EKINST*): The first indicator is a count of the number of methods/channels of external knowledge inflows a firm had used from science-based actors (Moilanen *et al.*, 2014). It has a possible range of 0–14. The second indicator is a count of whether the firms had used sources of ideas from science-based actors for their innovation activities (Veugelers and Cassiman, 1999). It has a possible range of 0–4. The third indicator is also a count of which science-based actors the firms collaborated with (Escribano *et al.*, 2009). It has a possible range of 0–8.

Mediating variables

Absorptive Capacity (*AC*) is a latent variable with six indicators. The six indicators are dummy variables that equal 1 if the firms had engineering or scientific skills (Grimpe and Sofka, 2009) or internal R&D activities, and zero otherwise.

Innovation Outcomes (*IO*) is a latent variable with four indicators. The four indicators are dummy variables that equal 1 if the firms undertake different types of innovations including product, operational, organisational and/or managerial methods, and marketing methods (ABS, 2003, 2005; OECD, 2005), and zero otherwise.

Dependent variable

Firm performance (*FP*) is a multidimensional construct (Huang and Rice, 2009; Lumpkin and Dess, 1996; Orser *et al.*, 2000; Wiklund and Shepherd, 2005). The first indicator is an ordinal variable which asks about firms' sales growth in 2005–06 compared with 2004–05. Sales growth scale items range from "negative" to "more than 35%". The second indicator is a (reverse coded) ordinal variable which asks about firms' market share in 2005–2006 compared with 2004–2005. The scale of the items ranged from -10% to 10%.

Analysis

Table 1 shows the means, standard deviations and correlations for all latent variables included in the analysis.

We use structural equation modelling (SEM) to analyse the measurement and structural models. SEM, a multivariate statistical technique, is appropriate for the analysis of our model as dependent variables for one regression become independent variables in the next one (Hair *et al.*, 2005). LISREL version 8.80

Table 1. Means, standard deviations and correlations.

Parameter path	Mean*	Std. dev.	EKMK	EKINST	AC	IO	FP
EKMK	2.723	2.242	1				
EKINST	0.897	1.634	0.6762	1			
AC	0.553	0.859	0.3891	0.5061	1		
IO	2.163	1.915	0.7248	0.396	0.456	1	
FP	4.287	2.781	0.3237	0.1918	0.242	0.4425	1

*Factor scores, and their means generated from the estimated equations:

(Jöreskog and Sörbom, 2006) was used in the analyses as it is able to analyse non-normal non-continuous data.

Structural model results

The analyses of the model were undertaken using the diagonally weighted least squares (DWLS) method based on the matrix of correlation (Pearson product moment, polychoric or polyserial as appropriate). DWLS provides more accurate parameter estimates where the assumption of "multivariate normality is violated and/or data are ordinal" (Mîndrilă, 2010).

The full structural model — incorporating external knowledge inflows, market- and science-based actors, absorptive capacity, innovation and firm performance — was evaluated. The model fit statistics suggest that the data fits the full structural model. The χ^2 was 458.81 and df 127, although χ^2/df is not in the generally acceptable range between 1 and 2 there is some agreement that χ^2 statistic is sensitive to sample size and in the case of samples with $N > 200$ the statistic has a tendency to indicate a poor model even when the model is satisfactory (Kelloway, 1998; Kenny, 2014). Thus, Kelloway (1998, p. 26) states, it is "highly unlikely that you will be able to obtain a non-significant test statistic with large sample sizes" and hence χ^2 may be an unrealistic statistic for finding well-fitting models. Other measures of model fit are used in our analysis such as RMSEA (root mean square error of approximation) = 0.056, AGFI (adjusted goodness-of-fit index) = 0.94 and CFI (comparative fit index) = 0.96. An RMSEA value less than or equal to 0.08 indicates an acceptable fit (Bollen, 1990). CFI is greater than 0.95 and AGFI is greater than 0.80 (Schreiber *et al.*, 2006), indicating a reasonable fit of the model. Overall, the fit indices suggest that the model fits the data satisfactorily and hence is useful for explaining the relationships between the latent variables.

Notwithstanding the non-significance of three paths between latent variables, they are maintained in the model. First, as outlined previously, theory supports the inclusion of these paths. Second, paths from external knowledge inflows from market-based actors to absorptive capacity and external knowledge inflows from science-based actors to innovation while non-significant at the 5% level are borderline significant (specifically, external knowledge from market-based actors to absorptive capacity has a p-value of 0.055; external knowledge inflows from science-based actors to innovation has a p-value of 0.105) and a conservative statistical approach could argue these two paths, at least, should be retained. Third, testing the nested model with the paths excluded resulted in models in which the absolute size of the remaining path coefficients changed somewhat but did not alter the conclusions that were to be drawn based on the reported model. Fourth, the reported model explains the theory rather than simply describe the data as it does in the alternative nested model.

The complete structural model relationships were examined. Table 2 presents the full structural model results. This shows the extent to which the latent variables (external knowledge inflows from market- and science-based actors, absorptive capacity and innovation) influence firm performance. Each of the individual hypotheses was tested and confirmed or rejected based on the significance of the relationships identified in the results in Table 2.

As shown in Table 2 and Fig. 2, the results indicate the following:

- H1a is supported as external knowledge inflows from market-based actors were positively related to innovation outcomes ($\gamma 11 = 0.82$, $p < 0.001$).

Table 2. Hypothesis testing.

Hypotheses	Parameter path	Direct	Indirect	Total	Hypothesis test
H1a/H3b	EKMK → (- ->) IO	0.82**	0.02	0.84**	Accepted/Rejected
H1b/H3c	EKINST → (- ->) IO	−0.30	0.13*	−0.17	Rejected/Accepted
H2a	EKMK → AC	0.09		0.09	Rejected
H2b	EKINST → AC	0.45**		0.45**	Accepted
H3a	AC → IO	0.29*		0.29*	Accepted
H4a/H4b	AC → (- ->) FP	0.05	0.12*	0.17*	Rejected/Accepted
H5	IO → FP	0.42**		0.42**	Accepted

**$P < 0.001$ and *$P < 0.01$. Non-significance at the 1% level or better (note that for 2 paths, there is significance at the 10% level.

- -> indicates indirect effects.

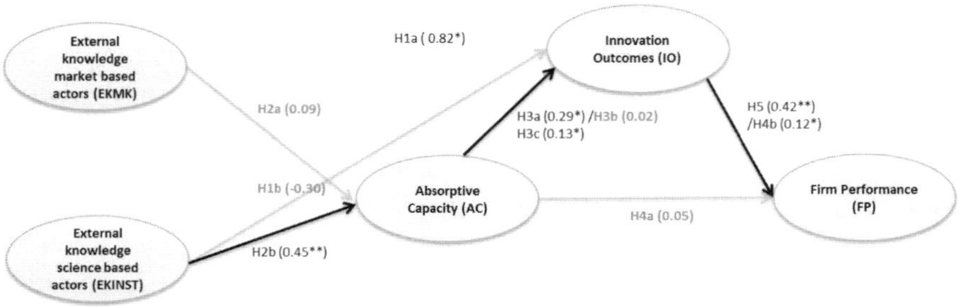

Note: For simplicity reasons, error variances are not shown in this diagram. $**P < 0.001$ and $*P < 0.01$. Bold H represents indirect (mediating) effect. Non-significance at the 1% level or better (note that for 2 paths, there is significance at the 10% level.

Fig. 2. Path diagram.

- H1b is rejected as external knowledge inflows from science-based actors had no significant influence on innovation outcomes ($\gamma = -0.30$, $p > 0.05$).
- H2a is rejected as external knowledge inflows from market-based actors had no significant influence on absorptive capacitive ($\gamma = 0.09$, $p > 0.05$).
- H2b is supported as external knowledge inflows from science-based actors were positively related to absorptive capacity ($\gamma = -0.45$, $p < 0.01$).
- H3c is supported as absorptive capacity was significantly positively related to innovation ($\beta = 0.29$, $p < 0.01$).
- H4a is rejected as absorptive capacity had no significant influence on firm performance ($\beta = 0.05$, $p > 0.05$).
- H5 is supported as innovation outcomes were significantly positively related to firm performance ($\beta = 0.42$, $p < 0.01$).

Mediation effects

The following hypotheses positing mediation effects were also tested: H3a states that absorptive capacity mediates the effects of external knowledge inflows from market-based actors on innovation, H3b proposes that absorptive capacity mediates the effects of external knowledge inflows from science-based actors on innovation, and H4b states that innovation mediates the effects of absorptive capacity on firm performance. To test for mediation effects we examine the direct and indirect effects in the path analysis following the related procedures in LISREL.

As shown in Table 2 and Fig. 2 the indirect effect of external knowledge inflows from market-based actors on innovation is non-significant ($\beta = 0.02$,

$p > 0.05$), rejecting H3b. In contrast, the indirect effect of external knowledge inflows from science-based actors on innovation is significant ($\beta = 0.13$, $p < 0.05$), while the corresponding direct effect is not (see H1b). This indicates that absorptive capacity mediates the effects of external knowledge inflows from science-based actors on innovation, providing support for H3c. Similarly, the indirect effects of absorptive capacity on firm performance is significant ($\beta = 0.12$, $p < 0.05$), while the direct effect is not (see H4a), suggesting that innovation mediates the relationship between absorptive capacity and firm performance, in line with hypothesis H4b.

Finally, additional analyses were performed in order to determine whether the total effects of paths between variables differed. No statistically significant differences between comparable paths were identified.

Discussion and Conclusion

The purpose of this study was to examine how, and the extent to which, external knowledge inflows from specific external actors influence absorptive capacity, innovation and firm performance in SMEs. Although theoretical models identify external knowledge inflows as antecedents of absorptive capacity and innovation and performance as outcomes (Lane *et al.*, 2006; Zahra and George, 2002) there is a dearth of empirical studies (Kostopoulos *et al.*, 2011), particularly in the SME context, that delineate the relationships between external knowledge inflows from science- and market-based actors, absorptive capacity, innovation and firm performance. The current study contributes to filling this gap in the extant literature in a number of ways.

First, the study investigated the direct and indirect effects of external knowledge inflows from market- and science-based actors on absorptive capacity, innovation and firm performance, thereby adding to our understanding of the mechanisms behind, and the effect of, external knowledge inflows from specific actors (Moilanen *et al.*, 2014). We find that the effects of knowledge inflows from specific external actors on innovation outcomes vary, with external knowledge inflows from market-based actors exerting a significant positive direct effect on innovation. In contrast, the direct effect of knowledge inflows from science-based actors on innovation is not significant, indicating that the direct effect of knowledge inflows from market-based actors on innovation is greater. This result is in line with a number of studies that indicate that SMEs are more likely to utilise market-based actors vis-à-vis science-based actors (Faems *et al.*, 2005; Huggins and Johnston, 2009; Hughes *et al.*, 2009; Lasagni, 2012; Laursen and

Salter, 2006; Mina *et al.*, 2014; West and Bogers, 2014). However, it goes against one of the central tenets of absorptive capacity (Cohen and Levinthal, 1989; Zahra and George, 2002) — that firms cannot benefit from external knowledge flows merely by being exposed to them (Cohen and Levinthal, 1989, 1990) and instead, must develop and utilise "absorptive capacity".

Thus external knowledge inflows from market-based actors influence innovation directly, or what Zahra and George (2002, p. 190) characterise as "serendipitously", and more recently, Love and Roper (2015) refer to as unanticipated knowledge flows. A possible explanation is that because SMEs are accessing knowledge from customers and suppliers who are part of their value chain the knowledge is "ready to use" — for example, knowledge accessed as part of the package, or as a by-product, with the acquisition of new equipment from suppliers.

In contrast, external knowledge inflows from science-based actors influence innovation indirectly via absorptive capacity, which is consistent with the fundamental tenet of absorptive capacity theory noted above (Cohen and Levinthal, 1989; Zahra and George, 2002) and some recent empirical research that demonstrates a firm's absorptive capacity mediates the relationship between external knowledge inflows in general and innovation (Kostopoulos *et al.*, 2011). But while theory suggests that firms derive innovation benefits from new external knowledge only if they recognise its value, internalise and exploit it (Cohen and Levinthal, 1990; Zahra and George, 2002) our results suggest that there is a need to delineate external knowledge flows by specific actors, as the circumstances in which absorptive capacity transforms external knowledge inflows into innovation are not universal, in line with recent work (Moilanen *et al.*, 2014).

Second, we find that external knowledge inflows from science-based actors are positively associated with absorptive capacity, in line with Moilanen *et al.* (2014) finding. In contrast, the relationship between external knowledge inflows from market-based actors and absorptive capacity was non-significant. This was unexpected, given the literature tends to indicate that SMEs are more likely to acquire and absorb knowledge from market-based actors (Spithoven *et al.*, 2013), who, as part of their value chain, are cognitively closer than their science-based counterparts (Knoben and Oerlemans, 2006).

Finally, the results reveal that the effect of absorptive capacity, in turn, is twofold: absorptive capacity has a positive direct effect on innovation outcomes, and an indirect effect on firm performance through innovation, consistent with theoretical expectations (Cohen and Levinthal, 1989; Zahra and George, 2002) and some empirical evidence (Kostopoulos *et al.*, 2011). These results, together with the indirect effect of knowledge inflows from science-based actors on innovation via absorptive capacity, underline the importance of absorptive capacity to innovation outcomes and ultimately, firm performance. We elucidate the implications of these findings next.

Implications, Limitations, and Future Research

The findings are of interest to managers because they contribute to a better understanding of the extent to which, and how, external knowledge inflows from science- and market-based external actors directly and indirectly influence absorptive capacity, innovation and firm performance differentially. The implication for managers is the need to invest in and develop the firm's absorptive capacity, in order to realise the benefits of knowledge inflows from science-based actors, since the study clearly shows that absorptive capacity is a precondition for external knowledge inflows from science-based actors influencing innovation. The direct effect of absorptive capacity on innovation, and indirect effect on firm performance via innovation, underscores the need for investment in and development of this capacity in SMEs. Conversely, we found that knowledge inflows from market-based actors influence innovation directly which underlines the importance of the supply chain as a source of valuable knowledge for innovation outcomes.

From a policy perspective, government agencies that focus on SME development should introduce programs or activities to facilitate SME access to knowledge from science-based actors, given its positive relationship to innovation outcomes. As noted, such SMEs would, of course, require absorptive capacity in order to make use of this type of knowledge. This finding also has research implications. Researchers should continue to delineate the role of absorptive capacity, in particular, under what circumstances absorptive capacity transforms external knowledge inflows into innovation outcomes, since the results of this study, together with recent empirical evidence (Moilanen *et al.*, 2014) suggests the need to make a clear distinction between knowledge inflows from particular external actors. As long as theory and empirical research does not make the distinction between such knowledge inflows, it may be difficult both for scholars and practitioners to fully understand the relationships between external knowledge inflows, absorptive capacity and innovation outcomes.

There are some limitations associated with the survey due to its cross-sectional nature which affects the explanatory power of the dependent variable which may have lag effects; this can be overcome by adopting longitudinal panel data in future research. We also acknowledge that there are sampling issues as the sample is taken from one Australian state, thereby raising issues of regional bias, that can also be addressed in future research by employing a national survey.

Acknowledgments

The authors appreciate the financial and other support from the South Australian Government's Department of Trade and Economic Development, and Marion, Onkaparinga, Salisbury, and Playford Municipal Councils which made this research possible. They are also appreciative of the access provided to them by the Australian Bureau of Statistics.

Appendix A. Variables Used in the Analysis

A.1. External knowledge

External knowledge inflows from market-based actors (EKMK) and External knowledge from science-based actors (EKINST. These are the variable labelled "EK" in the conceptual model and are subdivided into two groups from market-based actors (EKMK) and from institutional actors (EKINST).

External knowledge from market-based actors (EKMK): The first indicator is a count of the number of methods/channels of external knowledge inflows a firm had used from market-based actors (Moilanen *et al.*, 2014). It has a possible range of 0–15. This dimension reflects the types of methods/channels from market-based actors that the firms have used during a two-year period (2005–2006). Participants were asked which types of methods were used and the geographical location of them (see question 18 below). This construct is a behaviour variable and a list of items for selection was provided to the firms. Question 18 corresponds with these variables. Each type of method included a list where firms could select their options. The question was:

18. Did your business use any of the following methods to acquire knowledge or abilities for innovative purposes during 2005 and 2006?

The second indicator is a count of whether the firms had used sources of ideas from market-based actors for innovation activities (Veugelers and Cassiman, 1999). It has a possible range of 0–4. This dimension reflects the use of informal sources of ideas from market-based actors that the firms have used during a two-year period (2005–2006). Participants were asked if they used informal sources of ideas from market-based actors and for what type of innovation activities (see

Methods to Acquire Knowledge/ Abilities	From within the same state or territory	From elsewhere in Australia	From Overseas
Employed new skilled staff	❒	❒	❒
Interchanged staff with another business	❒	❒	❒
Used consultants (or other paid advisors)	❒	❒	❒
Acquired new equipment or technology for producing the business' goods or services	❒	❒	❒
Merger/takeover with/of another business (in whole or part)	❒	❒	❒

question 13 below). This construct is a behaviour variable and a list of items for selection was provided to the firms. Question 13 corresponds with these variables. Each type method included a list where firms could give their options (only one part of the question was used for this indicator). The question was:

13. Which of the following were the key sources of ideas or information that helped your business in 2005 and 2006 (in the indicated areas)?

	New Products	New Operational Processes	New Organisational/ Managerial Methods	New Marketing Methods
Marketing Sources • Clients or customers • Suppliers of equipment, materials, components or software • Consultants (including paid professional advice of all kinds) • Competitors and other businesses from the same industry	❒	❒	❒	❒

The third indicator is also a count of which market-based actors the firms collaborated with. It has a possible range of 0–6. This dimension reflects which actors had collaborated with the firms during a two-year period (2005–2006). Participants were asked the type of organisations with which they collaborated and their location (see question 21 below). This construct is a behaviour variable and a list of items for selection was provided to the firms. Question 21 corresponds with these variables. Each type of method included a list where firms could select their options (only one part of the question was used for this indicator). The question was:

21. Please indicate the types of organisation and their location with which your business collaborated in 2005 and 2006?

Collaboration means participation in shared projects between your business and other companies or organisations. It is not necessary that both parties obtain benefit from the collaboration

Type of Organisation	From Australia	From Overseas
Suppliers of equipment, materials, components, or software	❐	❐
Clients or customers	❐	❐
Competitors and other businesses from the same industry	❐	❐

***External Knowledge from Institutional Based Actors* (*EKINST*):** the first indicator is a count of the number of methods/channels of external knowledge inflows a firm had used from science-based actors. It has a possible range of 0–14. This dimension reflects the types of methods/channels from institutional based actors that the firms have used during a two-year period (2005–2006). Participants were asked about which types of methods were used and their geographical location (see question 19 below). This construct is a behaviour variable and a list of items for selection was provided to the firms. Question 19 corresponds with these variables. Each type of method included a list where firms could select their options. The question was:

19. Did your business use any of the following methods to acquire knowledge or abilities from higher education or research institutions for innovative purposes during 2005 and 2006?

Methods to Acquire Knowledge/Abilities	From Australia	From Overseas
Employed new graduate(s)	❐	❐
Employed academic or research staff	❐	❐
Used research results published by these institutions	❐	❐
Used research facilities of these institutions	❐	❐
Used patents, designs, or other intellectual property rights initially from these institutions	❐	❐
Used consultants from these institutions	❐	❐
Contracted out research and development to these institutions	❐	❐

The second indicator is a count of whether the firms had used sources of ideas from science-based actors for their innovation activities. It has a possible range of 0–4. This dimension reflects the use of informal sources of ideas from institutional based actors that the firms have used during a two-year period (2005–2006). Participants were asked if they used informal sources of ideas from institutional-based actors and for what type of innovation activities (see question 13 below). This construct is a behaviour variable and a list of items for selection was provided to the firms. Question 13 corresponds with these variables. Each type of method included a list where firms could select their options (only one part of the question was used for this indicator). The question was:

13. Which of the following were the key sources of ideas or information that helped your business in 2005 and 2006 (in the indicated areas)?

	New Products	New Operational Processes	New Organisational/ Managerial Methods	New Marketing Methods
Institutional Sources • Universities or other higher education institutions • Government agencies • Private non-profit research institutions • Commercial laboratories/ research and development enterprises	❐	❐	❐	❐

The third indicator is also a count of which science-based actors the firms collaborated with. It has a possible range of 0 to 8. This dimension reflects which actors had collaborated with the firms during a two-year period (2005–2006). Participants were asked the type of organisations with which they collaborated and their location (see question 21 below). This construct is a behaviour variable and a list of items for selection was provided to the firms. Question 21 corresponds with these variables. Each type of method included a list where firms could select their options (only one part of the question was used for this indicator). The question was:

21. Please indicate the types of organisation and their location with which your business collaborated in 2005 and 2006?

Collaboration means participation in shared projects between your business and other companies or organisations. It is not necessary that both parties obtain benefit from the collaboration

Type of Organisation	From Australia	From Overseas
Universities or other higher education institutions	❒	❒
Government agencies	❒	❒
Private non-profit research institutions	❒	❒
Commercial laboratories/research and development enterprises	❒	❒

A.2. Absorptive capacity (AC)

AC is a latent variable with six indicators. The six indicators are dummy variables that equal 1. The indicators used in this latent variable come from two different questions.

The first group is connected with specialised skills or capabilities (see question 9 below). This dimension reflects if the firms employed engineers and/or scientists during a two-year period (2005–2006). This construct is a behaviour variable and a list of items for selection was provided to the firms. Question 9 corresponds with these variables (only one part of the question was used for this indicator). The question was:

9. Which of the following skills and capabilities did your business look for when engaging people for innovation during the past two years?

❒ Engineering
❒ Scientific

The second group is connected to R&D intensity. This dimension reflects if the firms have undertaken R&D activities during a two-year period (2005–2006). Participants were asked about R&D activities (see question 10 below). This construct is a behaviour variable and a list of items for selection was provided to the firms. Question 10 corresponds with these variables. The question was:

10. Which of the following *formal* methods were used to protect your business' intellectual property during 2005 and 2006?

- ❐ Patents
- ❐ Registered design
- ❐ Copyright or trademark
- ❐ Other formal method (please specify)

A.3. Innovation Outcomes (IO)

OI involved four indicators — dummy variables that equal 1. The firms were asked if they undertook different types of innovations including product, operational, organisational and/or managerial methods, and marketing methods. This construct is a behaviour variable and a list of items for selection was provided to the firms. Question 6 corresponds with these variables. The question was:

6. During 2005 and/or 2006, has your business introduced any new or significantly improved innovations in the following areas?

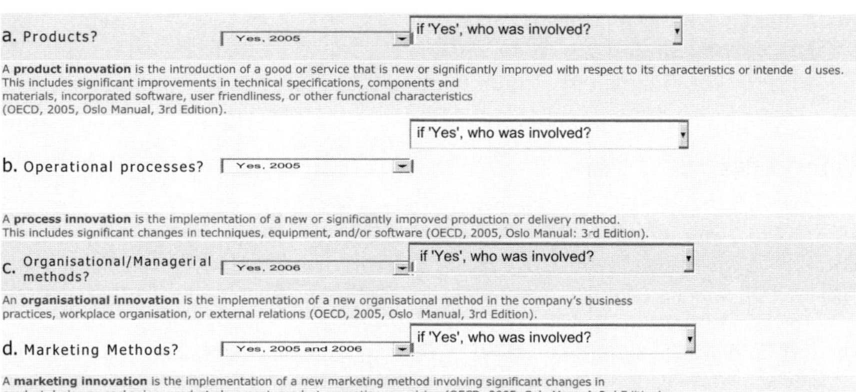

a. Products? Yes, 2005 if 'Yes', who was involved?

A **product innovation** is the introduction of a good or service that is new or significantly improved with respect to its characteristics or intende d uses. This includes significant improvements in technical specifications, components and materials, incorporated software, user friendliness, or other functional characteristics (OECD, 2005, Oslo Manual, 3rd Edition).

b. Operational processes? Yes, 2005 if 'Yes', who was involved?

A **process innovation** is the implementation of a new or significantly improved production or delivery method. This includes significant changes in techniques, equipment, and/or software (OECD, 2005, Oslo Manual: 3rd Edition).

c. Organisational/Manageri al methods? Yes, 2006 if 'Yes', who was involved?

An **organisational innovation** is the implementation of a new organisational method in the company's business practices, workplace organisation, or external relations (OECD, 2005, Oslo Manual, 3rd Edition).

d. Marketing Methods? Yes, 2005 and 2006 if 'Yes', who was involved?

A **marketing innovation** is the implementation of a new marketing method involving significant changes in product design or packaging, product placement, product promotion, or pricing (OECD, 2005, Oslo Manual, 3rd Edition).

A.4. Firm performance (FP)

The first indicator is an ordinal variable which asks about firms' sales growth in 2005–2006 compared with 2004–2005. Sales growth scale items range from "negative" to "more than 35%". The question was:

34. What was your business' sales growth in 2005–2006 (compared with 2004–2005)?

- ⦿ Negative
- ❍ No growth
- ❍ Under 5%
- ❍ 5% to 9%
- ❍ 10% to 19%
- ❍ 20% to 34%
- ❍ 35% to 50%
- ❍ More than 50%

The second indicator is a (reverse coded) ordinal variable which asks about firms' market share in 2005–2006 compared with 2004–2005. The scale of the items ranged from −10% to 10%. The question was:

33. Has the market share of your business in 2005–06 (compared with 2004–2005):

- ❍ Increased significantly (more than 10%)
- ❍ Increased slightly (1% to 10%)
- ❍ Stayed roughly the same
- ❍ Decreased slightly (−1% to −10%)
- ❍ Decreased significantly (more than −10%)

References

ABS (2003). *Innovation Survey 2003*. Canberra: Australian Bureau of Statistics'.

ABS (2005). *Innovation in Australian Business*. Canberra: Australian Bureau of Statistics.

ABS (2007). *Counts of Australian Businesses, Including Entries and Exits*. Canberra: Australian Bureau of Statistics.

Ahn, JM, T Minshall and L Mortara (2015). Open innovation: A new classification and its impact on firm performance in innovative SMEs. *Journal of Innovation Management*; 3(2), 33–54. Available at SSRN: http://ssrn.com/abstract=2431205 or http://dx.doi.org/10.2139/ssrn.2431205.

Barney, J (1991). Firm resources and sustained competitive advantage. *Journal of Management*, 17, 99–120.

Belderbos, R, M Carree and B Lokshin (2004). Cooperative R&D and firm performance. *Research Policy*, 33, 1477–1492.

Bianchi, M, A Cavaliere, D Chiaroni, F Frattini and V Chiesa (2011). Organisational modes for open innovation in the bio-pharmaceutical industry: An exploratory analysis. *Technovation*, 31, 22–33.

Bollen, KA (1990). Overall fit in covariance structure models: Two types of sample size effects. *Psychological Bulletin*, 107, 256–259.

Boschma, RA (2005). Proximity and innovation: A critical assessment. *Regional Studies*, 39, 61–74.

Chesbrough, H (2003). *Open Innovation: The New Imperative for Creating and Profiting from Technology*. Cambridge, MA: Harvard Business School Press.

Chesbrough, H (2006). Open innovation: A new paradigm for understanding industrial innovation. In: *Open Innovation. Researching a New Paradigm*, Chesbrough, H, W Vanhaverbeke and J West (eds.), pp. 1–14. Oxford: Oxford University.

Chesbrough, H and A Crowther (2006). Beyond high tech: Early adopters of open innovation in other industries. *R&D Management*, 36, 229–236.

Christensen, JF, MH Olesen and JS Kjær (2005). The industrial dynamics of Open innovation — evidence from the transformation of consumer electronics. *Research Policy*, 34, 1533–1549.

Cohen, W and D Levinthal (1989). Innovation and learning: The two faces of R&D. *The Economic Journal*, 99, 569–596.

Cohen, W and D Levinthal (1990). Absorptive Capacity: A new perspective on learning and innovation. *Administrative Science Quarterly*, 35, 128–152.

Cohen, WM, RR Nelson and JP Walsh (2002). Links and impacts: The influence of public research on industrial R&D. *Management Science*, 48, 1–23.

Crepon, B, E Duguet and J Mairesse (1998). Research innovation and productivity: An econometric analysis at the firm level. *Economics of Innovation and New Technology*, 7, 115–158.

Cutler, T (2008). The Cutler report and its implications for R&D in Australia. In Department of Innovation, I, Science and Research.

Danneels, E (2002). The dynamics of product innovation and firm competences. *Strategic Management Journal*, 23, 1095–1121.

Davenport, S (2005). Exploring the role of proximity in SME knowledge-acquisition. *Research Policy*, 34, 683–701.

Department of Industry (2014). Australian Innovation System Report 2014: Annual Series Office of the Chief Economist. Department of Industry.

Du, J, B Leten and W Vanhaverbeke (2012). Does open innovation improve the performance of R&D projects? In *Paper Presented at the 'Open Innovation: NewInsights and EvidenceUK-IRC Conference*. London: Imperial College London.

Enkel, E, O Gassmann and H Chesbrough (2009). Open R&D and open innovation: Exploring the phenomenon. *R&D Management*, 39, 311–316.

Escribano, A, A Fosturi and J Tribo (2009). Managing external knowledge flows: The moderating role of absorptive capacity. *Research Policy*, 38, 96–105.

Faems, D, B Van Looy and K Debackere (2005). Interorganizational collaboration and innovation: Toward a portfolio approach. *Journal of Product Innovation Management*, 22, 238–250.

Fernández-Esquinas, M, H Pinto, MP Yruela and TS Pereira (2015). Tracing the flows of knowledge transfer: Latent dimensions and determinants of university–industry interactions in peripheral innovation systems. *Technological Forecasting and Social Change*.

Flatten, T, A Engelen, SA Zahra and M Brettel (2011). A measure of absorptive capacity: Scale development and validation. *European Management Journal*, 29, 98–116.

Fosfuri, A and JA Tribó (2008). Exploring the antecedents of potential absorptive capacity and its impact on innovation performance. *Omega*, 36, 173–187.

Gassmann, O, E Enkel and H Chesbrough (2010). The future of open innovation. *R&D Management*, 40, 213–221.

Grant, RM (1996). Prospering in dynamically-competitive environments: Organizational capability as knowledge integration. *Organization Science*, 7, 375–387.

Grimpe, C and W Sofka (2009). Search patterns and absorptive capacity: Low-and high-technology sectors in European countries. *Research Policy*, 38, 495–506.

Gronum, S, ML Verreynne and T Kastelle (2012). The role of networks in small and medium-sized enterprise innovation and firm performance. *Journal of Small Business Management*, 50, 257–282.

Hair, JF, WC Black, BJ Babin, RE Anderson and RL Tathan (2005). *Multivariate Data Analysis*. Upper Saddle River, New Jersey: Prentice Hall.

Hargadon, A and RI Sutton (1997). Technology brokering and innovation in a product development firm. *Administrative Science Quarterly*, 42, 716–749.

Helfat, CEC (2006). Book review of open innovation: The new imperative for creating and profiting from technology. *Academy of Management Perspectives*, 20, 86.

Huang, F and J Rice (2009). The role of absorptive capacity in facilitating "open innovation" outcomes: A study of Australian SMEs in the manufacturing sector. *International Journal of Innovation Management*, 13, 201–220.

Huggins, R and A Johnston (2009). Knowledge networks in an uncompetitive region: SME innovation and growth. *Growth and Change*, 40, 227–259.

Hughes, T, N O'Regan and M Sims (2009). The effectiveness of knowledge networks: An investigation of manufacturing SMEs. *Education + Training*, 51, 665–681.

Jöreskog, KG and D Sörbom (2006). *LISREL 8.80 for Windows* [*Computer Software*].

Kang, K and J Kang (2009). How do firms source external knowledge for innovation? Analysing effects of different knowledge sourcing methods. *International Journal of Innovation Management*, 13, 1–17.

Kaufman, A, CH Wood and G Theyel (2000). Collaboration and technology linkages: A strategic supplier typology. *Strategic Management Journal*, 21, 649–663.

Kelloway, EK (1998). *Using LISREL for Structural Equation Modeling: A Researcher's Guide*: Sage.

Kenny, D (2014). Measuring model fit. Available at http://davidakenny.net/cm/fit.htm.

Klomp, L and G van Leeuwen (2001). Linking innovation and firm performance: A new approach. *International Journal of the Economics of Business*, 8, 343–364.

Knoben, J and LAG Oerlemans (2006). Proximity and inter-organizational collaboration: A literature review. *International Journal of Management Reviews*, 8, 71–89.

Kostopoulos, K, A Paplexandris, M Papachroni and G Ioannou (2011). Absorptive capacity, innovation, and financial performance. *Journal of Business Research*, 64, 1335–1343.

Lane, PJ, B Koka and S Pathak (2006). The reification of absorptive capacity: A critical review and rejuvenation of the contruct. *Academy of Management Review*, 31, 863–883.

Larsen, MT (2011). The implications of academic enterprise for public science: An overview of the empirical evidence. *Research Policy*, 40, 6–19.

Lasagni, A (2012). How can external relationships enhance innovation in SMEs? New evidence for Europe. *Journal of Small Business Management*, 50, 310–339.

Laursen, K and A Salter (2006). Open for innovation: The role of openness in explaining innovation performance among U.K. manufacturing firms. *Strategic Management Journal*, 27, 131–150.

Love, JH, S Roper and J Du (2009). Innovation, ownership and profitability. *International Journal of Industrial Organization*, 27, 424–434.

Love, JH and S Roper (2015). SME innovation, exporting and growth: A review of existing evidence. *International Small Business Journal*, 33, 28–48.

Lowik, S, D van Rossum, J Kraaijenbrink and A Groen (2012). Strong ties as sources of new knowledge: How small firms innovate through bridging capabilities. *Journal of Small Business Management*, 50, 239–256.

Lumpkin, GT and GG Dess (1996). Clarifying the entrepreneurial orientation construct and linking it to performance. *Academy of Management Review*, 21, 135–172.

Mackinnon, D, K Chapman and A Cumbers (2004). Networking, trust and embeddedness amongs SMEs in Aberdeen oil complex. *Entrepreneurship & Regional Development*, 16, 87–106.

Mina, A, E Bascavusoglu-Moreau and A Hughes (2014). Open service innovation and the firm's search for external knowledge. *Research Policy*, 43, 853–866.

Mîndrilă, D (2010). Maximum likelihood (ML) and diagonally weighted least squares (DWLS) estimation procedures: A comparison of estimation bias with ordinal and multivariate non-normal data. *International Journal of Digital Society*, 1, 60–66.

Moilanen, M, S Ostbye and K Woll (2014). Non-R&D SMEs: External knowledge, absoptive capacity and product innovation. *Small Business Economics*, 43, 447–462.

Morgan, R and C Turnell (2003). Market-based organizational learning and market performance gains. *British Journal of Management*, 14, 255–274.

Murray, S and J Peyrefitte (2007). Knowledge type and communication media choice in the knowledge transfer process. *Journal of Managerial Issues*, 19, 111–133.

Narula, R (2004). R&D collaboration by SMEs: New opportunities and limitations in the face of globalisation. *Technovation*, 24, 153–161.

Neuman, WL (2000). *Social Research Methods*. Needham Heights, MA: Allyn and Bacon.

Newby, R, J Watson and D Woodliff (2003). SME survey methodology: Response rates, data quality, and cost effectiveness. *Entrepreneurship Theory and Practice*, 28, 163–172.

Nooteboom, B (2000). *Learning and Innovation in Organizations and Economies*: OUP Oxford.

Nooteboom, B, W Van Haverbeke, G Duysters, V Gilsing and A Van den Oord (2007). Optimal cognitive distance and absorptive capacity. *Research Policy*, 36, 1016–1034.

OECD (2005). *Oslo Manual: Guidelines for Collecting and Interpreting Innovation Data*. Paris: OECD.

O'Regan, N, A Ghobadian and M Sims (2006). Fast tracking innovation in manufacturing SMEs. *Technovation*, 26, 251–261.

Orser, BJ, S Hogarth-Scott and AL Riding (2000). Performance, firm size, and management problem solving. *Journal of Small Business Management*, 38, 42–58.

Perkmann, M, V Tartari, M McKelvey, E Autio, A Broström, P D'Este, R Fini, A Geuna, R Grimaldi and A Hughes (2013). Academic engagement and commercialisation: A review of the literature on university–industry relations. *Research Policy*, 42, 423–442.

Pittaway, L, M Robertson, K Munir, D Denyer and A Neely (2004). Networking and innovation: A systematic review of the evidence. *International Journal of Management Reviews*, 5, 137–168.

Rosenbusch, N, J Brinckmann and A Bausch (2011). Is innovation always beneficial? A meta-analysis of the relationship between innovation and performance in SMEs. *Journal of Business Venturing*, 26, 441–457.

Saviotti, P (1998). On the dynamics of appropriability of tacit and of codified knowledge. *Research Policy*, 26, 843–856.

Schreiber, JB, A Norab, FK Stagec, EA Barlowb and J Kinga (2006). Reporting structural equation modeling and confirmatory factor analysis results: A review. *The Journal of Educational Research*, 99, 323–337.

Spithoven, A, W Vanhaverbeke and N Roijakkers (2013). Open innovation practices in SMEs and large enterprises. *Small Business Economics*, 1–26.

Teirlinck, P and A Spithoven (2013). Research collaboration and R&D outsourcing: Different R&D personnel requirements in SMEs. *Technovation*, 33, 142–153.

Terziovski, M (2010). Research notes and commentaries innovation practice and its performance implications in small and medium enterprises (SMEs) in the manufacturing sector: A resource-based view. *Strategic Management Journal*, 31, 892–902.

Theyel, N (2012). Extending open innovation throughout the value chain by small and medium-sized manufacturers. *International Small Business Journal*.

Todorova, G and B Durisin (2007). Absorptive capacity: Valuing a reconceptualization. *Academy of Management Review*, 32, 774–786.

Torkkeli, M, C Kock and P Salmi (2009). The "Open Innovation" paradigm: A contingency perspective. *Journal of Industrial Engineering and Management*, 326.

Tsai, W (2001). Knowledge transfer in intraorganizational networks: Effects of network position and absorptive capacity on business unit innovation and performance. *The Academy of Management Journal*, 44, 996–1004.

Uyarra, E (2010). Conceptualizing the regional roles of universities, implications and contradictions. *European Planning Studies*, 18, 1227–1246.

Veugelers, R and B Cassiman (1999). Make and buy in innovation strategies: Evidence from Belgian manufacturing firms. *Research Policy*, 28, 63–80.

Von Hippel, EA (2005). *Democratizing Innovation*. Cambridge, MA: MIT Press.

West, J and M Bogers (2014). Leveraging external sources of innovation: A review of research on open innovation. *Journal of Product Innovation Management*, 31, 814–831.

Wiklund, J and D Shepherd (2005). Entrepreneurial orientation and small business performance: A configurational approach. *Journal of Business Venturing*, 20, 71–91.

Zahra, SA and G George (2002). Absorptive capacity: A review, reconceptualization, and extension. *Academy of Management Review*, 27, 185–203.

Zikmund, WG (2000). *Business Research Methods*. Orlando: The Dryden Press.

Chapter 16

UNDERSTANDING RELATIONSHIPS BETWEEN UNIVERSITIES AND SMEs IN EMERGING HIGH TECHNOLOGY INDUSTRIES: THE CASE OF OPTO-ELECTRONICS

Chris Hendry

City University Business School, Barbican Centre

James Brown

City University Business School, Barbican Centre

Robert Defillippi

Suffolk University, Beacon Hill

It has long been recognised that the innovative and entrepreneurial capabilities of the small medium-sized enterprise (SME) sector can make an important contribution to the commercialisation of emerging technologies. In their role as centres of expertise and originators of new technical knowledge, universities are vital contributors to this process. Understanding the nature of relationships between universities and SMEs is therefore important, particularly in view of the fact that current theories on regional development suggest that concentrations of SMEs in certain regions, clustered around one or more university centres, can be effective locations for accelerating this process. As a counter to regional development theory, an alternative viewpoint is that the way emerging industries develop is affected more by the dynamics of industry life-cycles. The opto-electronics sector, which is characterised by regional clusters in the UK and USA, offers lessons for how SMEs and universities interact against a backdrop of these theories.

Keywords: Opto-electronics; technology transfer; universities, SMEs.

This paper was first published in *International Journal of Innovation Management*, Vol. 4, No. 1 (March 2000), pp. 51–75.

Introduction

Regional development policy in the UK over the last 30 years has shifted from an inward investment focus aimed at attracting multinationals into depressed areas, to the promotion of high-tech centres such as science parks, and efforts to create linkages between university research centres and local firms. As the shortcomings of the science park's simple linear model of technology transfer have become apparent (Massey *et al.*, 1992), regional strategy has embraced the idea of local clusters (Enright, 1995) with an emphasis on the formation and growth of high-technology-based firms operating in the same industry sector. The essence of the cluster concept is that it emphasises on how a concentration of small firms with a common and related set of technologies can build local skills and capabilities and stimulate the local economy through networked supply chain relationships (Porter and Sölvell, 1998).

Studies in innovation likewise stress the importance of external linkages and processes at all points along the technology transfer pathway (Tidd *et al.*, 1997). Innovation is seen increasingly as a multi-firm networking process (Rothwell, 1992), involving close collaboration between companies and a consequent linking of technology-push and market-pull factors. There is also a presumption that collaboration between universities and SMEs is desirable. However, Hoffman *et al.* (1998), echoing Oakey (1994), qualify this enthusiasm for networking with academics by pointing out that there are mixed views on the effectiveness of links with universities as sources of technology for SMEs. Tang *et al.* (1996) suggest this may be due to an inherent mismatch between the fundamental science research interests of universities and the market application needs of most SMEs.

The benefits of networking for knowledge diffusion and technology transfer are expected to be particularly strong when the participating firms are geographically concentrated. This is especially true at the early life-cycle stage of a new technology as at this point, the knowledge is tacit and best exchanged in face-to-face meetings (Swann *et al.*, 1998). However, recent empirical evidence has questioned the significance of regional clustering among new technology-based firms. Keeble (1994:205), for example, observes that large and small firms in high-tech industry have been "markedly more dispersed than clustered", and that this is particularly so for newly created (small) firms. Moreover, even where firms are concentrated, they may have more relationships outside their area than within it (Garnsey and Cannon-Brookes, 1993).

Evidence supporting the value of clustering for technologically based SMEs in the UK is thus somewhat contradictory. It is clear that there are a number of questions that need to be answered in the context of the UK in order to generate the type of clustering dynamic that has been seen in high-technology industries in the USA in such places as Silicon Valley (Saxenian, 1994). Temple (1998), for example, argues that although the UK has industrial concentrations or agglomerations, it lacks true clusters. He identifies three main drivers for the growth of clusters on which policy attention should be focused. These are: the life-cycle dynamics of clusters (in particular, achieving critical mass); the creation of a pool of innovative entrepreneurs and specialist supporting services; and the governance of clusters aimed at the creation of shared visions, based on trust and co-ordination regarding technological expectations. A key factor underlying these three ideas is the relationship between SMEs and universities. University research laboratories are a source of new technology, entrepreneurial talent and, in the early stages of the life-cycle, "untraded interdependencies" (Storper, 1995). As the industry enters later stages of the life-cycle, universities can also be a source of continued technological expertise.

In this paper, we examine the nature of SME–university relationships from an industry life-cycle perspective by analysing developments in one high technology sector, namely opto-electronics, in the UK and USA. Most life-cycle theories are derived from the original product life-cycle model, and in pointing this out, Malecki (1997:70) concludes that "while they do not provide universal applicability, they capture the skill and knowledge differences between economic activities and types of products." We interpret this as meaning that although there may be differences between product sectors, the general four-stage model of innovation, growth, maturity and decline, provides a framework for evaluating the respective contributions to technological innovation from industry and academia.

Malecki hedges his bets about the universal application the of life-cycle theory and Autio (1997) develops this point by showing that specifically, it may not apply to the majority of new technology-based firms. The evolution of such firms is not predominantly one of growth in size but is rather the accumulation and evolution of technological resources. Nevertheless, this model still requires such firms to be active in exploiting external sources of technology and information, even if this results in dominance in the technological niche markets rather than conventional product life-cycle growth.

Much of the research in this area of linkages between universities and SMEs has been biased towards the university viewpoint — that is, how university researchers can overcome institutional barriers that inhibit them from working across the university–firm divide (Lee, 1996; Rosenberg and Nelson, 1994).

A balancing view from the SME perspective is needed on what SMEs expect to gain from relationships with universities and what inhibits them from entering such engagements. In this treatment of SME–university relations, we therefore have the aim of illustrating the uses SMEs make of universities during the various stages of an industry life-cycle and to highlight particular problems and issues that arise in these relationships. The broad question, which this paper addresses as a contribution to this debate, is thus, "how do high-technology SMEs in regional clusters in the UK and USA relate to university research centres for the purposes of innovation?"

Overview of Opto-Electronics

Opto-electronics (also known as photonics) has been defined as "the integration of optical and electronic techniques in the acquisition, processing, communication, storage and display of information" (ACOST, 1988). As such, it is a prime example of "technology fusion" (Dubarle and Verie, 1993; Kodama, 1992), involving the interaction of photons (particles of light) with electrons (particles of electricity). Photons travel at the speed of light and interact very slightly with material environments, so they are ideal for the transmission of information. Electrons interact strongly with each other and with most materials, so they can be finely controlled in ways suitable for information processing. Managing these processes and creating these effects require a deep understanding of the physics of light transmission, electronics and the development of highly refined advanced materials. (Kaounides, 1995)

In a report entitled *Harnessing Light* and commissioned by the National Research Council in the USA, the assertion is made that opto-electronics (the term used in the report is optics and engineering) is an important technology for fostering the creation and growth of new businesses. Of equal if not greater significance than this, however, is the role of opto-electronics as an enabling technology. An investment of a few hundred million dollars in optical-fibre technology has made it possible for the trillion-dollar worldwide communications revolution to happen (National Research Council, 1998).

The optics part of opto-electronics has a long history. Mirrors were in use thousands of years ago and, by the early seventeenth century, lenses were being ground for microscopes and telescopes. In 1704, Isaac Newton published his classic text "Opticks", which set down the fundamental principles of reflection and refraction. But it was not until after the Second World War and the development of electronics and the semiconductor materials industry that the fusion effect took off. A major breakthrough came in 1960 with the invention of the laser (Townes, 1999). Lasers produce light with the property of coherence and this permits it to be directed, focused and propagated in ways that are impossible for incoherent light. Laser light has made possible fibre-optic communications, optical data storage, laser surgery and materials processing. Other key technologies were also developed at about the same time as the development of lasers, such as liquid crystal displays, which moved to Japan after the basic development work had been done in the UK, and infra-red sensing systems, which now form the basis for night-sight applications (ACOST, 1988).

Opto-electronics technologies emerged from a period of laboratory-based R&D in the 1960s and '70s into an applications and diffusion phase in the 1980s and '90s. At the early stages of the industry, much of the developmental activity was typically buried within large firms. As the technology evolved and branched out into new areas, a diverse set of product opportunities emerged. At this stage, divisions formed within large firms and independent firms began to appear through spin-off activity, either through employees leaving the parent firm to form their own enterprises, or individuals with an entrepreneurial streak coming out of the university research environment. Firms then became aware of their inability to encompass the complete set of skills required for successful commercialisation and began to create external linkages to gain leverage from retained core competencies (Rothwell, 1992).

At this point, the relationships between firms and infrastructure elements such as universities, the labour market, research institutions and government agencies become more critical, as R&D activity and skills are externalised into markets rather than hierarchies (Williamson, 1975), and as the industry becomes characterised by the presence of large numbers of SMEs, often in the form of agglomerated networks surrounding the original players. The industry now consists of large numbers of SMEs concentrated in industrial clusters and engaged in symbiotic relationships with multinational firms. The Photonics Spectra industry guide for 1995 lists over 4,000 firms world-wide in the industry (Photonics Spectra, 1995), most of which come into the SME category, while Roychoudhuri (1996) shows how this breaks down into clusters with particular reference to the USA.

Technologically, the industry can be analysed at three levels — a basic level of generic technologies and materials, a key components level, and products and systems with end-user applications (Miyazaki, 1995). End-products emerge from the combination of intermediate components and underlying generic technologies and this has given rise to a diversity of applications and product-markets, which remain somewhat fragmented. Some of these (such as optical lenses) are dependent on craft traditions whereas others (such as lasers) are based on "high science". (A detailed analysis of constituent technologies and markets can be found in ACOST (1988), Miyazaki (1995), Kaounides (1995) and the USA's National Research Council (1998) report, *Harnessing Light.*).

Life-Cycle Theory and its Application to University–SME Relationships in Opto-Electronics

Most of the literature on life-cycle patterns looks at the industry sector level and focuses on industry structure and innovation processes. Klepper (1996), for example, shows how regularities concerning entry, exit, market structure and innovation vary from the birth of technologically progressive industries through to maturity, with a prediction that, over time, firms devote more effort to process innovation and that the number of firms and the rate and diversity of product innovation eventually wither. This has its roots in Utterback (1994).

These regularities are interesting but the problem with the product life-cycle model is its generalisability at an industry level (Tether and Storey, 1998). Teece (1986) contends that the product life-cycle model is most suited to mass markets where consumer tastes are homogeneous. It may therefore be particularly applicable to the "Fordist" era of mass-production and may not prove as valid in the "post-Fordist" era of "flexible specialisation". Suarez and Utterback (1995) emphasise that the model may only hold for assembled manufactured goods. It is also clear that the definition of industry is central to the model and, nowadays, the notion of an industry is becoming somewhat difficult to define (Utterback and Suarez, 1993).

This raises particular problems for opto-electronics, which is characterised by the fusion of previously distinct scientific fields, and the enabling role it plays in the development of such diverse industries as telecommunications, consumer electronics, industrial and military imaging systems. In the diversity and pervasiveness of opto-electronics lies great strength, but these same qualities pose problems for anyone looking for concise assessments and simple prescriptions. However, while taking due account of the reservations expressed above, the general tenor of industry life-cycle theory is that companies will invest more resource in process rather than product innovation in the later phases of development. The approach we have adopted in this study is to examine the nature of university–industry relations from the perspective of the individual small firm to see whether there are substantial differences at the various life-cycle stages and what light this casts on the theoretical constructions.

In a paper that points to the diversity of growth strategies in small high technology firms, Tether (1997) builds on Pavitt's (1984) work to present an interpretation of the development of innovative small firms. Three broad types of innovative technology-based new firms are distinguished:

(i) "new generic technology-based firms";
(ii) "new, niche market, specialist technology-based firms"; and
(iii) "new opportunist design-concept-based firms".

Three phases in the development of the firms are also distinguished: "infancy", "adolescence" and "maturity".

It is arguable that examples of opto-electronics firms are to be found in each of Tether's categories. Designers of lighting and display systems are increasingly using lasers and light emitting diodes as part of the product without getting too involved with the underlying technology, while at the other end of the three-level model, manufacturers of generic materials are in evidence. In the main, though, our study concentrates on new niche-market opto-electronics companies, as these are the ones that are most commonly found and are the ones that are explicitly looking at transferring technology into marketable products.

Tether's thesis is that, at the infancy stage, innovative technology-based firms concentrate on establishing the credibility and viability of their technology for the chosen market application (or generic use). The extent to which they rely on external "community" sources of knowledge varies between "high", in the case of generic technology companies which are looking for consensus, and "low", in the niche-market companies which are intent on keeping their technology and market insights to themselves while working closely and sharing knowledge with potential users of the technologies.

At the adolescence stage, the main preoccupation for both these types of firms is how to achieve a sustainable market share. Tether suggests that with the growing definition of the market potential, generic technology firms become less willing to share knowledge and this may be reflected in an increasing emphasis on internal knowledge and capability generation. These activities can draw on both science-based knowledge (for new products) and engineering knowledge (for process improvements). On the other hand, niche technology firms tend to have a clear vision at the outset of the specific applications of their knowledge. Typically, these firms have specialised technological capabilities, which are focused on a relatively small but defined group of potential customers. Whilst scientific knowledge may be important, engineering and production knowledge and knowledge of users' needs are likely to be more prominent.

Having survived infancy and adolescence, Tether suggests that most innovative and technology-based smaller firms evolve into established niche-market technology-based SMEs (a few may break through into large company status). These companies are mature "specialist suppliers", their markets tend to be well-defined but limited, with the key to their strategic competencies being the ability to combine knowledge of users' changing needs with their own evolving capabilities and to develop new products through incremental innovation.

The implications of this theory for the nature of relationships between SMEs and universities are that, at the infancy stage, they would be expected to be at their most intense and focused on technological development but, thereafter, they diminish in intensity and firms look more towards process improvements and incremental product innovation for new markets. This implication is made on the assumption that SMEs regard universities as originators of "basic" science rather than solvers of practical problems under commercial business conditions.

Methodology and Findings

This paper draws on an international research study into the development and performance of opto-electronics (Hendry *et al.*, 1999). One hundred firms were interviewed in order to build up a picture of the dynamics of cluster development in six selected regions in the UK, USA and Germany. Additional case material is derived from follow-up interviews on firms' innovation processes.

In this paper, we focus specifically on the interaction between universities and manufacturing SMEs in the UK and USA — namely Wales, East Anglia, Massachusetts and Arizona. In doing so, we have excluded large firms with more than 500 employees and those firms whose principal activity is distributing or servicing new technology products. The result is 59 manufacturing SMEs concerned with bringing new products to the market. The rationale for selecting firms in regions of clustered opto-electronic activity is that the extent of university–SME interaction would be expected to be at its height here, and so this sample makes a good test case. To the extent that these regions are also the locations for some of the most active involvement in opto-electronics in the UK and USA, they also serve as indicators of national characteristics. Company cases are supplemented by interviews with government agencies and university researchers in order to understand SME–university relationships from a variety of perspectives.

In the interviews, company respondents were asked about their perception of the technology transfer process and how relationships with universities contributed to it. The interviews were all tape recorded and transcribed. Analysis of the interview data was carried out by all three researchers and the process aimed at identifying instances where a university connection was in place (for each of five reasons that emerged from this grounded theory approach) and an explanation for the connection was established in the interview. The six types of university–firm interconnections are listed in Table 1 in order of importance to SMEs, as interpreted by the number of firms claiming to make use of them.

Two limitations of this approach need to be mentioned. We did not attempt to get the respondents to make quantitative value judgements about the different forms of relationships (although opinions were expressed), nor did we attempt to assess the intensity of the relationship in terms of the amount of resource that was applied to it. Rather, in the first instance, we were concerned about understanding the reason for the connection and the link it had to the innovation process. In the second case, a proxy for the intensity of university relationships is derived by the simple expedient of summing the number of different types of connections to give an "intensity score". In this way, we are able to get a mixture of qualitative and quantitative data. The following sections discuss these, omitting the formation process at start-up.

Table 1. Proportion of companies in sample making use of different types of university–company connections listed in the order of occurrence.

Type of university connection	Percentage of 59 case study firms using this form of link
Use of university faculty or other research staff as advisers and consultants	56
A direct source of staff recruited at graduate or postgraduate level	39
Any form of funding connection on research to solve particular problems	32
A source of start-ups by students or staff leaving directly from a university[1]	25
A source of product ideas and proven technologies	29
Use of university facilities and equipment	24
No evidence of use of university resources	25

[1]The percentage shown here is the actual number of start-ups (i.e. 15 firms or 25% of sample).

Use of faculty and other research staff as advisers and consultants

Opto-electronics is a demanding scientific discipline with a crucial dependency on the fundamental physical properties of materials and components. Compounding this is the rapid rate of change in the technology, which gives rise to the need to keep a "technology watch". There is, therefore, an underlying need for SMEs to keep well informed about the scientific feasibility of new product ideas and future technology developments, and this goes some way to account for the widespread use of university faculty as advisers.

In most cases, the relationship is with a known and respected expert in the precise field of interest to the firm and, normally, the link extends over a period of time, suggesting that a form of trust builds up between the two. This is to be expected as, usually, the nature of the dialogue is one where the SME is looking for confirmation that confidential ideas for product or process development are well grounded in scientific or engineering theory. So, although it would be expected that the use of faculty will be at its height during infancy, the confidence and trust established at this stage may well carry through to adolescence and maturity. We also explored this connection at a more junior level in the use of graduate student placements. This happened in only a small number of cases and was used primarily as part of a recruitment process and as a means of developing some peripheral aspect of the company's product portfolio.

Though the use of faculty and graduate students is a common connection, its significance is often played down by respondents who describe it as a means of validating ideas that they themselves (or their customers) have initiated. Faculty advisers are seen as playing a useful but not essential risk evaluation role.

Recruitment direct from university at graduate or postgraduate level

Second in frequency of occurrence but widely stated as the most important reason for establishing a link with the education sector is the use made of a university as a direct source of staff. Almost all SMEs in opto-electronics have some staff with university qualifications in engineering or scientific disciplines. Our interest is in establishing the extent to which SMEs recruit new staff directly from university as a way of tapping into the most recent theoretical work in the subject. This could be at graduate level but is more likely to be of value at post-doctoral level, and there are examples of both kinds.

There is a dilemma here, however, that affects SMEs in many industries, in that SMEs need to recruit experienced people, since they do not have the time or resources to invest in training before making staff productive. This is expressed at its most extreme by the very fast-growing Company A:

> We look for research and development people who have a reasonable amount of industry experience in a research laboratory plus a first degree. Some of them have a Ph.D. Engineers are expected to be able to contribute to a project within one week of joining and be up to full speed within one month. They are expected to keep up-to-date with developments themselves by attending one or two technical conferences a year. (Company A, East Anglia)

Any funding connection on research to solve particular problems

Third in frequency of occurrence is the number of SMEs having funding connections with universities for the purpose of doing scientific research. This is surprisingly high. However, this includes joint projects where the funding comes from a third party (such as a government agency). Nevertheless, this represents a genuine resource commitment from the SME in terms of time. The following example is from one of the most active participants in externally funded research projects.

> We have had a number or research and development contracts with external agencies such as companies and universities but, as far as we are concerned, they are all aimed at "raising our technology base". In other words, we use these projects to improve our knowledge and scale in making epitaxy wafers as part of the overall research attempt to get higher end-user product performance and functionality. Sometimes, these projects will result in a published specification of a new epitaxy wafer for use in a new kind of product but, more often than not, the outcome for us is increased skill and knowledge that allows us to be more responsive to our customers' demands. (Company B, Wales)

SMEs do not ordinarily commission research or fund university posts directly. This latter category of funding was especially hard to find in our study and it accounted for only 3 instances out of 59 cases. It is only larger, more established companies that have the resources to fund directly. The examples that we did find in the UK were directed towards improvement of a company's existing products. In one case, this was to investigate the physical properties of material components in order to get more efficiency and performance out of a product and in the other, it was to support a process improvement programme.

A significant problem with SMEs funding universities as a source of new product ideas is the issue of confidentiality. This is particularly acute in a localised cluster, such as Arizona, where one university has relations with a number of firms:

> We do not fund any product-oriented research in the universities because of the need to keep technology information proprietary — especially in a locality where you have two direct competitors very close to each other and likely to use similar resources at the university. (Company C, Arizona)

Concerns were also expressed about a perceived lack of "business-like" approach from academics but this can be countered by using a local university.

> We regard universities as a good source of expertise but they do need to be put onto a strong commercial footing. Over the years, we have worked with the local university, both by feeding them with research contracts from our clients and by persuading people from industry to take up positions there, to build up a more business-like view of its equipment and skills. The local connection is a factor as we would probably not have done quite so much with a more distant university. (Company B, Wales)

Source of new product ideas

Oakey (1995) observes that the single largest reason for new high-technology-based firm formation is the existence of a product idea that founders could exploit. New firm founders are strongly influenced by their previous experience, so they largely develop new product ideas based on work they were performing in their previous employment or university. Consequently, most of the 22% of firms that attributed new product ideas to university origins were also SMEs that were formed directly out of a university environment. The few instances of firms not originating from a university and attributing new product ideas to university connections were usually looking for product and process improvements in the mature stages of the product life-cycle (Utterback, 1994). An exception is the rare case where a local university transferred product ideas to a favoured SME. Clearly though, this is built upon long-standing personal relationships.

> We have close links with the local college. Most of the people that work here have come through the college. There is a facility up at the college called the Product Development Centre (PDC), which is there to help develop new product ideas. The general idea is that if the PDC cannot help the originator of the idea, then the idea is passed on to us. We have facilities for prototype manufacturing and if we cannot manufacture from scratch, we can source parts and assemble equipment. (Company D, Wales)

Another exception occurs in Arizona where a company is developing products based on original research done at the University of Southampton. Significantly, this has an international rather then local perspective, implying the importance of "technology watch" and the global transfer possibilities of codified knowledge that exists in the form of research results.

The absence of university links as a source of new product ideas may not imply backwardness, though. The stage of development in basic opto-electronics technologies and the shift to an applications phase led by companies may explain this. As the head of one such small company said:

> In many ways, the frontier work being done on applications is being done by us. Doubtless there is research into basic science going on in universities but it is not changing significantly enough for it to have an impact on the application side in the immediate future. What happens is that technology gets developed and filters through to one or two high profile applications. Then it just sits there and rots until someone like us comes along and uses that technology in more mundane applications. (Company E, Wales)

Use of equipment and facilities

The number of firms making use of university equipment and facilities is quite small (although for some very small firms, it is quite significant). It is normally based upon the use of specialised calibration equipment that the university owns and the SME cannot justify purchasing for the limited use it is likely to get. This activity depends upon the existence of informal relationships (as usually the cost of using such equipment is not charged at its market rate, if at all) and on proximity:

> The Optical Sciences Center is key to us. They loan us equipment and they do work for us. Some pieces of equipment in the optics business are unique — we can take a piece of equipment into their place and they will measure it. We all know where the back door is and it's such a small community. I usually have lunch once a week with someone from the optics cluster or the university and we will talk about what's going on at their place or things we have read in reports. (Company F, Arizona)

Summary and life-cycle analysis

These patterns of usage may differ from that of large firms and from received wisdom. The use of faculty as advisers is high because this not only reflects the relatively small in-house R&D resources of SMEs but also does suggest an underlying need to maintain contact and a potential for more intensive relationships. Staff recruitment is understandably high for all firms and is seen by universities as a prime purpose. It is also the one that serves a dual purpose without any conflict of interests (Rosenberg *et al.*, 1994), in that industry clearly needs fresh inflows of technically trained staff while universities will be keen to establish reputations as centres of excellence in their chosen fields of teaching and research. The commissioning of research is perhaps surprisingly high in view of the difficulties over appropriation of research results (Teece, 1986) and may be accounted for by the increasing involvement by SMEs in multi-partner government-funded projects.

In order to analyse this data from a life-cycle perspective, we divided the sample of firms into three life-cycle categories, namely, infancy, adolescence and maturity. This was done using the simple expedient of defining firms of up to 5 years old as in "infancy", 6 to 17 as in "adolescence" and greater than 18 years as in "maturity" (the age being the age of the company at the time of interview in 1996). The rationale for these specific break points is that:

(i) the age frequency distribution (adjusted for isolated high age cases) has a mean age of 11.2 (median 11) and a standard distribution of 6.38, and
(ii) that the 25% and 75% percentiles came out at roughly these points.

This is a very simple construction but its validity is supported by the fact that most venture capital arrangements with new firms are for a period of five years duration (suggesting that this is the point at which companies move from infancy to adolescence) and that an analysis of failure rates for new high-technology companies in Cambridge (Garnsey *et al.*, 1994) indicates that 80% survive this period and presumably carry on until maturity. Further chi-square analyses were carried out with varying boundaries in a sensitivity analysis test that did not significantly alter the findings. The results are shown in Table 2.

Table 2 indicates that there is indeed a decline in the number of interactions with universities with age. The average intensity score (being the sum of different types of connections) falls from 2.00 through 1.86 to 1.47 as firms move from infancy to maturity.

In order to get better understanding of the underlying pattern, we looked at each type of interaction and counted the number of firms in each life-cycle category using the connection. Table 2 shows the actual and expected frequency patterns and chi-square significance. The results show that there is very little variation in the pattern of university relationships from the overall frequency distribution. The one exception is in the "use of university faculty as advisers and consultants", where there is the suggestion (chi-square 0.071, which is approaching significance) that firms in infancy make greater use of this type of connection than at adolescence and maturity. (The significance of the finding for "use of university facilities" is also interesting but the figures are too low to be of genuine significance.). The conclusion we draw from this data is that there is some evidence to suggest a heightened degree of interaction with university expertise in the early years of these companies' existence. This is supportive of Tether's thesis that this period is dominated by attempts to realise and validate the technology as a practical commercial proposition.

Other than that, the picture is one of gradual decline in the use of university knowledge resources; and the even distribution suggests that they are being utilised for a variety of different purposes. This is supported by evidence from the interviews where we found instances of firms looking both for process improvement and incremental adaptation of the existing technology to fit new market opportunities.

As the counter to this view, we looked to see if there were differences in the use of university resources between the UK and the USA. A similar chi-square analysis was carried out, this time by simply dividing companies into whether they are UK- or US-based. The results are shown in Table 3.

This time, it is notable that at least two significant patterns emerge. One is that UK firms are more likely to connect to a university in some form of funded

Table 2. Opto-electronics SMEs sampled in each age category categorised by type of connection with university infrastructure and start-up pattern.

	N Number of firms			
	Firms in "Infancy"	Firms in "Adolescence"	Firms in "Maturity"	Chi-Square Significance
Use of university faculty or other research staff as advisers and consultants	12(8.4)	15(16.2)	6(8.4)	0.071
A direct source of staff recruited at graduate or postgraduate level	5(5.8)	11(11.3)	7(5.8)	0.746
Any form of funding connection on research to solve particular problems	4(4.8)	10(9.3)	5(4.8)	0.866*
Firm founded by faculty or student leaving directly from university	4(3.8)	8(7.4)	3(3.8)	0.854*
A source of product ideas and proven technologies	4(4.3)	9(8.4)	4(4.3)	0.934*
Use of university facilities and equipment	5(3.6)	9(6.9)	0(3.6)	0.043*
Average score	2.00	1.86	1.47	
Median score	2.00	2.00	1.00	
	N 15	*N* 29	*N* 15	

The figure in bracket is the expected number of firms in each category derived from the overall distribution.
*Number of cells with expected frequency less than 5, is greater than 20%.

research activity US firms are more likely to be while start-ups coming from a university. This suggests that increased interaction with universities does not by itself lead to an increased number of companies being started.

These observations may not seem shattering but they do suggest an overall conclusion — that national innovation systems have a more differential impact on the innovation and technology transfer process than does the life-cycle factor. That is, the life-cycle factor is universal but the national innovation system does make a difference. This can only be a limited conclusion as we are just looking at one aspect of technology transfer — namely the relationship between industry and the academia.

Discussion and Conclusions

The aim of this paper has been to understand what use high-technology-based SMEs make of university resources and whether there is a differential pattern that can be explained in terms of classic industry life-cycle theory (Utterback, 1994).

Life-cycle conclusions

The first point is that not all high-technology-based SMEs make use of university resources. In this sample, some 25% did not. These companies may have good reasons for not engaging with university research centres. The issue of intellectual property rights and protecting proprietary knowledge is a significant barrier (Bowie, 1994). More generally, we need to recognise the circumstances which cause SMEs to seek to adapt and grow (Hendry *et al.*, 1995), and the specific motivations, triggers and decision processes which lead SMEs to engage with universities and technical centres.

The three key reasons for doing so are informal engagement with experts with relevant scientific and engineering knowledge, recruitment of scientific and engineering personnel, and collaborative research on both product and process improvements. There is a gradual diminution in the extent to which such firms work with university resources as they move into the more mature stages of the life-cycle. This suggests that they are becoming increasingly reliant on their own (or other non-university) sources of expertise and this can be interpreted as a greater focus on process innovation. Other evidence (Hendry *et al.*, 2000) shows that customers are far and away the most common source for product innovation and this is likely to increase as an SME becomes established, and becomes more visible and credible to larger end-user customers.

On the other hand, the pattern of usage analysed in terms of the different types of connections appears to be evenly spread across the three life-cycle stages. There is some suggestion that there is a greater tendency to use faculty expertise in the early years of a company's existence and this is to be expected in the light of the need to establish technical viability and credibility. But this is not strong enough to be statistically significant at this level of analysis. Rather, the fact that contacts with universities remain equally distributed over the years suggests that SMEs do have a continuing need for these relationships for a number of different reasons and this could include product as well as process innovations. To this extent, then, the evidence is more supportive of the Tether version of life-cycle theory (Tether, 1997).

This is summed up by Autio (1997). "The traditional life-cycle model of small firms may not apply well to the majority of new technology-based firms. The evolution of such firms is not predominantly one of growth in size. The evolution is rather the evolution of technological resources. Many such firms can be viewed as a unique bundle of technological resources, whose mission, motivated by survival, is to maintain technological leadership."

National innovation systems

Further light is thrown on the issue of SME–university relations by looking at the differences between the UK and US firms in the sample, bearing in mind that they all come from areas with a high university presence in opto-electronics and above average concentration of opto-electronics firms. The make-up in terms of age is broadly comparable, with the UK having a slightly greater presence in the "adolescence" category and the USA scoring more frequently with mature firms (Table 3). The evidence from Table 3 is that US firms are more likely to be formed from academics spinning off from a university UK while firms are more likely to be working with universities on funded research activity. US firms are overall less frequent users of university resources as shown by their "intensity score" being 1.4 compared to the UK value of 2.2.

It would appear, then, that pressures in the direction of increasing globalisation and industry standardisation are not reducing the importance of national and regional institutions and their impact in creating an environment in which firms can generate and accumulate technological competencies. Although the origins of technological change may be local and path-dependent, the way in which it takes shape and diffuses is moderated by sectoral, regional and national characteristics (Temple, 1998). Relationships between SMEs and university research centres are conditioned by the "national innovation system" (Lundvall, 1992; Nelson, 1993) in which they are embedded, and by circumstances which are unique to the way technology has developed in the particular locality (Archibugi *et al.*, 1999; Cooke *et al.*, 1997; Zysman, 1996).

Universities and industry clusters

One inference from this study is that the benefits which SMEs get from relationships with universities can, in theory, support a rich environment for

Table 3. Opto-electronics SMEs sampled in each country categorised by type of connection with university infrastructure and start-up pattern.

	N Number of firms		
	UK	USA	Chi-Square Significance
Use of university faculty or other research staff as advisers and consultants	20(16.8)	13(16.2)	0.091
A direct source of staff recruited at graduate or postgraduate level	13(11.7)	10(11.3)	0.486
Any form of funding connection on research to solve particular problems	14(9.7)	5(9.3)	0.016
Firm founded by faculty or student leaving directly from university	4(7.6)	11(7.4)	0.030
A source of product ideas and proven technologies	9(8.6)	8(8.4)	0.838
Use of university facilities and equipment	10(7.1)	4(6.9)	0.078
Firms in infancy	7(7.6)	8(7.4)	0.182
Firms in adolescence	18(14.7)	11(14.3)	0.182
Firms in maturity	5(7.6)	10(7.4)	0.182
	N 30	*N* 29	*N* 59

The figure in bracket is the expected number of firms in each category derived from the overall distribution.

innovation but, in practice, this must be moderated by taking into account SME views on intellectual property rights and their perspectives on the commercial discipline displayed by universities. Regional clustering around appropriate research centres provides a natural environment for innovation but the success of this depends on the research culture which pervades the university, and this in turn will derive much from the national culture in which universities operate.

The institutional character of individual universities, the national research culture, and the motivations and career incentives of university staff vary considerably. The "purist" view of the academic role among opto-electronics researchers is put most forcibly in the UK:

> While in industry, we all developed very strong views about what university research should be about and intend to be as academic

> as we can be. Our job is to be generating ideas, publishing papers, researching concepts, researching and evaluating the early stages of new possibilities. We would not purposely get involved in research if it was remotely near what we might call development. We should not be doing industry's job for them. At the point where it looks like the technology could be useful, our role stops and industry's starts. (UK University A)

At one of the major centres for opto-electronics research in the UK, however, they take a different line and are establishing a technology transfer company to take equity stakes in start-ups:

> The tendency is towards the "privatisation of research". If this is going to be the picture, you must have large integrated centres — you cannot hope to compete for research work on the basis of having small university departments operating on traditional lines combining teaching and research. The big new idea is corporate venturing, based on an alliance between a small contract research company and a large multinational. The multinational puts up funds to start a new company, takes a share of the equity and regards it as an incubator for new ideas. It is a way of outsourcing their R&D department. (UK University B)

This role for the university not only gets into the problems of intellectual property rights which others have run into and (some would say) the problem of universities keeping to deadlines. More importantly, it could shut out SMEs from ready access to research.

Ultimately, we suggest, the impact of the universities in a region is likely to depend on two factors:

(i) The extent to which university research is organised by discipline (Cambridge University) or applications-oriented (Massachusetts Institute of Technology — MIT), and
(ii) Whether university–industry collaborations have been planned with government leadership (Optical Sciences Center, Arizona) or evolved in a decentralised fashion through competition for funds (MIT) (DeFillippi, 1997).

An applications orientation, for example, is likely to mean multidisciplinary centres (important in an industry characterised by "technology fusion"), extensive collaborations not limited to contract research and incubation of start-up ventures. MIT exhibits all of these.

A planned approach will reflect regional economic priorities but the monopolistic role of the university may threaten commercial networks as a competitor. Decentralised, market-based evolution, on the other hand, may allow smaller, second-tier firms easier access to the university but fragment and duplicate research effort — what Sternberg (1992) has called "decentralised pork-barrel politics".

Networks to overcome university–SME barriers

SMEs have been shown to be more dependent than larger companies on external sources of scientific and technological information. However, as an OECD study argues, fundamental differences in the character of SMEs and universities give rise to a number of problems that make it difficult for SMEs to use universities:

> Because of these problems, university–SME relations remain marginal, both in comparison to university–industry relations as a whole and in relation to total scientific and technological information transfers to SMEs. (Estime *et al.*, 1993)

The same authors note efforts to overcome this barrier by using intermediaries rather than attempting to establish direct bilateral relations. However, they note that information tends to reach SMEs through personal, informal networks — an observation made by a number of others (Faulkner and Senker, 1995):

> These aspects explain why the regional and local dimensions were found to be important in SME networks even though SMEs may sometimes also make use of national or international networks, especially with regard to highly specialised scientific and technological information. Obtaining a better understanding of the type of networks used by SMEs, improving the effectiveness of these networks, and in particular, improving their ability to "distil" the expertise and know-how of universities are therefore important potential ways of promoting the competitiveness of SMEs. (Estime *et al.*, 1993)

However, the character and scope of personal networks in opto-electronics may be different from those in SMEs generally. Opto-electronics is high-tech and those who work in it include many trained scientists and engineers with Master's and PhD degrees. The best are accustomed to searching for information and dealing with experts nationally and internationally, and in the Internet, they have the means to communicate globally.

In the UK, the 1998 Competitiveness White Paper (Department of Trade and Industry, 1998) is an ambitious attempt to promote better clusters and networks, and improve SMEs' access to technological developments. In addition, the Technology Foresight Programme is sensitive to the need for cross-industry R&D (Kodama, 1992), which is important for an industry characterised by "technology fusion".

However, as one of our American respondents commented

> You'll never get research out of university labs. You have to take SMEs to them, like a trade mission, to take it out.

This implies a less "laissez-faire" attitude to the dissemination and diffusion of what are in effect "public goods" produced in university laboratories.

Acknowledgements

We are grateful to the Leverhulme Trust for initially funding this research, and to the Welsh Development Agency and the Anglo-German Foundation for the Study of Industrial Society for additional financial support.

References

ACOST (1988). *Opto Electronics: Building on our Investment*. London: HMSO.

Archibugi, D, J Howells and J Michie (1999). Innovation systems in a global economy. *Technology Analysis & Strategic Management*, 11(4), 527–539.

Autio, E (1997). New, technology-based firms in innovation networks symplectic and generative impacts. *Research Policy*, 26, 263–281.

Bowie, NE (1994). *University-Business Partnerships: An Assessment (Issues in Academic Ethics)*. Lanham, MD: Rowman & Littlefield.

Cooke, P, MG Uranga and G Etxebarria (1997). Regional innovation systems: Institutional and organisational dimensions. *Research Policy*, 26(4–5), 475–491.

DeFillippi, RJ (1997). *Patterns of Development in US Regional Clusters. Managing in a Global World*. Dublin: Eastern Academy of Management.

Department of Trade and Industry (1998). *Our Competitive Future: Building the Knowledge Driven Economy*. London: The Stationery Office.

Dubarle, P and C Verie (1993). *Technology Fusion: A Path to Innovation. The Case of Opto-Electronics*. Paris: Organisation for Economic Cooperation and Development.

Enright, M.J. (1995). *Regional Clusters and Economic Development: A Research Agenda*. Working Paper 94-02, Harvard Business School, Cambridge, MA.

Estime, M, G Drilhon and P Julien (1993). *Small and Medium-Sized Enterprises: Technology and Competitiveness*. Paris: Organisation for Economic Cooperation and Development.

Faulkner, W and J Senker (1995). *Knowledge Frontiers: Public Sector Research and Industrial Innovation in Biotechnology, Engineering Ceramics and Parallel Computing*. Oxford: Oxford University Press.

Garnsey, E and A Cannon-Brookes (1993). The "Cambridge Phenomenon" revisited: Aggregate change among Cambridge high-technology companies since 1985. *Entrepreneurship & Regional Development*, 5(2), 179–207.

Garnsey, EW, SC Galloway and SH Mathisen (1994). Flexibility and specialization in question; birth, growth and death rates of Cambridge new technology-based firms 1988–1992. *Entrepreneurship & Regional Development*, 6, 81–107.

Hendry, C, MB Arthur and AM Jones (1995). *Strategy Through People: Adaptation and Learning in the Small-Medium Firm*. London: Routledge.

Hendry, C, J Brown and R DeFillippi (2000). Regional clustering of high technology-based firms: Opto-electronics in three countries. *Regional Studies* (Forthcoming).

Hendry, C, J Brown, R DeFillippi and R Hassink (1999). Industry clusters as commercial, knowledge and institutional networks: Opto-electronics in six regions in the UK, USA and Germany. In *Interfirm Networks: Negotiated Order and Industrial Competitiveness*, ed. A Grandori, pp. 151–184. London: Routledge.

Hoffman, K, M Parejo, J Bessant and L Perren (1998). Small firms, R&D, technology and innovation in the UK: A literature review. *Technovation*, 18(1), 39–55.

Kaounides, LC (1995). *Advanced Materials*. London: Financial Times.

Keeble, DE (1994). Regional influences and policy in new technology-based firm creation and growth. In *New Technology-Based Firms in the 1990s*, ed. R Oakey, pp. 204–218. London: Paul Chapman.

Klepper, S (1996). Entry, exit, growth, and innovation over the product life cycle. *American Economic Review*, 86(3), 562–583.

Kodama, F (1992). Technology fusion and the new R&D. *Harvard Business Review*, 70(4), 70–78.

Lee, YS (1996). Technology-transfer and the research university — a search for the boundaries of university–industry collaboration. *Research Policy*, 25(6), 843–863.

Lundvall, BA (1992). *National Systems of Innovation: Towards a Theory of Innovation and Interactive Learning*. London: Pinter.

Malecki, EJ (1997). *Technology and Economic Development: The Dynamics of Local, Regional and National Competitiveness*. Harlow: Addison Wesley/Longman.

Massey, D, P Quintas and D Wield (1992). *High-Tech Fantasies: Science Parks in Society, Science and Space*. London: Routledge.

Miyazaki, K (1995). *Building Competences in the Firm: Lessons from Japanese and European Opto-Electronics*. London: MacMillan Press Ltd.

National Research Council (1998). *Harnessing Light: Optical Science and Engineering for the 21st Century*. Washington, DC: National Research Council.

Nelson, RR (Ed.) (1993). *National Innovation Systems: A Comparative Analysis*. Oxford: Oxford University Press.

Oakey, R (1994). *New Technology-Based Firms in the 1990s*. London: Paul Chapman.

Oakey, R (1995). *High-Technology New Firms: Variable Barriers to Growth*. London: Paul Chapman.

Pavitt, K (1984). Sectoral patterns of technological change: Towards a taxonomy and a theory. *Research Policy*, 13(6), 343–373.

Photonics Spectra (1995). *The Photonics Corporate Guide to Profiles and Addresses*. Pittsfield, MA: Laurin Publishing Company.

Porter, ME and Ö Sölvell (1998). The role of geography in the process of innovation and the sustainable competitive advantage of firms. In *The Dynamic Firm; The Role of Technology, Strategy, Organization, and Regions*, ed. AD Chandler, P Hagstrom and Ö Sölvell. pp. 440–457. Oxford: Oxford University Press.

Rosenberg, N and RR Nelson (1994). American universities and technical advance in industry. *Research Policy*, 23(3), 323–348.

Rothwell, R (1992). Successful industrial innovation: Critical factors for the 1990s. *R&D Management*, 22(3), 221–239.

Roychoudhuri, C, R Becherer, D Young and A DeMaria (1996). *Review of the CONNECT Program Progress and the Connecticut Photonics Cluster. Global Networking of Regional Optics Clusters*. Denver: SPIE.

Saxenian, A (1994). *Regional Advantage: Culture and Competition in Silicon Valley and Route 128*. Cambridge, Massachusetts: Harvard University Press.

Sternberg, E (1992). *Photonic Technology and Industrial Policy: US Responses to Technological Change*. New York: State University of New York Press.

Storper, M (1995). The resurgence of regional economies, ten years later: The region as a nexus of untraded interdependencies. *European Urban and Regional Studies*, 2(3), 191–221.

Suarez, FF and JM Utterback (1995). Dominant designs and the survival of firms. *Strategic Management Journal*, 16, 415–430.

Swann, GMP, M Prevezer and D Stout (1998). *The Dynamics of Industrial Clustering*. Oxford: Oxford University Press.

Tang, NKH, A Agnew and O Jones (1996). Technological alliances between HEIs and SMFs: Examining the current evidence. *Small Business and Enterprise Development*, 3(1), 48–56.

Teece, DJ (1986). Profiting from technological innovation — Implications for integration, collaboration, licensing and public-policy. *Research Policy*, 15(6), 285–306.

Temple, P (1998). Clusters and competitiveness: A policy perspective. In *The Dynamics of Industrial Clustering*, ed. GMP Swann, M Prevezer and D Stout. pp. 257–297. Oxford: Oxford University Press.

Tether, BS (1997). Growth diversity amongst innovative and technology-based new and small firms: An interpretation. *New Technology, Work and Employment*, 12(2), 91–107.

Tether, BS and DJ Storey (1998). Smaller firms and Europe's high technology sectors: A framework for analysis and some statistical evidence. *Research Policy*, 26(9), 947–971.

Tidd, J, J Bessant and K Pavitt (1997). *Managing Innovation: Integrating Technological, Market and Organizational Change*. Chichester: John Wiley & Sons.

Townes, CH (1999). *How the Laser Happened: Adventures of a Scientist*. Oxford: Oxford University Press.

Utterback, JM and FF Suarez (1993). Innovation, competition and industry structure. *Research Policy*, 22, 1–21.

Utterback, JU (1994). *Mastering the Dynamics of Innovation*. Boston, MA: Harvard Business School Press.

Williamson, O (1975). *Markets and Hierarchies*. New York: Free Press.

Zysman, J (1996). Nations, institutions, and technological development. *International Journal of Technology Management*, 12(5/6), 651–678.

Chapter 17

DO SMES CLUSTER AROUND INNOVATION ACTIVITIES? DISCOVERING ACTIVE, INCREMENTAL AND OPPORTUNISTIC INNOVATORS

Gian Luca Casali, Roxanne Zolin and Sukanlaya Sawang

QUT Business School

Queensland University of Technology, Brisbane, Australia

Do SMEs cluster around different types of innovation activities? Are there patterns of SME innovation activities? To investigate we develop a taxonomy of innovation activities in SMEs using a qualitative study, followed by a survey. First, based upon our qualitative research and literature review we develop a comprehensive list of innovation activities SMEs typically engage in. We then conduct a factor analysis to determine if these activities can be combined into factors. We identify three innovation activity factors: R&D activities, incremental innovation activities and cost innovation activities. We use these factors to identify three clusters of firms engaging in similar innovation activities: active innovators, incremental innovators and opportunistic innovators. The clusters are enriched by validating that they also exhibit significant internal similarities and external differences in their innovation skills, demographics, industry segments and family business ownership. This research contributes to innovation and SME theory and practice by identifying SME clusters based upon their innovation activities.

Keywords: Innovation; innovation activities; taxonomy; SME; factor analysis; cluster analysis.

Introduction

Innovative products and services make important contributions to the economy, making innovation a widely discussed topic in business, information technology,

---

This paper was first published in *International Journal of Innovation Management*, Vol. 20, No. 7 (October 2016) 1650062 (24 pages). © World Scientific Publishing Europe Ltd. DOI: 10.1142/S1363919616500626

engineering, and public development contexts (Branzei and Vertinsky, 2006; Sawang and Unsworth, 2011). Despite the fact that many innovations come from small and medium-sized enterprises (SMEs), much of the innovation research has been conducted in large firms, leaving SMEs largely unstudied (Hoffman *et al.*, 1998). Recent research calls for investigation of the impact of innovation on performance among SMEs, because they likely differ from large firms (Rosenbusch *et al.*, 2011). Studies of large firms measure innovation based upon their number of patents and dollars invested in research and development (R&D), yet many SMEs still manage to innovate without applying for patents or investing in R&D.

But not all SMEs engage in innovation. It could be misleading to make the assumption that all SMEs are the same, particularly when it comes to innovation. There might be differences between SMEs in the activities they engage in based upon differences in size, differences between SMEs in new high growth market segments or mature market segments and there may be differences between young SMEs and mature firms.

Researchers examining innovation by SMEs have identified factors that contribute to innovation, and concluded that SMEs tend to be innovative and innovation improves performance (Shobha *et al.*, 2008), but these studies do not acknowledge that there might be some clusters of SMEs that have high levels of engagement in innovation activities and other clusters with low engagement in innovation activities.

Yu *et al.* (2012) categorised outcome variables of family businesses to develop a numerical taxonomy with seven clusters (performance, strategy, social and economic impact, governance, succession, family business roles, and family dynamics) but they did not consider the innovation activities of SMES.

Similarly, Rotaba and Beaudry (2012) look at how low-tech and high-tech firms differ in "leading "and "catch-up" economies but their study does not distinguish between large firms and SMEs. Lazoi *et al.* (2011) note the importance of innovation networks in an aerospace SME cluster, but do not investigate how SMEs might cluster around a broader range of innovation activities.

Our research asks whether SMEs cluster around different types of innovation activities. To investigate this phenomenon we collected qualitative and quantitative data from the Spatial Information (SI) Industry in Australia to develop a taxonomy of innovation activities in SMEs.

This is an ideal data source because, while being largely representative of SMEs the industry SI has a few larger firms for comparison. Finally, SI is also a good industry for our study because it has two clear segments: the traditional surveying industry using mature technology (our *low-tech* segment), and SI firms using new technologies, such as global positioning systems (GPS) technology (our *high-tech* segment).

This paper makes important contributions by identifying the factors of innovation activities and by identifying clusters of firms with similar levels of engagement in these innovation activities. In addition to engaging in different innovation activities, these groups also differ on other parameters such as internal development of different innovation skills sets. More research is needed to evaluate these factors and clusters in different industries and countries.

Literature review of SME innovation activities

Innovation is a widely discussed topic, especially in business, information technology, engineering, and public development contexts (Unsworth *et al.*, 2012). We define innovation broadly as ideas, systems, technologies, products, processes, services, or policies that are new to the adopting organisation (Sawang and Unsworth, 2011; Zaltman *et al.*, 1973). Opportunistic recognition plays an essential role in innovative change (Ardichvili *et al.*, 2003). It is the interpretive role of business owners/managers that allows us to understand organisational innovation (Eggers and Kaplan, 2009).

Identifying opportunities has been recognised as one of the most critical antecedents of organisational innovation and consequently has become a central element of the study of entrepreneurship and innovation management (Alvarez and Busenitz, 2001). Generally, prior to developing new innovation activities, firms must go through a discovery process, which includes search and recognition. As a result, scanning behaviour can drive innovative opportunity recognition. A discovery process can lead to answers about unsolved problems within organisations (Hsieh *et al.*, 2007).

But not all SMEs innovate, or innovate in the same ways. To understand how SMEs innovate, it would be useful to answer this question using a taxonomy of innovation activities.

One place to start in teasing out how SMEs cluster in their innovation activities is by looking at the innovation activities they engage in.

Empirical Taxonomy

It is important to clarify what taxonomy is, and is not, especially in light of the debate about the difference between taxonomy and typology.

Taxonomy is defined as "the numerical evaluation of the affinity or similarity between taxonomic units and the ordering of these units into taxa on the basis of their affinities" (Sokal and Sneath, 1963). Hence, a taxonomy of SMEs' innovation activities is a grouping of the SMEs into taxa, also called clusters, based upon the similarities and differences between their innovation activities.

Despite a number of scholars suggesting that there is little difference between taxonomies and typologies (Burns, 1967), the literature clearly defines typology as an invention of individual creativity, and taxonomy as an empirical tool for building complex filing systems that allow both the ordering and the retrieval of large amounts of data (Davidsson, 1989; Rich, 1992). In other words, a typology is designed by forming priori classes/groups and subsequently placing cases into these groups, whereas in taxonomy classes/groups emerge from the empirical procedures used to sort cases on the basis of similarity or contrast (Rich, 1992). It has been widely argued that developing empirical taxonomies can reduce the complexity of empirical phenomena to a few constructs (Archibugi, 2001; De Jong and Marsili, 2006; Ketcher and Shook, 1996). Additionally, constructing a usable taxonomy plays a fundamental role in the development of a scientific body of knowledge (Sabherwal and Robey, 1995). Based upon our qualitative study, a number of potential innovative activities have been used in this study to develop a usable taxonomy that can assist in understanding the differences between organisations and their approaches toward organisational innovative activities. Those innovation activities adopted in this study vary from improving efficiency in the provision of existing products and/or services to experimenting with new products and/or services in our local market and from regularly implementing small adaptations to existing products and/or services to frequently utilising new opportunities in new markets.

Methodology

Our research question is "do SMEs cluster around different types of innovation activities?" We do not start with the assumption that all SMEs are the same when it comes to innovation. We question and probe for differences between SMEs in the activities they engage in, differences in size, differences between SMEs in new high growth market segments and mature market segments and differences between young SMEs and mature firms.

Research design

Based on the uniqueness and size of the industry studied in this paper, we used a combination of qualitative and quantitative research methodologies to better address the research question.

First, prior to our survey study, we interviewed our case organisations to understand a preliminary question of "*How do firms capture innovative opportunity in their current environment*?" Then, we employed a quantitative survey of innovation activities to develop the taxonomy.

A guided, theoretically based, sample was created from the list of organisations in the spatial industry business association (SIBA) members list, including surveying and other spatial firms with different sizes on both the east and west coast of Australia. This list represents the majority of the population of firms in Australia using SI. Firms in the industry are largely described as being Surveyors or Other Spatial Firms.

Surveyors' primary activities are measuring, assembling and assessing land and geographic-related information to be used for land planning and implementing the efficient administration of the land and the structures thereon (e.g., engineering and mining surveyors or boundary surveyors). Other spatial firms consist of SI users and IT firms that manage and analyse data that has geographic, temporal and/or a spatial context. This category also includes development and management of related IT tools, such as aerial and satellite remote sensing imagery, GPS and computerised GIS.

SI is a rapidly growing industry that consists of companies offering a wide range of geographic-related services such as surveying, remote sensing, location-based services, photogrammetry, mapping, aerial imagery, land development, environmental management, geographic information systems, web services, and GPS amongst others.[1] The SI industry contributes up to $12.5 billion annually to Australia's gross domestic product.[2] This industry includes diverse firms; some with a history of surveying, and other spatial service firms more focused on application of information technology. Some small firms are family businesses while medium-sized firms are partners in international collaborations.

For our pilot interviews, we used theoretical sampling that included four firms in Queensland, four in New South Wales, five in Western Australia, six in Victoria, and one in the Australian capital territory. We began with a list of companies obtained from our industry partner, SIBA and their characteristics, and contacted companies to request their participation in face-to-face semi-structured interviews with one member of the research team. We updated existing information from information provided and ascertained through company information on the web and phone conversations, who would be the appropriate personnel

---

[1] http://www.spatialbusiness.org/aus/Position-On-Issues.

[2] http://www.spatialbusiness.org/.

with direct involvement with innovation. We used the qualitative data to develop the survey instrument and to provide a richer profile of firms in the clusters. To provide the richer profile the firms that completed the interviews were also invited to do the survey, which allowed us to allocate them to a specific cluster.

Survey instrument development

For our survey study, the survey instrument was designed specifically to target a larger group of firms in order to gain insights into innovation activities in as many firms as possible across the Spatial Industry sector, the circumstances in which innovation appears and the implementation processes used. Survey items identified for inclusion were then assessed through a literature study to identify innovation attributes that were revealed by previous research and our pilot interviews. Subsequently, the survey was tested by the research team and representatives from the spatial industry. Their comments were considered in the finalisation of questions for the whole survey, more than 100 items covering a number of different constructs. For the purpose of this study, only the items measuring the different types of innovative activities (14) and the innovation skills (13) employed by the firm were used.

Development of the research instrument

For our pilot interview, semi-structured face-to-face interviews that on average lasted approximately 1 to 1.5 hours were employed to explore the activities and orientation of 20 firms. All interviews were recorded and transcribed and interviewers also took field notes. Both transcribed interviews and the field notes were used in the data analysis. In addition to the face-to-face interview and company observations with each company during the site visits, follow-up phone calls or emails were used to seek clarification or greater depth in particular areas. In our survey study, we contacted senior management from 20 firms from pilot interview phase to participate in the survey as well as to recommend additional firms.

Data analysis overview

In total, 64 firms participated in our survey study, but due to missing data only 59 qualified for analysis. The 59 survey respondents represent 20.4% from a total of 289 Australian industry association spatial organisations. Figure 1 illustrates the

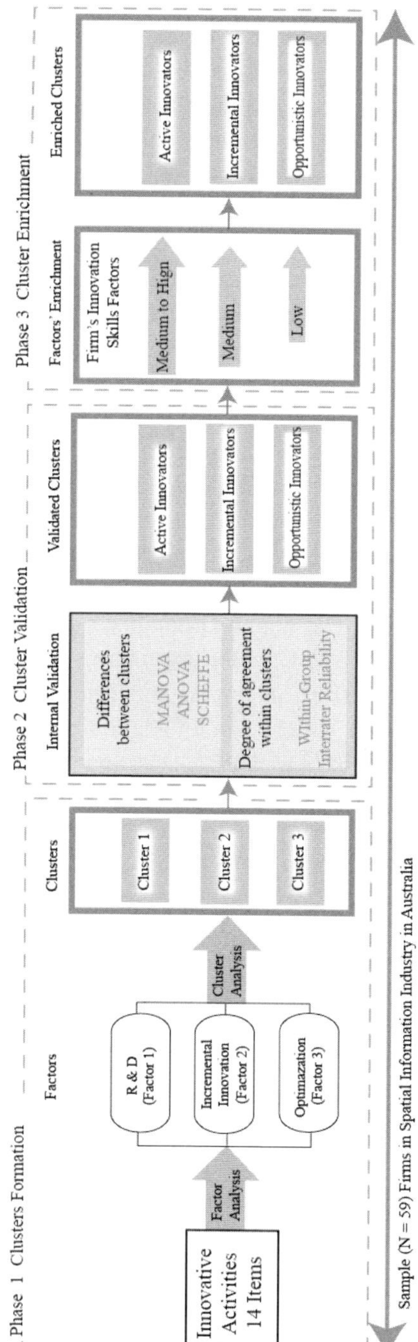

Fig. 1. Cluster formation, validation, and enrichment process.

process undertaken to develop and validate the cluster taxonomy. The survey data is analysed in several ways. First, in Phase 1 the 14 items reflecting the Innovation Activities are factor analysed in order to identify possible factors that summarise the main possible types of innovative activities that a firm could engage in. Once the factors are identified, the computed results of each factor are used in a cluster analysis in order to cluster firms based on their preferences around the different types of innovative activities. As this is an exploratory study no *a priori* number of clusters are expected, different cluster solutions are tested empirically and theoretically. The three clusters structure is chosen due to its strong internal consistency and significant external differences. In Phase 2, this paper uses two main tools to validate the three clusters: the first is evidence of high external heterogeneity and the second is high internal homogeneity (Hair *et al.*, 2003). External heterogeneity involves measuring the differences between the five clusters in relation to the eight innovative activity factors and proving that there are significant differences between them. Manova, Anova and Scheffe analyses are used for this purpose. Internal homogeneity shows the degree of agreement between managers within each cluster in relation to the degree of influence that the eight factors play on their decision making process. The Within-Group Interrater Reliability test is used for this purpose (James *et al.*, 1993). Once the taxonomy is developed, in Phase 3 we run a factor analysis of firm skills to further enrich the clusters by explaining significant differences between the clusters. In addition a multinomial regression analysis was performed to further enrich the clusters by assessing the relationship between the three clusters and the ownership of the firm as family and non-family owned.

Sample demographics

Based on the information provided in the survey and on firms' websites, the firms are divided into two different categories. The first category, Surveyors, consists of predominantly surveying firms. The second category, Other Spatial firms, consists of predominantly spatial sciences firms. Surveyors' primary activities are measuring, assembling and assessing land and geographic-related information to be used for land planning and implementing the efficient administration of the land and the structures thereon (e.g., engineering and mining surveyors or boundary surveyors).[3] Other Spatial firms consist of SI users and IT firms that manage and analyse data that has geographic, temporal and/or a spatial context. This category

[3]http://www.sssi.org.au/details/cat/8/sub/9.html.

also includes development and management of related IT tools, such as aerial and satellite remote sensing imagery, GPS and computerised GIS.[4] The group of Other Spatial firms is further subdivided into 35% IT firms, 30% consultancy firms, 15% geology firms, 10% wholesale firms, and 10% unclassified firms.

A total of 49 respondents mentioned their firm ages, which is on average 35 years. However, the spatial firms are younger than the surveying firms. The surveying firms are on average 36 years old, while the Other Spatial firms are on average 25 years old. The majority of the firm' owners are 51 years old or older.

Further, a large proportion (45%) of respondents indicated their firms are SMEs with fewer than 10 FTEs. A large majority (69%) of the respondents further indicate that their main market is local, whereas 14% of the respondents serve a national and 17% an international market. Over a third (36%) of firms in the research sample are family-owned.

Results

The study results are divided into:

(1) Table 1 shows the exploratory factor analysis of the innovative activities, which tells us how the innovation activities align into three different factors.
(2) Table 2 shows the cluster analysis and validation, i.e., how the firms align with the three factors.
(3) Table 3 shows the firm demographics by cluster.
(4) Table 4 shows the cluster naming, indicating how each cluster maps onto the three factors.
(5) Table 5 shows the factor analysis of firms' innovation skills.
(6) Table 6 shows the Anova between firm clusters and firms' innovation skills factors.
(7) Table 7 shows the multinomial regression between clusters and firm ownership (family and non-family owned).

Results of the exploratory factor analysis

Principal component analysis and varimax rotation method is used in the factor analysis of the 14 innovation activity items as shown in Table 1. Three factors

[4]http://www.crcsi.com.au/About/What-is-Spatial-Information.

Table 1. Factor analysis results of innovation activities.

Items	*Factor* 1 R&D activities	*Factor* 2 Incremental changes	*Factor* 3 cost optimisation
1. Our enterprise accepts demands that go beyond existing products and/or services	**0.510**	0.156	0.361
2. We invent new products and/or services	**0.791**	0.120	−0.078
3. We experiment with new products and/or services in our local market	**0.797**	0.251	0.242
4. We commercialise products and/or services that are completely new to our enterprise	**0.837**	0.179	0.091
5. We frequently utilise new opportunities in new markets	**0.637**	0.210	0.415
6. Our enterprise regularly uses new distribution channels	**0.647**	0.054	0.382
7. Our enterprise expands services for existing clients	**0.607**	0.500	0.110
8. We frequently refine the provision of existing products and/or services	0.473	**0.677**	0.181
9. We regularly implement small adaptations to existing products and/or services	0.452	**0.703**	−0.081
10. We introduce improved, but existing, products and/or services for our local market	0.082	**0.841**	0.147
11. We improve efficiency in the provision of existing products and/or services	0.059	**0.878**	0.209
12. We regularly search for and approach new clients in new markets	0.479	0.269	**0.631**
13. We increase economies of scales in existing markets	0.180	0.400	**0.690**
14. Lowering costs of internal processes is an important objective	0.027	−0.031	**0.679**
Cronbach's Alpha	**0.86 (6 items)**	**0.86 (4 items)**	**0.64 (3 items)**

Note: The Cronbach alpha of three items is lower than the other three cluster due to short scales. However, the optimal range for the inter-item correlation of 0.2 to 0.4, which is recommended by the Briggs and Cheek (1986).

were found to have an eigenvalue over 1.0, explaining 65.6 % of the total variance. We label these three factors based upon the type of innovation activities involved in each factor as following:

(1) R&D factor.
(2) Incremental innovation factor.
(3) Cost optimisation factor.

Results of the cluster analysis

Table 2 shows the results of the two-step cluster analysis performed on three desired clusters.

Internal validity: Manova, Anova, Scheffe, and within-group interrater reliability

The first part of the internal validity analysis about the three clusters is to validate the significance of the degree of external heterogeneity between them. Manova shows the three clusters are significantly different (Table 2, $F = 15.787$, $p < 0.0001$). Variations in dimensions of the three factors among the three clusters were gleaned from Anova, as summarised in Table 2. Scheffe tests also highlighted the distinguishing qualities of the three clusters. In reviewing the profiles of the three clusters (Table 1), the reader should know that because the number and content of clusters were inseparable from the classification criteria used, the results of Anova were presented only to illustrate where the greatest differences existed among the clusters. The second part of the validation is to

Table 2. Two-step cluster analysis performed on three clusters.

Factors	Cluster 1 ($N = 21$)	Cluster 2 ($N = 29$)	Cluster 3 ($N = 9$)	Scheffe results (factor differences)	F	Sig.
Factor 1 R&D	4.16	3.42	2.40	1&2, 2&3, and 1&3	20.831	0.000
Factor 2 incremental innovation	4.69	4.15	2.81	1&2, 2&3, and 1&3	29.700	0.000
Factor 3 cost optimisation	4.75	3.55	2.41	1&2, 2&3, and 1&3	74.000	0.000

measure the degree of agreement (internal homogeneity) inside each cluster. The results of the application of the degree of interrater agreement within each of the three clusters are as follows: Cluster 1 (0.94), Cluster 2 (0.95), Cluster 3 (0.98). James *et al.* (1984) measure the degree of agreement between each member within a cluster 0 (no agreement) and 1 (total agreement). The results of the three clusters is between 0.94 and 0.98, which indicates very high internal agreement results. It can be argued that firms within each of the three clusters have very strong agreement about their choice respectively to the different types of innovative activities they pursue.

The results from the cluster analysis confirm the existence of three well-defined clusters. In general, this study has found that a three-cluster structure exhibits high internal homogeneity or a strong degree of agreement in within each cluster (James *et al.*, 1984). The results suggest that each individual firm included in any of the three clusters displays very strong similarities in terms of the type of innovative activities they pursue. In addition to that, each cluster exhibits a high external heterogeneity in terms of keeping consistent, significant differences between each cluster, and in particular, each cluster represents a unique mix of preferences with respect to the three types of innovative activities (factors). In developing these three clusters, each cluster has been analysed individually, and inferences have been based on examination of the internal relations between the scores from the three factors. Each cluster can be treated as a unique pattern about the degree of influence that different innovative activities are likely to have on the decision making of SI firms.

Cluster naming and definition

Once the taxonomy has been empirically tested and validated, the next step is to discuss the clusters solution in a way that managers/owners can identify with and use in a beneficial way in their daily activities. Therefore, each cluster was triangulated with our pilot interview with an additional 20 firms, in order to provide the meaningful cluster names. Semi-structured face-to-face interviews that on average lasted approximately 1 to 1.5 hours were employed to explore the activities and firms' orientation. We used semi-structured interviews to have comparability across firms and used these interviews to obtain better ideas about innovation activities that were carried across firms. By using the results of the interviews and three clusters, three taxonomies were chosen that encapsulated the essence of each profile (Table 3).

Table 3. Three emerging taxonomies of innovation activity by firm size and industry type.

Clusters	Cluster 1 active innovators	Cluster 2 incremental innovators	Cluster 3 opportunistic innovators
By firm size			
Small	6	8	9
Medium	8	12	0
Large	3	4	0
By industry type			
Surveyor	11	3	2
Spatial sciences firms	10	25	7
By firm ownership			
Family owned	6	9	7
Non-family owned	13	18	2

After we identified the innovative firm taxonomy (active innovator, incremental innovator, and opportunistic innovator), we further explored how these taxonomies are mapped on the innovation drivers (i.e., R&D, incremental changes, and optimisation) (see Table 4). The results are discussed below.

In order to enhance our definition of each cluster additional data was used to assess the skill set that firms in each of those clusters were perceived to possess in relation to their competitors (see Table 5 for a list of these skill sets). The results of factor analysis yielded three (factors) sets of skills, which are: innovation & technology, marketing & advertisement, and human resources.

Once those three sets of skills are identified, Anova is used to identify significant differences between the three clusters and the level of development of each of those skills (Table 6). As a result of this further analysis, it is possible to

Table 4. Innovation taxonomies by innovation activity factors cluster naming based upon how clusters map onto factors.

Innovation activity factors	Cluster 1 active innovators ($N = 21$)	Cluster 2 incremental innovators ($N = 29$)	Cluster 3 opportunistic innovators ($N = 9$)
Factor 1 R&D Factor	High (4.16)	Moderate (3.42)	Low (2.40)
Factor 2 incremental changes	High (4.69)	High (4.15)	Low (2.81)
Factor 3 optimisation	High (4.75)	Moderate (3.55)	Low (2.41)

Table 5. Factor analysis of firms' innovation skills.

Items	*Factor* 1 innovation & technology	*Factor* 2 marketing & advertisement	*Factor* 3 human resources
1. Assessing the feasibility of new technology	**0.893**	0.120	0.164
2. Learning about technology it has not used before	**0.865**	0.160	0.081
3. Identifying promising new technologies	**0.770**	0.305	0.340
4. Assessing the potential of new markets	**0.614**	0.438	0.419
5. Implementing new types of production processes	**0.603**	0.379	0.252
6. Developing new pricing strategies	−0.004	**0.886**	0.029
7. Developing new advertising or promotion strategies	0.414	**0.776**	0.185
8. Researching new competitors and new customers	0.415	**0.693**	0.242
9. Leveraging its brand reputation or company image to new markets	0.339	**0.681**	0.445
10. Recruiting specialised staff in technical areas it is not familiar with	−0.032	0.030	**0.813**
11. Setting up a new sales force	0.113	0.311	**0.799**
12. Setting up new distribution channel.	0.485	0.189	**0.672**
13. Building relationships in new markets	0.407	0.385	**0.670**
Cronbach's Alpha	**0.912 (5 items)**	**0.875 (4 items)**	**0.833 (4 items)**

argue that the active innovators perceive themselves very high skilled in the areas of innovation and technology and marketing and reasonably high in terms of HRM. The Incremental Innovator firms seem to position themselves around the average or just above average in all three set of skills. Firms in opportunistic innovator cluster perceived themselves as less skilled in all the three areas than their competitors.

Cluster 1 ($N = 21$) active innovators: "*We're good at finding new products, services and markets!*"

Table 6. Anova between firm clusters and firms' innovation skills factors.

Factors	Cluster 1 active innovators ($N = 21$)	Cluster 2 incremental innovators ($N = 29$)	Cluster 3 opportunistic innovators ($N = 9$)	Sig.
Factor 1 innovation & technology skills	High (4.40)	Moderate (3.85)	Low (2.80)	0.000
Factor 2 marketing and advertising skills	High (4.47)	Moderate (3.45)	Low (2.71)	0.000
Factor 3 human resources skills	Moderate (3.75)	Moderate (3.26)	Low (2.25)	0.000

Those firms consistently rate all of the three types of innovative activities highly; R&D, incremental innovation and cost optimisation. They endeavour to maximise their competitive advantage through actively pursuing innovative solutions. A firm in this cluster invents, experiments, commercialises new products and services, in new markets and new distribution channels.

In order to enhance our definition of each cluster additional data was used to assess the skill set that firms in each of those clusters were perceived to possess in relation to their competitors (see Table 5 for a list of these skill sets). The results of factor analysis yielded three (factors) sets of skills, which are: innovation & technology, marketing & advertisement and human resources. Active innovators were a high on the innovation & technology, and marketing & advertisement, skill sets. They were moderate on human resources skills.

The following observations and comments are from the qualitative data collected in interviews.

The Cluster 1, Active Innovator firms were all highly technical. All three Cluster 1 firms in our qualitative study use spatial sciences. Of the three, one worked with satellite GPA, another developed terrestrial laser scanners for the mining industry and the third was a survey services firm with about 60% survey, 20% planning and 20% spatial using GPS across all three areas of the business.

The two spatial sciences firms claim to have developed and introduced radically new "World's First" innovations and the third survey services firm claims to use radically new technology. The two development firms have international operations, while the services firm was very large in one state of Australia.

The two development firms held patents, and invest in R&D. They have highly technical employees, who are also entrepreneurial in their thinking, highly motivated, for example one manager reported his people working until 4 am.

They also reported having an entrepreneurial culture and spending time identifying new product ideas through brainstorming.

> "We've got an awesome team, fantastic group of guys, you know, working until four o'clock in the morning. I don't have a life, in fact I have a famous saying is that I've never worked a day in my life, because I've never done anything I don't want to do." — Small Spatial Development Firm

One of the barriers to development reported by the smaller spatial development firm (30 employees) was trying to balance time spent on innovation against time spent on generating revenue. Size was mentioned as a barrier to innovation, but that motivated the firm to apply for R&D grants, which they were successful in getting.

The services firm does not have a formal R&D program but they do embark on R&D projects. They report allocating staff to such projects during slow periods to avoid having to lay people off. They did report buying new software and hardware. Like the development firms the service firm reports brainstorming, internal needs analysis, looking for new opportunities, improvements and market diversification. Their research activities included fact finding, looking for new products and new ways to use them. They reported spending 1% of turnover on R&D, another 1% on experimenting with new technologies and 5% on purchase of the new software. They attended conferences and were members of industry organisations, which gave them access to new technologies.

> "... this is where we actually had some spare capacity during the GFC which we were able to deploy into R and D projects so that when we came out of the GFC we had some new things to put to the market." — Survey Services Firm

Firm size was seen as a barrier to innovation by the services firm due to the need to spread the high cost of new technologies across their relatively small workforce (150 employees). The Services firm measured innovation in the same the way they would justify a purchase to the Board, in terms of productivity increases:

> "That job I did two years was very similar to this job, that took two days, this took a day and a half, so it is measured by improvements in productivity, which affects the bottom line, but I know that the board look at it and say, yeah, the return on investment. They spent $100K on a piece of equipment, they want to attribute $200K worth of work on that equipment in that year." — Large Survey Firm

Cluster 2 ($N = 29$) incremental innovator: "*We're good at maintaining a continuous process of small improvements*"

Firms in this cluster scored very high incremental innovation factors (Factors 2) and moderately on the other two, suggesting that incremental innovation is the main focus. They concentrate on frequently refining the provision of existing products and/or services, implementing small adaptations to existing products and/or services, introducing improved, but existing, products and/or services for their local market, and improving their efficiency in the provision of existing products and/or services.

Incremental innovators were a moderate on all three skill sets: innovation & technology, marketing & advertisement and human resources.

The following observations and comments are from the qualitative data collected in interviews.

One Incremental Innovator firm was a small firm with 45 employees started 42 years ago in the 1980s by the current Managing Director's Father, importing radar guns and survey equipment and providing mapping and spatial services to clients. The firm serves the Australian market, with only 5% overseas sales in the Asia Pacific. The major customer is government (60%), utilities firms and people who want to change land use, for example from a farm to a subdivision, golf course, or mine, or prepare the property for sale. The Managing Director had been with the firm 12 years, who took over from his father about eight years ago. Recently, five years ago, they purchased another company, which had provided them with photography capture services. The firm is experiencing greater competition from Europe, where there is not much work during the winter. And there is a constant change in technology with new cameras available all the time. This firm estimates that 40% of their sales come from products or services introduced in the last five years.

This firm defines innovation as "improving current practices, whether they're internal practices or external practices." They measure innovation by the speed of processes, the amount of time to do a job takes compared to how long it would have taken in the past. They rely upon innovation because they have to move with the times, for example with hardware and software. The innovation activities they carry out includes being open to new ideas and encouraging the people actually completing the work to improve the methods they use. So if they are doing a pilot program they might improve the process by using a new piece of software to a small part of the business before they roll it out to the client. This incremental innovation also applies to marketing with new online tools.

They describe their business strategy as evolving and growing the firm from a one manager firm to managers in different areas, who cross pollinate with different ideas they have learned in their area to help each other and to improve what is happening in the various areas. So in addition to the incremental innovation is internal collaboration.

This firm does not develop radically new innovations, but operates more in the area of using new tools and finding new applications for existing tools. For example, with their digital camera they got the same result with less flying time, which helped them reduce the price for the customers, who were quite pleased with the results. They got help from the vendor of the equipment, which gave them the opportunity to trial the innovation. This firm also uses trial and error to identify innovations.

The company seems to have a dense network of clients, and other companies in the industry, who give them support for overflow of work; "if they had too much work and needed some help they could come to us, and vice versa, you know we could support each other to increase that capacity... so that we sort of had an informal network there. But we also kept in touch with each other to keep up with what they're doing and they keep us with what we're doing." This firm also has contacts with universities, government and industry groups.

The Incremental Innovator firm has no formal R&D budget, patents, licencing or other intellectual property. They do spend an estimated $100,000 per year on innovation associated costs, such as sending people on training courses, and buying new software and hardware, plus allowing time for trial and error testing.

The incremental innovator can be discouraged from innovation by short-time frames: "you know people want things yesterday... so the time constraints can, I guess you stick to what you know in order to get it done... but when you get the opportunity to innovate then sometimes it can save you time in the end."

They were also discouraged by clients who are quite prescriptive about how they want things done, which does not give much leeway to improve the process: "Yes, instead of saying we want the final product and we want it to be this, they say well you need to do this and this and this and then get the final product."

Government programs to encourage innovation were avoided due to the paperwork involved.

Cluster 3 ($N = 9$) opportunistic innovator: "*Innovation... is changing your work practices to suit the technology that is available.*"

Firms in this cluster scored low across all the three factors, R&D, incremental innovation and cost optimisation, compared with the two previous clustering firms. Nonetheless their score for incremental innovation was highest among the three factors.

Opportunistic innovators were a low on all three skill sets.

The following observations and comments are from the qualitative data collected in interviews.

The Opportunistic Innovator firm was a small survey practice, which was started in 1975. The firm previously had two partners and 22 staff but one partner left and the decision was made to focus on the more profitable business so over time the firm reduced to 10 employees. Their major clients include a power utility company and City Councils. The industry is perceived to be very competitive. Their innovation activity was described as a "toolbox talk":

> "We've got this thing where we are going to be setting out a block of units, five multi storey, it is not technical, it is just mundane, but there are all these new instruments and ways of doing it, and so we have developed, and gone to the instrument factory and said we've heard so and so are doing this, so we've brought this, what they call a robotic instrument, so instead of the picture of the surveyor standing there and the man standing down the road, now you can get away with just the man standing there, the instrument is all electronic, it's got little motors in it, so it is telling you via a board what to do and where to go.
>
> But then there is all the technical, it is very simple to say yeah that tells you that, but you know there is a lot of work going on in the program for it to tell you that. So you are going through this innovation, change for doing it." — Small Survey Firm

The Small Survey firm in Cluster 3 perceived innovation as adjusting work practices to the technology:

> "Innovation to me is changing your work practices to suit the technology that is available." — Small Survey Firm

It appears as though the segment of the industry is also more responsive to technology change, rather than involved in active innovation:

> "I wouldn't say that we innovate ourselves, but more that we will hear about something and then adapt it to our particular, more an adaption of what is available, to suit our needs." — Small Survey Firm

For them innovation is learning how to use the new tools available:

> "Yeah and keeping up to date with the latest software programs which is always, yes you buy a new program, but then you've got to change something in your system, to get it to work, it is never just loaded up and keep going sort of thing." — Small Survey Firm

This group learns about innovation from their industry networks:

> "Yeah and because they've got this panel of surveyors, so they're dealing with, they will say oh do you know that Joe has gone and done this, or have you thought about doing that, and that has led us, they have said to us we will give you this particular job if you can do it in a certain way, so we do create a three dimensional model for them, it was nothing, it was different for us, but it was all just making more use of the functions in the CAD systems or in the systems, yeah." — Small Survey Firm

In this cluster change is perceived to be incremental and client driven:

> "So just this incremental step changing all the time, but it's brought about by these things coming onto the market and you finding a way of using it, and you sort of come up like, yeah you provide more information then that information gets used by the architects or engineers and then they ask for more and you've got to." — Small Survey Firm

Like the Survey Services firm in Cluster 1, the Small Survey Firm in Cluster 3 also saw innovation as measured by increases in productivity. This could be typical of services firms. In this cluster the respondent indicated that an innovation should be paid for by the first job in which it is needed:

> "I'm thinking about that, if someone came to you with a job that was a $200K job, let's say and you needed $100K, you worked out that if you could get an instrument that could do it for $100K I would say yes, we will go down that path because we will then use that somewhere else." — Small Survey Firm

Table 7. Multinomial regression between clusters and firm ownership (family and non-family owned).

	b	*z*	$P > \|z\|$	e^b	$e^{b^{StDY}}$	%	%StdX
Active versus opportunistic	−2.0260	−2.152	0.031	0.132	0.367	−86.8	−63.3
Opportunistic versus active	2.0260	2.152	0.031	7.583	2.723	658.3	172.3
Opportunistic versus incremental	1.9459	2.163	0.031	7.000	2.617	600.0	161.7
Incremental versus opportunistic	−1.9459	−2.163	0.031	0.143	0.382	−85.7	−61.8

One of the benefits of innovation was engaging and retaining staff:

> "Well better productivity and I think it is a way of engaging the staff, by having the best sort of stuff is a way of engaging them and keeping them." — Small Survey Firm

In summary the opportunistic innovation cluster perform a lower level of innovation, mainly to keep up with the industry, achieve productivity gains and retain staff. They are opportunistic because they only engage in innovation when the opportunity presents, such as when a client will pay for the innovation or it will save money on the job.

Table 7 shows the multinomial regression between clusters and firm ownership (family and non-family owned).

In addition we enriched our understandings of the clusters by investigating if the clusters differ in terms of their family business ownership due to the recent literature pointing to potential differences in innovation between family firms and other firms (De Massis *et al.*, 2015).

The main results of the multinomial regression are around the odds of having an Opportunistic Innovation firm versus having an Active Innovation firm increase by 7.58 times (658%) for family-owned companies. The odds of having an opportunistic innovation versus an Incremental Innovation firm increase by seven times (600%) for family-owned companies

Discussion

First, using exploratory factor analysis, we find that SME innovation activities group into three factors: R&D, incremental innovation and cost optimisation (see Table 1). Then we find that SME firms cluster around these factors into three groups based upon the innovation activities they engage in. The first group is the "active innovator", who are actively engaged in all three types of innovative activities (i.e., they are relatively high on all three innovation activity factors). This suggests that firms in this category can be seen as active innovators, as they are happy to commercialise their own knowledge and to use externally generated knowledge (Lichtenthaler, 2009; Robertson *et al.*, 2012). The second cluster, "Incremental Innovators", represents those firms that believe in continuous improvement, mostly from internal effort. In contrast, firms in the third cluster, "Opportunistic Innovator" scored low in all the three types of innovative activities, suggesting that they are adverse to innovation practice. It is interesting that the three clusters scored at largely the same level across all three of the innovation factors, rather than each clustering forming around a different factor, i.e., The active innovators are high on all factors, incremental innovators are medium to high on all factors and opportunistic innovators are low on all factors. Thus, it is the level of activity that mainly differentiates these innovation activity clusters. In addition in an enrichment analysis, we find that the different clusters develop different innovation skills. Active Innovators are high in innovation & technology skills and marketing & advertising skills and moderate in human resources skills. Incremental innovators are moderate in all skills and opportunistic innovators are low on all. Thus we show that these clusters have different behaviours, and so they can be used to predict more than just the innovation activities of the firm. More research is needed to investigate the behaviours of these clusters.

In addition we discovered that the odds of having an Opportunistic Innovation firm versus having an Active Innovation firm increase by 7.58 times (658%) for family-owned companies and the odds of having an Opportunistic Innovation versus an Incremental Innovation firm increase by seven times (600%) for family-owned companies. This means that family-owned businesses are much more likely to be opportunistic than active or incremental innovators. This finding greatly enriches our understanding of these firm clusters and justifies them based upon demographic differences, as well as the behavioural differences we found. Further research is needed to reconcile these findings with other studies of innovation in family business, such as De Massis *et al.* (2015), who found that family firms engage in innovation processes aimed at developing and bringing to market incremental new products, which are carried out relying on a functional organisation, with high levels of decisional autonomy given to the project leader.

As a result this research makes three very important contributions to innovation theory and practice. The first contribution is the identification of three SMEs' innovation activity factors, which have been shown to be useful in distinguishing different types of firms, based upon their innovation activities. These innovation activity factors will be very helpful for future researchers and practitioners to better understand SMEs and assess their engagement in innovation. The management implications are that SMEs, possibly at the recommendation of their consultants or advisors, can perform a simple survey to identify their scoring on these three innovation activity factors. Future research is needed to test the theory that it is mainly the amount of innovation activities the SME engages in that differentiates firms.

The second major contribution is the use of the three factors to identify three distinct clusters of firms, with each cluster having a different profile in terms of its practice of innovation activities and the skills they develop. Profiling firms based upon their innovation activities will allow future researchers to better select firms for qualitative studies, ensuring that they do not miss one of the clusters in their sample selection. Understanding these clusters will also help SME practitioners, including SME innovation instructors, consultants, advisors and mentors, to develop more targeted and appropriate materials and interventions. By understanding that there are three distinct types of firms with different innovation practices practitioners and their advisors can avoid the "one size fits all" approach to innovation advice and practice. Some SMEs likely fall into the category of active innovators, incremental innovators and others will be opportunistic innovators. Neither strategy is likely to be right or wrong, but can be judged based upon the context of the firm. This will allow for more nuanced practice and advice for SMEs. Additionally, if the firm is not clearly situated in a cluster, or they are not satisfied with their cluster as per the assessment, the survey provides an easy mechanism to indicate what actions the firm can take to transition into the more desirable cluster.

The third contribution is the enrichment of the clusters with the identification of their skill levels in innovation & technology, marketing and advertising, and human resources. Identification of these skill levels contributes to the theory of innovation activity clusters and provides further guidance to firms seeking to change their cluster positioning.

Like all research, this study has its limitations, which create opportunities for future research. First, this study is conducted in one country creating the opportunity to replicate this methodology in other countries and conduct across country comparative analysis. Secondly, the study was conducted in the SI industry, which provided both mature and growth industry segments and reduced the complexity of the data. This provides the opportunity to replicate this study in other industries and conduct an across industry analysis. Future research could also look for the possible external or internal factors, which differentiate clusters and investigate the performance across clusters.

Conclusion

How do SMEs innovate and what innovation activities do they perform? To investigate this question we collected qualitative and quantitative data from the SI Industry in Australia to develop a taxonomy of innovation activities in SMEs.

This paper has demonstrated that SMEs can be clustered around factors of innovation activities such as R&D, incremental innovation, and cost innovation. As a result a taxonomy of innovation activities in SMEs in the SI Industry in Australia has been developed based on qualitative and quantitative primary data. The taxonomy provided three clusters of firms engaging in similar innovation activities which were identified: active innovators, incremental innovators, and opportunistic innovators. These clusters also exhibit significant differences in other parameters such as internal development of particular skills sets and family business ownership.

References

Alvarez, S and L Busenitz (2001). The entrepreneurship of resource-based theory. *Journal of Management*, 27(6), 755–775.

Archibugi, D (2001). Pavitt's taxonomy sixteen years on: A review article. *Economics of Innovation and New Technology*, 10(5), 415–425.

Ardichvili, A, R Cardozo and S Ray (2003). A theory of entrepreneurial opportunity identification and development. *Journal of Business Venturing*, 18(1), 105–123.

Branzei, O and I Vertinsky (2006). Strategic pathways to product innovation capabilities in SMEs. *Journal of Business Venturing*, 21(1), 75–105.

Burns, T (1967). The comparative study of organizations. In *Methods of Organizational Research*, VH Vroom (Ed.), pp. 113–170. Pittsburgh, PA: University of Pittsburgh Press.

Davidsson, P (1989). Entrepreneurship — and after? A study of growth willingness in small firms. *Journal of Business Venturing*, 4(3), 211–226.

Briggs, SR and JM Cheek (1986). The role of factor analysis in the evaluation of personality scales. *Journal of Personality*, 54, 106–148.

De Jong, JP and O Marsili (2006). The fruit flies of innovations: A taxonomy of innovative small firms. *Research Policy*, 35(2), 213–229.

De Massis, A, F Frattini, E Pizzurno and L Cassia (2015). Product innovation in family versus nonfamily firms: An exploratory analysis. *Journal of Small Business Management*, 53(1), 1–36.

Eggers, JP and S Kaplan (2009). Cognition and renewal: Comparing CEO and organizational effects on incumbent adaptation to technical change. *Organization Science*, 20(2), 461–477.

Hair, JF Jr, B Babin, AH Money and P Samouel (2003). *Essential of Business Research Methods*. USA: Wiley.

Hoffman, K, M Parejo, J Bessant and L Perren (1998). Small firms, R&D, technology and innovation in the UK: A literature review. *Technovation*, 18(1), 39–55.

Hsieh, C, JA Nickerson and TR Zenger (2007). Opportunity discovery, problem solving, and the theory of the entrepreneurial firm. *Journal of Management Studies*, 44(7), 1255–1277.

James, LR, RG Demaree and G Wolf (1993). RWG: An assessment of within-group interrater agreement. *Journal of Applied Psychology*, 78(2), 306–309.

Ketcher, DJ and CL Shook (1996). The application of cluster analysis in strategic management research: An analysis and critique. *Strategic Management Journal*, 17(6), 441–458.

Lazoi, D, F Ceci and A Corallo (2011). Collaboration in an aerospace SMEs cluster: Innovation and ICT dynamics. *Journal of Innovation and Technology Management*, 8(3), 393.

Lichtenthaler, U (2009). The role of corporate technology strategy and patent portfolios in low-, medium- and high-technology firms. *Research Policy*, 38(3), 559–569.

Rich, V (1992). Code on medical ethics divides Poland's doctors. *New Scientist* (1971) 133(1811), 13.

Robertson, PL, G Casali and D Jacobson (2012). Managing open incremental process innovation: Absorptive capacity and distributed learning. *Research Policy*, 41(5), 822–832.

Rosenbusch, N, J Brinckmann and A Bausch (2011). Is innovation always beneficial? A meta-analysis of the relationship between innovation and performance in SMEs. *Journal of Business Venturing*, 26(4), 441–457.

Rotaba, Z and C Beaudry (2012). How do high, medium, and low tech firms innovate? A system of innovation (Si) approach. *International Journal of Innovation and Technology Management*, 09(5), 1250034. DOI: 10.1142/S0219877012500344.

Sabherwal, R and D Robey (1995). Reconciling variance and process strategies for studying information system development. *Information Systems Research*, 6(4), 303–327.

Sawang, S and K Unsworth (2011). Why adopt now? Multiple case studies and survey studies comparing small, medium and large firms. *Technovation*, 31(10), 554–559.

Sokal, RR and PHA Sneath (1963). *Principles of Numerical Taxonomy*. San Francisco: W. H. Freeman & Co.

Shobha S, DAS and H Zheng (2008). Innovation in high-technology SMEs: Insights from Singapore. *International Journal of Innovation and Technology Management*, 05, 475. DOI: 10.1142/S0219877008001473.

Unsworth, K, S Sawang, J Murray, P Norman and T Sorbello (2012). Understanding innovation adoption: Effects of orientation, pressure and control on adoption intentions. *International Journal of Innovation Management*, 16(01), 1250004–1250039.

Yu, A, GT Lumpkin, RL Sorenson and KH Brigham (2012). The landscape of family business outcomes a summary and numerical taxonomy of dependent variables. *Family Business Review*, 25(1), 33–57.

Zaltman, G, R Duncan and J Holbek (1973). *Innovations and Organizations*. New York: John Wiley.

Chapter 18

THE IMPACT OF THE PERCEIVED USEFULNESS OF WORKPLACE SOCIAL NETWORKS UPON THE INNOVATIVE BEHAVIOUR OF SME EMPLOYEES A SOCIAL CAPITAL PERSPECTIVE

Matthew J. Xerri and Yvonne Brunetto
Southern Cross University

This research includes an examination about the impact of three organisational factors upon the perceived usefulness of workplace social networks for problem solving in engineering SMEs. As well this research examines the impact of the perceived usefulness of workplace social networks upon the innovative behaviour of engineering SME employees. More specifically, the dimensions of Social Capital Theory are applied as a lens to develop an understanding into the effect of the strength of workplace social network ties, sociability and organisational culture upon the perceived usefulness of workplace social networks for problem solving. This study examines the proposed model by applying mixed methods. The research method includes a survey with engineering employees' and interviews with senior management. The findings confirm that the organisational factors tested (tie strength, sociability and organisational culture) impact upon the perceived usefulness of workplace social networks for problem solving. Furthermore, the findings also confirmed that the perceived usefulness of workplace social networks affects the innovative behaviour of engineering SME employees. Therefore, this research adds to the current body of literature by providing insight into the usefulness of workplace social networks for problem solving and the impact this has on the innovative behaviour of engineering SME employees.

Keywords: Network content; social networks; Social Capital Theory (SCT); innovative behaviour; small to-medium-sized enterprises (SMEs); tie strength; sociability; knowledge-based firms.

This paper was first published in *International Journal of Innovation Management*, Vol. 15, No. 5 (October 2011), pp. 959–987. © Imperial College Press. DOI: 10.1142/S1363919611003350

Introduction

Developing innovative business practices has become an imperative function in today's organisations. The constantly changing global economy, within which firms operate, is forcing organisations to develop competitive practices that can provide innovative solutions to work based problems. The transfer of knowledge is a key factor in providing employees the resources to solve work based problems innovatively. Considering that innovation is a vital factor in the development of a competitive advantage, there is a lack of research examining the workplace network factors that impact upon a small-to-medium sized enterprise (SME) employees innovative behaviour and their ability to produce innovative solutions to work based problems.

Current literature describes innovative behaviour as a three-step process. The first step is to identify a work based problem, the second step is to create new ideas and solutions for problems, and the last step is to create support for the new ideas and solutions for integration into the firm (Carmelli *et al.*, 2006). However, this process assumes an easy flow from one step to another. Instead, the process of fostering innovative behaviour may be quite the opposite. This paper examines the role of workplace social networks in affecting the second step, solving workplace problems. The significance of workplace social networks becomes apparent because in spite of the advancement of decision making tools and decision support systems, it is proposed that in an organisational context employees' still prefer to communicate with others who form their workplace social network as a way of gathering important information for decision making (Nebus, 2006). Therefore, the aim of this study is to examine the content of the workplace network and how sociability, organisational culture and the strength of workplace social network ties impact upon the perceived usefulness of workplace social networks for problem solving.

The development of effective workplace social networks is believed to impact on the perceived usefulness of such networks for problem solving (Adler and Kwon, 2002; Lin, 2001). Furthermore, the ability of an employee to solve a problem efficiently and effectively and the support they are given will have a direct impact upon the innovative behaviour of SME employees (Janssen, 2005). So far, there have been several studies that examine the impact of organisational factors upon the innovative behaviour of employees (Carmelli *et al.*, 2006; Janssen, 2005; Maqsood *et al.*, 2007; Scott and Bruce, 1994). However, there has been a paucity of research about the innovative behaviour of employees in SME's, particularly in the service sector (Scozzi *et al.*, 2005). As such, this study aims to provide insight into SMEs and the service sector by examining a small-to-medium sized engineering consultancy. Moreover, a review of the literature suggests no studies that examined the impact of the perceived usefulness of workplace social networks upon the innovative behaviour of SME employees. Therefore, the examination of the relationship between these two factors provides new information about the connection between workplace social networks and innovative behaviour. Innovation is a popular topic amongst practitioners and researchers alike, because SMEs currently require the development of innovative business practices if they are to compete with larger organisations, especially when operating within competitive markets (Brunetto and Farr-Wharton, 2007; Huang and Rice, 2009). Therefore, this study contributes new knowledge for academics and practitioners alike.

Social Capital Theory (SCT) will be utilised as a lens for examining the direct impact of some organisational factors upon the perceived usefulness of workplace social networks for problem solving and the indirect impact upon the innovative behaviour of SME employees. The premise of SCT is that the quality and quantity of workplace social network relationships impacts upon an employees' ability to be able to source appropriate information and resources (Adler and Kwon, 2002; Lin, 2001; Tsai and Ghoshal, 1998). Past literature suggests that the quality of network relationships that are used to tie together two or more employees is reliant on whether there is a culture developed around obligations behaviour, mutual trust and reciprocity (Akdere, 2005; Al-Alwai *et al.*, 2007; Dobni, 2008; Koeberg and Chusmir, 1987). Such a culture is the platform upon which relationships are developed because this is how members determine whether they are exhibiting appropriate workplace actions and behaviour (Scott and Bruce, 1994). This research is imperative for organisations' seeking to gain a competitive advantage through the development of workplace social networks. Such networks are intended to directly facilitate the development of workplace social networks, which are designed to support the solving of work based problems and provide a platform for the development of the innovative behaviour of employees.

There are several facets that emphasise the significance of carrying out this research. These facets include the deficient amount of empirical research about SMEs regarding how to generate, maintain and transmit innovative knowledge (Johnson, 2002; Nelson *et al.*, 2007; Scozzi *et al.*, 2005). This study is designed to examine the impact of three organisational factors upon the perceived usefulness of workplace social networks and the impact the perceived usefulness has upon the innovative behaviour of SME employees. More specifically, the research is conducted to examine workplace social networks functionally by analysing the role of particular organisational factors in relation to the transfer of important work-based information. For that reason, the subsequent primary research questions were developed to guide the direction of the study, in particular the data collection.

> *"What is the impact of the tie strength, sociability and culture upon the perceived usefulness of workplace social networks in SMEs?"*
>
> *"What is the impact of the perceived usefulness of workplace social networks upon the innovative behaviour of employees in SMEs?"*

This paper is written in three parts. The first part explains the theory and concepts from which the hypotheses emerge. The next part outlines the methods utilised. Then finally the results and discussion are used to test and support the hypotheses.

Background

Social capital theory

Past research suggests that there are three organisational dimensions that can be used as a lens for examining the level of social capital within an organisation (Nahapiet and Ghoshal, 1998). They include the structural dimension (examining strength and content of network ties), relational dimension (examining sociability) and the cognitive dimension (examining culture) (Adler and Kwon, 2002; Akdere, 2005; Nahapiet and Ghoshal, 1998; Tsai and Ghoshal, 1998). Firstly, such structures include the quality of the bonds between workplace social network members (tie strength), as well as the quantity of members within a workplace social network (network content). Therefore, to promote tie strength and the development of effective workplace social networks organisations must embed mechanisms (structures) as support. Secondly, the relational dimension provides a lens for examining the impact of relational factors such the sociability of employees (the social skill required to develop effective workplace social networks) upon the perceived usefulness of workplace social networks. Taylor *et al.* (2004) suggest that to develop a supportive culture and successfully leverage the knowledge of network members, social relations must be effectively facilitated by the organisation. Lastly, the cognitive dimension provides a theoretical lens for examining the influence of cultural resources (organisational culture) in developing a common perception of the organisation's vision, values and protocol within a social system. The effective development of organisational culture can be facilitated by embedding structures and policies that clearly outline the organisation's vision and values (Akdere, 2005). However, in practice developing a culture that supports the development of workplace social networks has proved to be a major organisational challenge.

Examining the structural dimension of social capital theory

Tie strength

The structural dimension of SCT is usually examined using tools from networking literature. Current literature suggests that the information and resources that could be derived from a social network are dependent on the strength of ties amongst workplace social network members (Levin and Cross, 2004). Social capital is said to be a result of social relations because of the quantity and quality of relationships between dyads, the cultural support mechanism and the structure of the network (Adler and Kwon, 2002; Nahapiet and Ghoshal, 1998). A dyad is the development of a relationship between two people and can be analysed by examining the strength of the relationship or tie. Tie strength can be defined as the level of proximity between two workplace social network members. Granovetter (1973) suggests four determinants of tie strength the amount of time, the emotional intensity, the intimacy and the reciprocal services which characterise the tie. Therefore, increased contact over a period of time will allow greater opportunity for workplace social network members to form bonds and to develop a rapport.

Social network theory is a growing body of knowledge that argues that social interaction enhances or constrains access to valued resources (Sparrowe *et al.*, 2001). Current literature suggests that social network ties support collaborative thinking and the sharing of knowledge (Dooley and O'Sullivan, 2007; Fliaster and Spiess, 2008). In addition, such collaboration plays a pivotal role in developing an effective workplace social network to support problem solving. A network with strong ties means that network members have formed close relationships, beneficial for problem solving. In contrast, a network with weak ties provides network members with access to a range of diverse perspectives and new information (Levin and Cross, 2004).

Organisations that facilitate a culture that supports the transfer of knowledge and resources between workplace social network members are more likely to develop useful network ties for problem solving. Developing large numbers of interpersonal linkages by focusing on structure, relationships and culture is by no means an easy task for organisations (Granovetter, 1973; Li, 2007). Allen *et al.* (2007) suggest that there are two main types of workplace networks that should be develop within an organisation, formal and informal. Formal networks are a result of organisational structure and are considered beneficial for solving work based problems. In contrast, informal networks are the social interactions that form in the workplace and are considered to be facilitated by the formal networks constructed in an organisation. Therefore, an effective social network will consist of both formal and informal relationships, which are designed to disperse information and reduce the propensity of ambiguity in the workplace. Therefore, shared knowledge generates a support platform, and increases the usefulness of workplace networks for problem solving. However, what organisational factors will ensure the development and sustainability of an effective knowledge sharing environment that will promote the development of both strong and weak ties, which in turn will impact on the usefulness of workplace social networks for problem solving? Therefore, the first hypothesis is:

H1: *The strength of ties between workplace social network members will be positively and significantly associated with the perceived usefulness of workplace social networks for problem solving in SMEs.*

Network content

As previously stated, social capital is derived from the formation of effective workplace social networks. Past literature suggests that the structural dimension of SCT is a combination of both the form (structure, appearance) and content of workplace social network relationships, which are developed as intangible ties are formed between employees and are based on several interactions over a period of time. One way of examining the usefulness of workplace social networks is by analysing their network content (Hoang and Antoncic, 2003). Hite, Williams and Baugh (2007) suggest that network content is the information or resources that flows through the network of relationships embedded within firms. However, this study refers to network content as the actual workplace social network ties (relationships) that either facilitate or prevent the transfer of important work based information (Allen *et al.*, 2007). This information is captured by identifying how many people contacted a colleague when faced with a workplace problem. Subsequently, accessing network content also depicts the informal social network and identifies the path an individual takes when faced with an initially unresolvable work based problem. Therefore, this supports the notion that the usefulness of a workplace social network is dependent on the content of the network and the strength of the ties that binds together members within a social unit.

To examine the content of the workplace social network, it is essential to examine the advice network that has formed informally within the organisation. An advice network is referred to as an internal network of professional relationships, which are developed through social interactions when employees seek information and advice from colleagues to aid their problem solving (Gibbons, 2004). Lazega *et al.* (2006) suggest that an advice network represents a set of connections through which information is disseminated among network members. Furthermore, an appropriate advice network can aid the development of mentoring relationships within the organisation. Mentoring can be defined as a process where an influential individual with advanced knowledge and experience, provides support and advice to their mentee or protégé (Hetty Vanemmerick, 2004). Furthermore, Hetty Vanemmerick (2004) suggests that it is increasingly acknowledged that it is not only important to have appropriate mentoring relationships within the organisation, but to have developmental networking relationships between employees and/or management. Therefore, to effectively examine the workplace social network, it is necessary to ascertain who employees approach for help when problem solving. Thus, the secondary research question of this report is:

SRQ 1: *Who do engineers' approach for advice, what is the content of their workplace social network for solving work based problems?*

Examining the relational dimension of social capital theory

Sociability

SCT literature suggests that relations need to be effectively developed and supported if the organisation is to develop social capital. The relational dimension of SCT refers to examining the development of social capital by analysing the role of the factors affecting the quality of relationships. Sociability is the social skill, competence, intelligence or interdependence necessary to develop effective social ties and workplace social networks (Ferris *et al.*, 2001; Lizardo, 2006). When examining the factors that influence the perceived usefulness of workplace social networks, it is imperative to examine how well employees are able to develop friendships and workplace networks. Nie (2001) suggests that each employee will possess a different level of sociability, thus some employees may be better than others at developing relationships or gaining network ties. More specifically, an employee with a high level of social skill will not only possess an advantage when forming friendships and network ties, but will also be able to sustain these relationships over a period of time more effectively than someone with a lower sociability (Ferris *et al.*, 2001). Therefore, current literature supports the theory that the sociability of employees will impact on the perceived usefulness of the workplace social network.

Literature has explained social skills as the ability to effectively predict, understand, and control social interactions. Social skill literature can be attributed to the early work on social intelligence by (Thorndike, 1920). Ferris *et al.* (2001) suggest that workers who demonstrate high levels of social skills are more likely to be able to influence others to cooperate than those with low social skills. In addition, social skills include the ability to be able to clearly articulate to another in conversation, while at the same time being well aware of the different social situations, roles and norms (Riggio and Reichard, 2008). Social competence is a person's ability to be able to effectively interact with others based on their social skills (Baron and Markman, 2000). When examining sociability within a workplace it is important to determine if employees have the skills or competence required for effective social interaction and if the organisation has provided appropriate resources to support the social interaction between employees (Totterdell *et al.*, 2008).

Moreover, while it is important that employees develop the skills to communicate and develop networks, it is also vital that employees have the sociability to develop and sustain social network ties. Furthermore, the organisation can also assist in developing social networks, by outlining formal relationships in policies, job descriptions and organisational charts (Allen *et al.*, 2007). Formal relationships provide the contact between employees, facilitating the development of informal ties or social networks. It is also important there be a mix of strong and weak ties, although weak ties provide new information to aid creativity and innovative behaviour, strong ties are important, because they provide relationships that are beneficial for problem solving (Li, 2007). Therefore, it is important that the organisation provides an appropriate organisational structure and communication tools. As well as, ensuring employees have an appropriate level of sociability, to support the development of social networks and support problem solving and innovative behaviour. For this reason, it is important to examine the impact of sociability upon engineers' perceptions of the usefulness of workplace networks, and therefore the following hypothesis is proposed:

H2: *The sociability of SME employees' is positively and significantly correlated with the perception of the usefulness of their workplace social networks.*

Examining the cognitive dimension of social capital theory

Innovative culture

Past literature defines organisational culture in a plethora of different ways. (Schien, 1990) suggest that organisational culture is a construction of a collective values, beliefs, attitudes and assumptions that generates the standard of behaviour in the organisation. Another way to describe organisational culture is a shared perception of thoughts, feelings and bonds that motivates employees and it manages the way employees interpret information and values (McAleese and Hargie, 2004). To summarise the definitions it can be seen that organisational culture is related to how the employees perceive the organisations collective goals and etiquette within a social system. Additionally, Dobni (2008) suggests that an organisation with an innovative culture will have an intention to be innovative, the appropriate infrastructure to support innovative problem solving and the operational behaviours and environment to be able to support innovative behaviour. Dombrowski *et al.* (2007) proposes several elements associated with innovative cultures including: innovative vision and mission statements, democratic communication, flexibility, teams or units with resources to support facilitation, collaboration, incentives and leadership that support innovative behaviour.

To show employees that an organisation values innovative problem solving, it is important that incentives and rewards are associated with innovative behaviour. Moreover, management can indirectly encourage innovative problem solving and behaviour by being flexible and treating mistakes as a knowledge building process (Scott and Bruce, 1994). Also it is important to support the desired behaviour by ensuring an appropriate amount of time and resources are available. Furthermore, there are several factors of organisational culture that can support the innovative behaviour of employees, including strategy, structure, support mechanisms, particular behaviours and communications (Martins and Treblanche, 2003). This study will focus on outlining some of the determinants of organisational culture that impacts upon the perceived usefulness of workplace networks for problem solving and the innovative behaviour of employees.

Following a review of current literature, it was found that organisational culture can both encourage and support the transfer of knowledge for problem solving (Dodgson *et al.*, 2002). Firstly, the strategy of the organisation should support and encourage a culture of innovation and change, which should be communicated through the mission and vision statements (Dobni, 2008). Moreover, communication is not enough, it is also imperative to ensure that employees and the organisation itself support the organisational mission and vision

(Kotter, 1995). In addition, the support of employees can be achieved by developing a supportive structure, mechanisms and exhibiting behaviour that encourages innovative behaviour and the use of workplace networks for problem solving. The culture of the organisation should be one of flexibility and freedom that is poised for change, which can be achieved by sharing information and decision making with employees, leaving them with a sense of empowerment and autonomy (Johlke and Duhan, 2000). Furthermore, mechanisms should also be developed that are supportive of an innovative culture. Such mechanisms include rewarding and recognising the innovative behaviour of employees and by making resources available to employees through a workplace social network (Martins and Treblanche, 2003). However, even with flexibility and support it is still important that leaders exhibit behaviours that encourage an innovative culture. This can be achieved through the effective handling of employee mistakes, risk taking and conflict. In conclusion, there are several factors that can support the innovative culture of an organisation which sustains and encourages innovative problem solving. For the purpose of this study the importance of an innovative culture and its impact upon the perceived usefulness of workplace social networks will be examined.

H3: *An organisation's innovative culture is positively and significantly correlated to the perceived usefulness of workplace social networks to solve work based problems in SMEs.*

The perceived usefulness of workplace social networks

Knowledge sharing through informal workplace networks is not only a platform for the development of the innovative behaviour of employees, but can also be used for everyday problem solving. Empirical research suggests that networks are pivotal in the problem solving process and that some employees have high centrality and function as connectors within the network (Chang and Harrington, 2007). The importance of the social network concept becomes evident particularly in KBFs, because creating an environment where knowledge is transferred helps employees solve complex and often ambiguous problems (Cross *et al.*, 2001). Access to such knowledge supports an employee during the decision making process by providing them with the appropriate information to solve complex work based problems, empowering the employee to perform beyond their own personal capabilities (Borgatti and Cross, 2003; Branchos *et al.*, 2007; Marouf, 2007). In summary, access to resources and knowledge provides workplace social network members with the information they require to solve work based problems; however, it is their perception of the usefulness of the network that is imperative to its success.

While the value of social networks for decision making cannot be disputed, its value as perceived by employees will determine the effectiveness and sustainability of a workplace social network for solving problems. For example, if employees consider that the process of attaining new knowledge to solve a problem is too difficult, time consuming or will not provide appropriate help they may choose to use an inefficient, but tried and tested method to solve a problem (Gibbons, 2004; Hoang and Antoncic, 2003). This study will therefore examine amongst other things how social capital enablers such as sociability, tie strength and organisational culture impact upon the perceived usefulness of workplace social networks for solving work based problems. Therefore, the justification for examining the perceived usefulness of workplace social networks is highlighted by the requirement of knowledge based SMEs to develop the innovative behaviour of employees if they are to develop sustainable business practices (Alvesson, 2000; Brunetto and Farr-Wharton, 2007; Grant, 1996).

Innovative behaviour

Innovative behaviour is defined as the process of bringing new problem solving ideas into use, thereby enhancing a product, service or process. Furthermore, Innovation diffusion is the process by which, over a period of time, the innovation is communicated throughout a social system linked by a network. Subramaniam and Youndt (2005) emphasise that, innovation, like many business functions, is a management process that requires specific tools, rules, discipline and management support. Innovation begins with individuals in an organisation or team, working on a specific project mainly through the generation of creative ideas. However, these ideas need to make a positive change in a product or service for the innovation cycle to be completed. Furthermore, none of this can happen without organisational support, such as the facilitation of appropriate workplace social network ties and an organisational culture focused towards supporting innovative behaviour.

Bratton and Gold (2003) define innovative behaviour in KBFs as the development of what is known or introduction of something new by valuing the collection, dissemination and utilisation of new knowledge. Luecke and Katz (2003) suggest that innovative behaviour is generally understood as the successful introduction of a new thing or method. Innovative behaviour in a KBF is the embodiment, combination, or synthesis of knowledge in original, relevant, valued new products, processes, or a service which gives an organisation a competitive edge. However, following a review of current literature it is quite clear that most of the literature is either related to big business, the manufacturing sector or very generic. Hence, literature on innovation specifically associated with SME KBFs is quite scarce. Nevertheless, due to some of the similarities with other services and manufacturing sectors it might be possible to generalise some of the activities (and associated indicators) from these associated sectors. However, before this can be achieved it is proposed that there is a need to look at innovative behaviour from the lens of social capital enablers (tie strength, network content, sociability, organisational culture) associated with it, and the knowledge transfer required to support the development of social capital.

Brokel and Binder (2007) suggest that the use of face-to-face contact to transfer tacit knowledge is commonly thought to be the source of a number of opportunities including learning and innovative behaviour. Current literature explains that tacit knowledge cannot be articulated easily as it exists only in people's minds and bodies and is displayed by their actions (Perounce, 2007). Furthermore, people are more likely to engage in innovative behaviour (creating innovative solutions or transferring complex innovative ideas) to the point that they believe they will be listened to and will share in the accumulated benefits (Brokel and Binder, 2007; Clegg *et al.*, 2002). Although innovative behaviour is widely known to improve organisational effectiveness, the high risk for an employee who takes an innovative approach is often overlooked (Janssen, 2005). This can be attributed to the fact that innovative behaviour deviates from traditional linear behaviour, bringing uncertainty and a risky learning process (Kriegsman *et al.*, 2007). Furthermore, past literature suggests that only in a culture where mistakes can be made without fear of retribution will risky innovative behaviour increase to bring order to a chaotic business environment.

H4: *The perceived usefulness of workplace social networks will be positively and significantly correlated with the innovative behaviour of SME employees.*

Time in company moderator

The relationship between sociability and usefulness of workplace social networks is thought to be moderated by the amount of tenure an employee (social network member) has had within the company. More specifically, it is hypothesised that an employee's ability to develop workplace social networks would improve as an employee's time in the company increases. Such a concept was formulated from Granovetter (1973) conclusion that over a period of time and through numerous encounters employees will develop ties between one another. The moderating relationship suggests that as an employee's time in the company increases this will increase their leverage and ability to form social networks that are useful for solving work based problems. Therefore, this study seeks to examine amongst other things the impact of an employee's time in the company upon the relationship between sociability and the perceived usefulness of workplace social networks.

H5: *Time in the company will be positively and significantly correlated to sociability and the perceived usefulness of workplace social networks in SMEs.*

Methods

Sample and procedure

This study uses mixed methods to examine the usefulness of networks for problem solving and innovative behaviour for one national organisation. The mixed methods research design is used to gather data from professional engineers' working within an SME. Further, the findings are compared with current literature about other engineers' working within SMEs in the service sector. The hypotheses were tested using a sample of 150 engineers selected randomly from an engineering SME in Australia. Surveys were emailed to participants and 85 engineers returned useable surveys, concluding a response rate of 56.67%. The sample was selected based on industry and employee typicality to ensure the sample was representative of Australian engineering SMEs.

An engineering SME was selected for this study because this is an example of the type of firm that requires the continual generation of new knowledge, so as to maintain its competitiveness. Grant (1996) suggests that KBFs add value through the development of knowledge, especially where this knowledge is generated by employees within the organisation. As such, this type of firm was selected because it is a SME that states in its mission that it actively promotes innovative behaviour in its employees. An engineering SME was chosen for this study because they are required to develop innovative business practices if they are to remain competitive, especially during a period of intense competition (Brunetto and Farr-Wharton, 2007). That is both engineering firms and SMEs require the generation and transfer of knowledge, to facilitate innovative behaviour which contributes to gaining a competitive advantage. Therefore, past research suggests that the employee is of utmost importance because without them there is no knowledge and there would be no competitive edge for the organisation (Herling and Provo, 2000). In particular, past literature suggests that KBFs have the following attributes: they have a problem solving capability and may in some cases work in non-standardised production areas; employees are highly creative; employees in most cases are highly educated; the central assets for delivering value are not the tangible assets but instead the employees, client networks, customer relationships, tacit and explicit knowledge (Alvesson, 2000; Grant, 1996; Herling and Provo, 2000). Furthermore, Alvesson (2000) considers a KBF to be one where knowledge is related to individuals/employees rather than the organisation, machines and technologies and where the majority of the workforce is well educated.

Instruments used

The **Tie Strength** instrument (Levin and Cross, 2004) was selected to examine the strength of workplace social network ties and their impact on the usefulness of workplace social networks. This instrument includes three closed ended questions examining the tie strength between engineers in a workplace social network. The **network content** instrument was constructed based on the work of Granovetter (1973). The **Sociability** instrument included four sub-scales (propensity to make friends, propensity to make acquaintances, propensity to join others and propensity to belong to groups) to examine the sociability of workplace social network members. The instrument used for the survey was developed from Totterdell *et al.* (2008). This instrument examines whether engineers have the socially embedded skills or sociability to be able to develop social networks rich in strong and weak ties. The **Innovative organisational culture** instrument was developed by Scott and Bruce (1994). This instrument examines whether engineers surveyed perceive their organisation to value an innovative culture. Furthermore, it is suggested that organisations that value an innovative culture should have a greater amount of innovative employees than other organisations who do not value an innovative culture (Scott and Bruce, 1994). The perceived **usefulness of workplace social networks** instrument was developed from Brunetto, Farr-Wharton and Shacklock (2010). The **Innovative behaviour** instrument was developed by Janssen (2005) which was a modified version of Scott and Bruce (1994) measure of innovative behaviour. This instrument contains nine items and has been selected to examine how innovative behaviour is affected by the perceived usefulness of a workplace social network.

Analysis

A standard regression analysis was conducted to test the hypotheses. The regression analysed the relationship between three organisational factors (tie strength, sociability and organisational culture) and the perceived usefulness of workplace social networks for problems solving. Additionally, a regression analysis was also used to examine the relationship between the perceived usefulness of workplace social networks and the innovative behaviour of SME employees.

Results

Quantitative results

The sample selected consisted mostly of males (69.4%), with a wide spread of age up until fifty years (see Table 1). With regards to employment length, 55.3% of employees have been with the company for 2 years or less. Also, 70.5% of employees have at least an undergraduate degree. In summary, the demographic results portray diverse group of ages with high levels of education and low levels of employment length.

Descriptive statistics

Table 2 depicts the means, standard deviations, coefficient alphas and inter-correlations for all variables included within this study. All variables are significantly correlated to the perceived usefulness of workplace social networks which is significantly correlated to the innovative behaviour of engineering employees. The only variable that not significantly correlated was the age of employees (control variable).

Secondary research question

To examine who engineers' approach for help and therefore identify the content of their workplace social network for problem solving, engineers were sought to respond to the following question:

> *Who do you normally approach with frustrating work-based problems? This may be different for each specific problem you have, however, please place a (1) next to the person you approach the most and a (2) next to the position/person you approach the next most and so on.*

Table 1. Demographics.

Surveys	Engineering employees ($N = 85$)	%
Gender		
Male	59	69.4
Female	26	30.6
Age		
16–23	26	30.6
24–30	28	32.9
30–50	23	27.1
50+	8	9.4
Employment Length		
< 1	13	15.3
1–2	34	40.0
2–3	16	18.8
3+	22	25.9
Education		
High School	9	10.6
TAFE	19	18.8
Undergraduate	42	52.9
Postgraduate	15	17.6
Employment status		
Full-Time	81	96.4
Part-Time	2	2.3
Casual	2	2.3
Office Location		
City 1, State 1	44	51.8
City 2, State 2	15	17.6
City 3, State 3	8	9.4
City 4, State 4	18	21.2

The results of the network content of engineers suggests that engineers approach mostly their state managers and principals when they require support in solving work based problems. Additionally, only 22% of engineers had 2 or more network contacts that they could approach with work based problems and 8% approached no contacts when faced with a frustrating work based problem (see Table 4). Therefore, the findings depict that the majority of engineer's

Table 2. Means, standard deviations and correlations of the organisational factors tested.

Variable	Mean	SD	Usefulness of networks	Innovative behaviour	Tie strength	Sociability	Innovative culture	Time in company	Age
Usefulness of network	4.15	0.77	(0.895)						
Innovative behaviour	3.55	0.90	0.666**	(0.924)					
Tie strength	2.77	0.96	0.564**	−0.153	(0.913)				
Sociability	2.99	0.86	0.654**	0.946**	−0.196	(0.873)			
Innovative culture	3.79	0.56	0.769**	0.978**	0.164	0.938**	(0.812)		
Time in company	2.55	1.04	0.114*	0.275*	−0.229*	−0.095	0.100	1.000	
Age (control)	2.0706	0.25	−0.076	−0.198	−0.014	−0.168	0.202	0.222*	1.000

Note: $N = 85$. Numbers in parentheses on the diagonal are the Cronbach's Alpha in coefficients of the composite scales *Correlation is significant at the 0.05 level (2-tailed). **Correlation is significant at the 0.01 level (2-tailed).

Table 3. Advice network.

Position/person	Approach 1(%)	Approach 2(%)
Managing director	5.9	5.9
Principal	31.2	29.4
National manager	21.7	8.26
State manager	11.7	5.9
Associates	5.9	23.5
Senior engineers	11.7	15.1
Engineers	3.5	7.26
Graduate engineers	2.4	1.18
Designers	0	3.5
Drafters	0	0

Table 4. Network content.

Number of contacts	Amount of employee's	Percentage
0	7/85	8.24%
1	23/85	27.06%
2	36/85	42.35
3	10/85	11.76%
4	6/85	7.06%
5	3/85	3.53%

posses' minimal network contacts, this is likely to impact negatively on their ability to seek necessary information for problem solving.

Hypothesis 1

The coefficient value of 0.913 suggests that the tie strength variable has a high reliability rating. Hypothesis 1 declared that the strength of ties between workplace social network members will be positively and significantly associated with the perceived usefulness of workplace social networks for problem solving in SMEs. The findings provide full support for this hypothesis, the linear regression indicated that the strength of workplace social network ties is significantly and positively associated with the usefulness of workplace social networks ($F = 32.66, R^2 = 27\%$, $p < 0.000$). Additionally, the R^2 value suggests that tie strength contributes to 27 percent of the variance related to the usefulness of workplace social networks.

Hypothesis 2

The coefficient value of 0.873 suggests that the sociability variable has a high reliability rating. Hypothesis 2 declared that the sociability of SME employees' is positively and significantly correlated with the perception of the usefulness of their workplace social networks. This hypothesis is supported by the findings because the linear regression analysis indicates that the sociability of employees' is significantly and positively associated with the perceived usefulness of workplace social networks ($F = 120.02$, $R^2 = 58.6\%$, $p < 0.000$). Additionally, the R^2 value suggests that the sociability of an employee contributes to 58.6 percent of the variance in the usefulness of workplace social networks. The results suggest that the perceived usefulness of workplace social networks for problem solving is highly dependent upon the sociability of the employee.

Hypothesis 3

The coefficient value of 0.812 suggests that the innovative culture instrument has a high reliability rating. Hypothesis 3 stated that an organisation's innovative culture is positively and significantly correlated to the perceived usefulness of workplace social networks to solve work based problems in SMEs. This hypothesis should be accepted because the findings from the linear regression analysis indicate that an innovative organisational culture is significantly and positively associated with the usefulness of workplace social networks for problems solving ($F = 39.45$, $R^2 = 32.2\%$, $p < 0.000$). Additionally, the R^2 value suggests that innovative culture contributes to 32.2 percent of the variance in the usefulness of workplace social networks for problem solving. The results suggest that an organisation that supports innovative behaviour will inevitably be supporting the usefulness of social networks for problem solving. The support mechanism supports both innovation and workplace social networks because the two are highly interrelated.

Hypothesis 4

The coefficient value was 0.924 suggesting that the innovative behaviour instrument had a high reliability rating. Hypothesis 4 stated that the perceived usefulness of workplace social networks will be positively and significantly correlated with the innovative behaviour of SME employees. The hypothesis received full support because findings from the linear regression indicate that there is a significant association between the usefulness of workplace social networks and the innovative behaviour of SME employees ($F = 66, R^2 = 44.3\%, p < 0.000$). Additionally, the R^2 value suggests that the usefulness of workplace social networks on their own contributes to 44.3 percent of the variance in innovative behaviour of employees. The results infer that a useful workplace social network geared towards supporting employees when solving work based problems is also likely to facilitate the development of innovative behaviour.

Hypothesis 5

Hypothesis 5 stated that time in the company would be positively and significantly correlated to sociability and the perceived usefulness of workplace social networks in SMEs. The examination of this hypothesis suggests that an employee's time within the company would moderate the relationship between sociability and the perceived usefulness of workplace social networks for problem solving. An independent sample *t*-test was conducted to compare the usefulness of workplace social networks for employee's that have been with the company for 2–3 years and more than 3 years. Additionally, a second independent *t*-test was conducted comparing the sociability of employee's that have been with the company for 2–3 years and for more than 3 years. There was a significant difference when examining the usefulness of a workplace social network when comparing employee's that had been with the company for 2–3 years and employee's that had been with the company for more than 3 years. In addition, there was not a significant difference when examining the sociability of employee's when comparing those that have been with the company for 2–3 years and to those that have had more than 3 years with the company. The results from the *t*-test

Table 5. Independent *t*-test of sociability and usefulness of workplace social networks.

Variable	Mean score		Degrees of freedom		Two tail probability
	2–3 years	3+ years	*t*-value	df	*P*-value
Usefulness of workplace social networks	4.33	3.98	1.51	36	0.048
Sociability	4.19	3.88	0.989	36	0.165

proposes a contrasting view to current literature and suggests that employee's that have been with company for more than 3 years found the workplace social networks less useful than those that had been with the company for 2–3 years.

Qualitative results

Semi-structured interviews were conducted with fifteen engineering managers to provide insight and a greater understanding into the quantitative results. The interview participants were asked a total of seven questions (see Appendix 1). The first three questions addressed whether the organisational culture is supportive of innovative behaviour. In addition, questions four and five examine the workplace social network and its perceived importance. Moreover, questions six and seven address managements' perception of employee social skill and the best thing about working in the organisation respectively.

The first question asked interview participants to describe in their own words the values, vision and mission of the organisation. A review of the responses found that while all participants were aware and able to recite the mission and vision; five participants suggested that current practice is not aligned with organisational policy (mission and vision). This means that a third of the engineers identified a difference between the organisational policy and practice of the firm. The second question addressed how management supports the creativity of their employees. In response, 20% of participants suggested that support is provided by the allocation of time, 27% suggested that support is provided by resources, incentive and rewards, and public recognition; while 53% of participants suggested that support was not adequate to develop a supportive workplace social network or to foster innovative behaviour. This means that the majority of employees experienced poor levels of support from management. The third question addressed how much time is available to support creative ideas and whether or not this time is adequate. In response, 40% of participants responded that not enough time is provided to support creative ideas; 33.3% suggested the time provided was below average; while 6.7% suggested the time provided was both adequate and above average if compared to other engineering SMEs. In addition, 13.3% of participants suggested that employees' needed to be motivated and trained in time management skills in order to support innovative behaviour. In summary, the findings suggest that participants perceive that not enough support or time is allocated to support and facilitate innovative behaviour in the workplace.

The fourth question examined how often participants used their workplace social network to solve work based problems innovatively. In response, 73% of participants suggested that they used their workplace social network one to five times per week; 20% suggested they used their workplace social network six to ten times per week; while only 7% suggested they used their workplace social network more than eleven times per week. The fifth question addressed how participants would rate the importance of their workplace social network in assisting them to solve work based problems innovatively. In response, 13.10% of participants suggested that the workplace social network was somewhat important; 13.57% suggested that the workplace social network was important; while 53.33% of participants suggested that their workplace social network was essential for assisting them to solve work based problems innovatively. What the findings from question four and five suggest is that the majority of participants only used their workplace social networks a limited number of times per week (1–5), even though the majority of participants suggested that their workplace social network was essential for assisting them in solving work based problems.

The sixth question examined whether participants perceive that engineers possess the necessary skills to develop a vast workplace social network for which to attain and transfer knowledge. In response, 60% of participants suggested that engineers do not possess the necessary social skills to be able to develop a vast workplace social network themselves; while 30% of participants suggested they think that engineers do possess the necessary social skills to be able to develop vast workplace social networks themselves. The seventh question examined what participants perceive is the best thing about working for the organisation under examination. In response, 16% of participants suggested that the best thing was the people; also 16% of participants suggested that the best thing was the organisation's culture; 24% of respondents suggested that the best thing was the opportunity for career development; while 52% suggested that the best thing about working for the organisation under examination was the projects they were able to undertake. Overall the qualitative findings suggest that while the organisation is winning projects that require problems to be solved innovatively, support for creativity by management is not provided, there is not enough time to be innovative and while there is a workplace social network that is considered essential, it is not used enough to assist in solving work based problems.

Discussion

This study used the three dimensions of SCT as a lens to examine the relationship between some organisational factors and the perceived usefulness of workplace social networks for problem solving and the impact of this network upon the innovative behaviour of SME employees. Using a social capital lens, it was expected that those employees that are embedded in effective workplace relationships are the same employees that would have access to more information and resources for solving work-based problems innovatively. The aim of this study was first to examine the content of the workplace social networks for problem solving and then to examine the impact of tie strength, sociability and organisational culture upon the perceived usefulness of workplace social networks. As such, the study tested and confirmed the proposed model about the impact of tie strength, sociability and organisational culture upon the perceived usefulness of workplace social networks. The study also tested and confirmed the impact of the perceived usefulness of workplace social networks for problem solving upon the innovative behaviour of SME employees. Therefore, following the premise of structural social capital, it was expected that engineers' would be part of a workplace social network that contain network contacts which include both strong and weak ties.

However, the findings from this study suggest that the majority of ties between engineers' are strong ties and that most engineers' usually only rely on one or two workplace social network contacts when faced with an unresolved work based problem. In addition, current literature suggests that relying on a small group of network contacts will limit the perspectives from which network members can gain knowledge, information and ideas from which to solve work based problems (Cole *et al.*, 2002) (Gibbons, 2004; Vanemmerick, 2004). Therefore, based on the findings of this study the usefulness of workplace social networks is dependent upon the strength and number of workplace social network ties. In summary, based on the theory and the findings from this research, by increasing the number of workplace social network ties that engineers' rely when solving work based problems should increase the perceived usefulness of workplace social networks for problem solving.

In addition, based on an analysis of Table 3; the findings suggest that the two hierarchical positions most approached by engineers' when faced with a work based problem were the principal and national manager, whom can be categorised as middle to senior level management. There are two interpretations of the data. Using one theory, the findings suggest that engineers do at least have effective networks in place for solving problems. On the other hand, using another theory the findings present a problem, because the same people are being approached each time. This could mean that these people are very knowledgeable and have all the answers. Although, there seems to be two main problems with this approach.

Firstly, gathering information from the same people is not providing engineers with access to diverse perspectives on the one problem. Secondly, relying on one or two people can pose a serious problem for engineers if help is not available when they need it. Moreover, another interpretation of the same data is that the engineers' workplace social network was limited in number and therefore the capacity for innovative solutions was probably reduced by the fact that engineers had such a limited network from which to gather information, knowledge, resources and support. Additionally, such a problem is becoming more frequent with the increasing work load of engineering managers. Therefore, a gap has emerged between the conditions required for the development of social capital and the current practice of engineering SMEs.

Previous research findings identified that weak ties or weaker relationships are typically beneficial for creativity and innovation (Gummeson, 2007; Levin and Cross, 2004; Perry-Smith, 2006). In contrast, the results from this study suggest that employees currently use their intra-office networks to solve problems and gather information and from the quantitative findings it has already been established that these networks are dense in strong ties. What this means is that while the organisation already has strong ties for problem solving, the networks are small and this infers that the central network contacts (advisors) may not always be available to assist in solving work based problems. Furthermore, the development of weak ties throughout the organisation will facilitate the transfer of new and unique knowledge and will promote the innovative work behaviour of employees.

Moreover, the development of network ties will only support the effective problem solving and innovative behaviour of professionals if they are part of a professional's culture that has embedded a process that passes on information and improves the profession. In contrast, the findings from this study contradict current literature suggesting that the usefulness of workplace social network ties is dependent on the strength of network ties both strong and weak. Also the development of inter office network ties are imperative to engineers as each office and division can learn from one another and benefit from new and unique knowledge within their network (Marouf, 2007). Additionally, formal networks can be utilised to facilitate the development of network contacts for engineers between divisions and offices. The findings from this study suggest that the usefulness of workplace social networks are dependent upon the support function of the organisation's culture, strength of workplace social network ties and the sociability of network members.

Under ideal conditions SCT would suggest that both strong and weak ties can be developed between offices by developing effective online communication channels and ensuring that employees have the socially embedded skills required to form informal social networks rich in weak and strong ties (Nie, 2001). For example, what this means is that to develop relationships across state borders it is important that the organisation has in place communication technologies that allow employees to communicate easily and in this case all over Australia. However, having the right type of communication technologies will not improve communication unless employees have the social skills to be able to interact with one another and to build networks in an online setting. Additionally, past literature suggests that the development of formal networks will assist in facilitating the development of informal workplace social networks. Therefore, under ideal conditions it is expected that the structures embedded within the organisation will support high quality relationships that will promote the development of both strong and weak ties that are appropriate for problem solving and the disseminating of knowledge between workplace social network members.

This study also aimed to examine the relationship between relational factors such as sociability and the perceived usefulness of workplace social networks for engineers. This relationship was examined through the lens of the relational dimension of SCT. The relational dimension theory suggests that under ideal conditions workplace social networks are developed and sustained depending on the social skill (sociability) of engineers' in the workplace. Moreover, the results suggest that employees' perceive that they have an appropriate level of social skill to be able to effectively develop workplace social network relationships. However, in contrast the results suggest that the development of inter-division or inter-office networks is almost not existent. Subsequently, while engineers perceive their level of sociability to be high, the results of developing vast workplace social networks rich in both strong and weak ties has yet to be achieved.

Moreover, to develop social capital (innovative and effective problem solving) it is imperative that organisations embed structures, facilitate the development of relationships and develop a supporting culture. Additionally, under ideal conditions the cognitive dimension, suggests that to develop an organisation that supports innovation and the continuous exchange of knowledge, innovation will be embedded into the culture of the organisation. It is important that an organisation have in place appropriate polices and organisational documents that supports their vision to be innovative through effective problem solving and, to effectively transfer knowledge between employees (Dobni, 2008). However, the results suggest that while innovative problem solving seems to be promoted amongst engineers it is not appropriately supported. More specifically, the results suggest that there is not an appropriate amount of time, resources and rewards to support the development of effective workplace social networks for problem solving. Therefore, if employees' perceive that that the organisation does not demonstrate that they value the use of workplace social networks for problem solving, there is no incentive for employees to develop workplace social networks.

Conclusion

This study has made several contributions to current literature including SCT, innovation and SME literature. The contribution to SCT was conducted by applying three social capital dimensions to examine the propensity to develop useful workplace social networks to assist in solving work based problems. The second contribution was to innovation literature by linking the role of social networks to the second step of developing innovative behaviour, solving work based problems. The third and final contribution was to SME literature examining the workplace social networks of engineers' impacts upon their ability to solve work based problems within an SME. In summary, this study made three separate contributions to current literature.

This study also contributed to the implication that it is management's responsibility to embed practices that facilitate and support the development of workplace social networks in any organisation and not simply SMEs and KBFs. In addition, it is imperative that management embed workplace social networks in their organisation so that employees' have access to people resources (support, information and knowledge) and can use this to build on their own capability and creatively address work based problems. As such, it is in the firm's best interest to assist engineers in developing effective workplace social networks, because it improves the capacity of individual engineer's to address workplace problems.

The results of this study open up a number of areas that are recommended for further research and outline some limitations that may impact on the generalisability of the study. For example, although the model proved advantageous in this study, further research using a larger sample of employees and a larger sample of organisations, will increase the generalisability of the results and provide further clarification and refinement of the model. Moreover, greater in-depth research into the organisational factors that impact on an employee's innovative behaviour is also recommended due to the post-positivist methodology undertaken in this study. More specifically, although this study was rich in quantitative findings there was a lack of in-depth qualitative findings that an interpretivist style study may obtain. In addition, this study was only investigated from an organisational perspective. Therefore, an examination into the innovative behaviour within several projects might provide an alternative explanation of the factors that impact on an employee's innovative behaviour.

References

Adler, P and S Kwon (2002). Social capital: Prospects for a new concept. *Academy of Management Review*, 27(1), 17–40.

Akdere, M (2005). Social capital theory and implications for human resource development. *Singapore Management Review*, 27(2), 1–27.

Al-Alwai, A, N Al-Marzooqi and Y Mohammed (2007). Organisational culture and knowledge sharing: Critical success factors. *Journal of Knowledge Management*, 11(2), 22–42.

Allen, J, A James and P Gamlen (2007). Formal versus informal knowledge networks in R&D: A case study using social network analysis. *R&D Management*, 37(3), 179–196.

Alvesson, M (2000). Social identity and the problem of loyalty in knowledge-intensive companies. *Journal of Management Studies*, 37(8), 1101–1124.

Baron, R and G Markman (2000). Beyond social capital: How social skills can enhance an entrepreneur's success. *Academy of Management Executive*, 14(1), 106–116.

Borgatti, S and R Cross (2003). A relational view of information seeking and learning in social networks. *Management Science*, 49(4), 432–445.

Branchos, D, K Kostopoulos, K Soderquist and G Prastacos (2007). Knowledge effectiveness, social context and innovation. *Journal of Knowledge Management*, 11(5), 31–44.

Bratton, J and J Gold (2003). *The Dynamics of Professionalization: Whither the HRM Profession*? Basingstoke: Palgrave Macmillian: Human Resource Management Phenomena — HRM & Beyond.

Brokel, T and M Binder (2007). The regional dimension of knowledge transfers — A behavioral approach. *Industry and Innovation*, 14(2), 151–175.

Brunetto, Y and R Farr-Wharton (2007). The moderating role of trust in SME owners/managers' decision-making about collaboration. *Journal of Small Business Management*, 45(3), 362–387.

Brunetto, Y, R Farr-Wharton and K Shacklock (2010). The impact of supervisor-subordinate relationships on morale: Implications for public and private sector nurses' commitment. *Human Resource Management Journal*, 20(2), 206–225.

Carmelli, A, R Meitar and J Weisberg (2006). Self-leadership skills and innovative behaviour at work. *International Journal of Manpower*, 27(1), 75–90.

Chang, M and J Harrington (2007). Innovators, imitators, and the evolving architecture of problem-solving networks. *Organization Science*, 18(4), 648–666.

Clegg, C, K Unsworth, O Epitropaki and G Parker (2002). Implicating trust in the innovation process. *Journal of Occupational and Organizational Psychology*, 4(75), 409–422.

Cole, M, W Schaninger and S Harris (2002). The workplace social exchange network. *Group and Organization Management*, 27(1), 142–167.

Cross, R, A Parker, L Prusak and S Borgatti (2001). Supporting knowledge creation and sharing in social networks. *Organizational Dynamics*, 30(2), 100–120.

Dobni, B (2008). Measuring innovation culture in organizations, the development of a generalized innovation culture construct using exploratory factor analysis. *European Journal of Innovation Management*, 11(4), 539–559.

Dodgson, M, D Gann and A Salter (2002). The intensification of innovation. *International Journal of Innovation Management*, 6, 53–84.

Dombrowski, C, J Kim, K Desouza, A Braganza, S Papagari and P Baloh (2007). Elements of innovative cultures. *Knowledge and Process Management*, 14(3), 190–202.

Dooley, L and D O'Sullivan (2007). Managing within distributed innovation networks. *International Journal of Innovation Management*, 11(3), 397–416.

Ferris, G, L Witt and W Hochwarter (2001). Interaction of social skill and general mental ability on job performance and salary. *Journal of Applied Psychology*, 86(6), 1075–1082.

Fliaster, A and J Spiess (2008). Knowledge mobilization through social ties: The cost benefit analysis. *Schmalenbach Business Review: ZFBF*, 60, 99–127.

Gibbons, D (2004). Friendship and advice networks in the context of changing professional values. *Administrative Science Quarterly*, 49, 238–262.

Granovetter, MS (1973). The strength of weak ties. *The American Journal of Sociology*, 78(6), 1360–1380.

Grant, R (1996). Towards a knowledge-based theory of the firm. *Strategic Management Journal*, 17, 109–122.

Herling, R and J Provo (2000). Knowledge, competence, and expertise in organisations. *Advances in Developing Human Resources*, 2(1), 1–7.

Hetty Vanemmerick, I (2004). The more you can get the better: Mentoring constellations and intrinsic career success. *Career Development International*, 9(6/7), 578–594.

Hite, JM, EJ Williams, SC Hilton and SC Baugh (2006). The role of administrator characteristics on perceptions of innovativeness among a network of public school administrators. *Education and Urban Society*, 38(2), 160–187.

Hoang, H and B Antoncic (2003). Network-based research in entrepreneurship: A critical review. *Journal of Business Venturing*, 18(2), 165–187.

Huang, F and J Rice (2009). The role of abosrptive capacity in facilitating "open innovation" outcomes: A study of Australian SMEs in the manufacturing sector. *International Journal of Innovation Management*, 13(2), 201–220.

Janssen, O (2005). The impact of perceived influence and supervisor supportiveness on employee innovative behaviour. *Journal of Occupational and Organisational Psychology*, 78, 573–579.

Johlke, M and D Duhan (2000). Supervisor communication practices and service employee outcomes. *Journal of Service Research*, 3(2), 154–165.

Johnson, S (2002). Lifelong learning and SMEs: Issues for research and policy. *Journal of Small Business and Enterprise Development*, 9(2), 285–295.

Koeberg, C and L Chusmir (1987). Organizational culture relationships with creativity and other job related variables. *Journal of Business Research*, 15(5), 397–409.

Kotter, P (1995). Leading change: Why transformation efforts fail. *Harvard Business Review*, 73(2), 59–67.

Kriegsman, B, T Kley and M Schwering (2007). Making organizational learning happen: The value of creative failures. *Business Strategy Series*, 8(4), 270–276.

Lazega, E, C Lemercier and C Mounier (2006). A spinning top model of formal organisation and informal behaviour: Dynamics of advice networks among judges in a commercial court. *European Management Journal*, 3, 113–122.

Levin, D and R Cross (2004). The strength of weak ties you can trust: The mediating role of trust in effective knowledge transfer. *Management Science*, 50(11), 1477–1490.

Li, P (2007). Social tie, social capital, and social behavior: Toward an integrative model of informal exchange. *Asia Pacific Journal of Management*, 24, 227–246.

Lin, N (ed.) (2001). *Social Capital: A Theory of Social Structure and Action*. Cambridge, NY: Cambridge University Press.

Lizardo, O (2006). How cultural tastes shape personal networks. *American Sociology Review*, 71, 778–807.

Luecke, R and R Katz (2003). *Managing Creativity and Innovation*. Boston, MA: Harvard Business School Press.

Maqsood, T, D Walker and A Finegan (2007). Facilitating knowledge pull to deliver innovation through knowledge management: A case study. *Engineering, Construction and Architectural Management*, 14(1), 94–109.

Marouf, L (2007). Networks and knowledge sharing in organizations: A case study. *Journal of Knowledge Management*, 11(6), 110–125.

Martins, E and F Treblanche (2003). Building organisational culture that stimulates creativity and innovation. *European Journal of Innovation Management*, 6(1), 64–74.

McAleese, D and O Hargie (2004). Five guiding principles of culture management: A synthesis of best practice. *Journal of Communication Management*, 9(2), 155–170.

Nahapiet, J and S Ghoshal (1998). Social capital, intellect capital and organizational advantage. *Academy of Management Review*, 23(2), 242–285.

Nebus, J (2006). Building Collegial information networks: A theory of advice network generation. *Academy of Management Review*, 31(3), 615–637.

Nelson, S, Y Brunetto, R Farr-Wharton and S Ramsay (2007). Organisational effectiveness of Australian fast growing small to medium enterprises (SMEs). *Management Decision*, 45(7), 1143–1164.

Nie, N (2001). Sociability, interpersonal relations and the internet: Reconciling conflicting findings. *American Behavioural Scientist*, 45, 420–435.

Perounce, D (2007). Tacit knowledge in the workplace: The facilitating role of peer relationships. *Journal of European Industrial Training*, 31(4), 244–258.

Perry-Smith, J (2006). Social yet creative: The role of social relationships in facilitating individual creativity. *Academy of Management Journal*, 49(1), 85–101.

Riggio, R and R Reichard (2008). The emotional and social intelligences of effective leadership: An emotional and social skill approach. *Journal of Managerial Psychology*, 23(2), 169–185.

Schien, E (1990). Organisational culture. *American Psychologist*, 45(2), 109–119.

Scott, S and R Bruce (1994). Determinants of innovative behaviour: A path model of individual innovation in the workplace. *Academy of Management Journal*, 37(3), 580–607.
Scozzi, M, C Carvelli and K Crowston (2005). Methods for modelling and supporting innovation processes in SMEs. *European Journal of Innovation Management*, 8(1), 120–137.
Sparrowe, R, R Liden, S Wayne and M Kraimer (2001). Social networks and the performance of individuals and groups. *Academy of Management Journal*, 44(2), 316–325.
Subramaniam, M and M Youndt (2005). The influence of intellectual capital on the types of innovative capabilities. *Academy of Management Journal*, 48(3), 450–463.
Taylor, D, O Jones and K Boles (2004). Building social capital through action learning: An insight into the entrepreneur. *Education and Training*, 46(4/5), 226–235.
Thorndike, E (1920). Intelligence and its use. *Haper's Magazine*, 140, 227–235.
Totterdell, P, D Holman and A Hukin (2008). Social networkers: Measuring and examining individual differences in propensity to connect with others. *Social Networks*, 30, 283–296.
Tsai, W and S Ghoshal (1998). Social capital and value creation: The role of intrafirm networks. *Academy of Management Journal*, 41(4), 464–476.

Appendix 1: Interview Themes

	Questions	Responses
1	Describe in your own words the organisations values, vision and mission? Are they currently being met?	All fifteen participants were able to recite the current vision and mission. However, five participants stated the current values and practices do not support the current mission and vision.
2	How does management support the creativity of employees?	Support with time allocation = 20% Support with resources = 6.7% Incentives and rewards = 13.3% Public recognition = 6.7% 53% of participants suggested there needs to be more support for innovation and creativity.
3	How much time is available to support creative ideas? Is this adequate?	Not enough = 40% Above average = 6.7% Below average = 33.33% Too much = 0.0% Adequate = 6.7% 13.3% of participants suggested employees needed to be motivated and trained in time management skills to support innovative behaviour.

Appendix 1. (*Continued*)

	Questions	Responses
4	How often would you use your social network during a week to solve work based problems innovatively?	0 = 0% 6–10 = 20% 1–5 = 73% 11+ = 7%
5	How would you rate the importance of your workplace social network in assisting you to solve work-based problems innovatively?	Not important = 0% Somewhat important = 13.10% Important = 13.57% Essential = 53.33%
6	Do you think that engineers possess the necessary social skills to develop a vast workplace social network from which to attain and transfer knowledge?	*No* = 60% *Yes* = 30%
7	What is the best thing about working at this organisation? Please name a few.	The people = 16% The projects = 52% The culture = 16% Opportunity for career development = 24%

Chapter 19

THE ROLE OF ABSORPTIVE CAPACITY IN FACILITATING "OPEN INNOVATION" OUTCOMES: A STUDY OF AUSTRALIAN SMEs IN THE MANUFACTURING SECTOR

Fang Huang and John Rice

The University of Adelaide Business School

The open innovation approach emphasizes porous knowledge boundaries between firms and upstream suppliers, but tends to ignore questions of transformative efficiency and effectiveness once the knowledge reaches the focal organization. In this paper, we test for the significance of interaction effects between open innovation strategies and absorptive capacity, finding support for the idea that effective knowledge absorption capabilities are of vital importance in the facilitation of innovation effectiveness.

Keywords: Open innovation; absorptive capacity; manufacturing; SMEs; Australia.

Introduction

Traditionally, innovation has been viewed as the outcome of effective internal research and development processes, with appropriation of rents flowing from effective tight control of the product or service developed. Open innovation models (Chesbrough, 2003) depart from these "closed innovation" assumptions at all stages, with development and appropriation regimes essentially undertaken within inter-firm network arrangements. In addition to Chesbrough's (2003, 2006) recent contributions, many other scholars are contributing to the literature relating to the transitional processes from the closed innovation to an open innovation paradigm (Lichtenthaler, 2006; Dodgson *et al.*, 2006; Gassmann, 2006; West and

---

This paper was first published in *International Journal of Innovation Management*, Vol. 13, No. 2 (June 2009), pp. 202–220

Gallagher, 2006). This literature in essence builds on a strong body of knowledge that has contextualized R&D externalization within the study of outsourcing, collaboration, licensing and inter-firm networks, etc. (e.g. Carr, 1995; Pisano, 1990; Freeman, 1991).

As the theoretical model (although perhaps not the actual practice) is quite novel, theoretical and empirical efforts are being made to explore the various multivariate relationships involved in the application of open innovation approaches.

This paper focuses on issues relating to the important role of "absorptive capacity" (Cohen and Levinthal, 1990) in creating the organizational climate to facilitate open innovation effectiveness. We then discuss issues related to the nature of the empirical data used, and undertake analysis and present the results of this analysis investigating the important role of firm level absorptive capacity in generating the climate for effective innovation.

Theoretical Background

The open innovation paradigm

According to Chesbrough *et al.* (2006: 1), open innovation entails the "use of purposive inflows and outflows of knowledge to accelerate internal innovation and expand markets for external use of innovation." This model fundamentally demonstrates the porosity of a firm's transactional and knowledge boundary in accordance with the logic of open system theory, and the steady loosening of constraints on the interactions between firms and their external environment (McGee *et al.*, 1995; Pisano, 1990).

Open innovation strategies are viewed as a source of further and potentially sustainable competitive advantage in that the successful integration of externally sourced knowledge with in-house capabilities can create a socially complex and "causally ambiguous" advantage that cannot be easily imitated by competitors (Sen and Rubenstein, 1989; Kodama, 2005a, 2005b). Other literature has compared open innovation's benefits in negating the inwardly-focused culture of closed innovation and also extending the potential of vertical and horizontal integration (West and Gallagher, 2006).

In essence, there are three crucial components of the open innovation paradigm. First, the acquisition of knowledge must be considered, whereby a wide range of knowledge sources for innovation should be created, including developing customer, supplier, competitor and research institution linkages (Laursen and Salter, 2006; Katila and Ahuja, 2002). Second, the development of effective appropriation strategies should be investigated whereby a broad range of internal and external commercialization channels should potentially be developed, involving not only traditional distribution approaches but also other forms such as licensing out, the creation of spin-off ventures or the establishment of joint ventures (Chesbrough, 2003). Finally, the development of the firm's systems and capabilities to identify, assimilate, and exploit external know-how from its strategic environment and successfully integrate it into its internal knowledge base must be considered.

The absorption of acquired knowledge — absorptive capacity

This final element of open innovation — namely the effective integration of the acquired knowledge into organizational value-adding processes — is akin to Cohen and Levinthal's (1990) "absorptive capacity" construct, a processual capability that builds on a stock of prior knowledge to facilitate the effective absorption of externally gained knowledge. Indeed, it has been suggested that external knowledge can only be assimilated when firms manage to change their organizational structure and culture to facilitate open innovation processes (Dahlander and Gann, 2007).

The identification and acquisition of external know-how does not necessarily mean that firms can incorporate this knowledge into their existing innovation systems to gain economic returns. Indeed, it is proposed in this paper that innovation outcomes will be elusive in the absence of absorptive capacity, even if valuable technology has been successfully obtained from outside. In analogy, both a sponge and a sieve can "attract" fluids, but only a sponge, with its strong absorptive nature, can retain fluids for later use.

When absorptive capacity is present, external research activities can act complimentarily to in-house research, achieving potentially strong synergies and yielding the best results in terms of innovation and innovation appropriation (Arora and Gambardella, 1990; Macpherson, 1997). Such synergies ensure that in-house R&D need not necessarily decline or become obsolete as the openness may stimulate new appropriation benefits for the internal R&D (Howells, 1999; Veugelers, 1997). As we have observed, this in-house R&D may be viewed as an essential catalyst for the development of a firm's absorptive capacity, improving the overall status of knowledge-based skills within the focal firm (Cohen and Levinthal, 1989; Lane *et al.*, 2006).

Basic open innovation approaches — networking and technology purchase

While there are potentially many forms of potential knowledge exchange between firms, Cassiman and Veugelers (2002) note that firms typically engage either in the transaction-based acquisition of knowledge on the technology market or acquire knowledge through collaboration with other firms. As such, inter-organizational networking and technology purchase could be regarded as the basic practices of a firm's open strategy.

Networking involves a set of relationships that binds a group of independent organizations together (Gulati, 1998). It is essentially a cooperative pattern for R&D to obtain complementary know-how. In addition to enhancing the availability of external knowledge, networking also provides an interactive platform for knowledge transfer with a high degree of reciprocity (Almeida and Kogut, 1999). Other than formal inter-organizational networking, sometimes, in order to avoid problems associated with costs and restrictions in contracting or licensing agreements, firms also depend on less formal inter-organizational collaborations in the form of opportunistic and temporary alliances and networks (Vanhaverbeke, 2006).

Acquisition of external technology can also take transactional forms such as the purchase of patents, trademarks or licenses (Sen and Rubenstein, 1989). The role of technology purchasing by contractual agreements can be regarded as the "buy" decision in the context of "make or buy" strategic choices which refer to conducting in-house R&D or commercially buying in technology respectively (Kurokawa, 1997). However, the role of a "buy-in" strategy is not always substitutive; it can also complement internal research to form a combinative means of R&D (Veugelers, 1997).

Limitations of the extant literature

Some evident problems exist with the extant open innovation literature, some of which we seek to address in this paper. First, the majority of previous studies of this issue, including Chesbrough's own seminal work, have largely been undertaken at a conceptual level, on the basis of axiomatic assumptions relating to knowledge synergy, rather than positivistic and statistical research based on large scale samples. Where empirical work has been undertaken (Laursen and Salter, 2006), the results have been ambiguous.

Secondly, Chesbrough's comparison between closed and open innovation models mainly focused on the specific cases of large or leading industrial corporations such as AT&T, IBM, Microsoft and Procter & Gamble (Chesbrough, 2003). The generalizability of this new paradigm to small companies has yet to be fully explored.

Thirdly, while many other studies in R&D externalization have investigated the implications of discrete inter-organizational activities, such as outsourcing, networking, licensing and R&D intensity, the integration of these modes to internal characteristics has not been empirically examined. The potential benefits provided by absorptive capacity, with regards to the costs of knowledge search and the transactional costs associated with external sourcing, clearly drive a complex transactional model. Testing some of the complex and interacting effects associated with open innovation and absorptive capacity, investigated in this paper has not, as of yet, been adequately undertaken.

Fourthly, the vast majority of empirical literature relating to the practice of open innovation is drawn from North American or European Union data. These jurisdictions tend to have elaborate knowledge diffusion systems (government agencies, strong universities, closely collated firms and industries). It is important to assess if the situation might be different outside these two domains.

Finally, in terms of industry context, more attention has been paid to high-technology and knowledge-intensive industries (e.g. chemicals, pharmaceuticals, semiconductors and electronics) than other sectors of the economy. In general, these sectors exhibit higher overall levels of innovation and greater external knowledge sourcing activities than other industries (Bee, 2003; Grindley and Teece, 1998). This raises questions on the applicability of this construct outside technology-based sectors (Simard and West, 2006).

Research Model

Drechsler and Natter (2008) note that there is no available model yet to specifically analyze the open innovation paradigm. Empirical studies often develop different theoretical models, with different determinants and consequences examined. The means by which firms have tended to adopt "openness" have included:

- acquisitions,
- the purchase of technology rights through licensing,
- the contracting out of internal R&D to external agents, other firms or research institutions, and
- the use of formal and informal inter-organizational networks

Our research utilizes the following variables and tests the following relational hypotheses.

Interorganizational networking and innovation performance

Much research has claimed that the role of inter-organizational networking has a significant effect in enhancing innovation output of firms (Monsted, 1994; Chaston, 2000; Porter and Ketels, 2003). Other research has showed that those firms that exclude such options, whether by engaging in formal or informal networking, or through the development of effective co-operative strategies, might gradually limit their knowledge base over the long run (Shaw, 1998).

The contribution of networking to innovation is largely based on the potential benefits accrued through the access to external technology and know-how that the firms themselves currently lack (Cooke, 1996; Powell *et al.*, 1996; Ahuja, 2000); the sharing of potentially complementary skills and resources (Hagedoorn and Duysters, 2002); the sharing of risks-rewards propositions (and hence commitment) among networked firms (Grandori, 1997); and the establishment of strengthened competitive advantage in terms of leading-edge technology (Human and Provan, 1996).

H1: *Interorganizational networking will have a significant effect on innovation performance (as measured by the internal commitment of R&D resources) of SMEs.*

Technology purchase and innovation performance

It is surprising that although the theoretical studies discussed previously have demonstrated the many benefits of technology acquisition, the purchasing or licensing of such knowledge has often been shown to have ambivalent effects (Jones *et al.*, 2001). There may be complex reasons behind this outcome, including the coincidence of such acquisitions with knowledge stock deficiencies and also the resource drain that such transactions create. Additionally, the firm might suffer the potential threat that the external technology acquisition might "crowd out" in-house research as R&D staff realize that the preferred technology in use is introduced from outside rather than from their own research efforts. This might lead to a reduced level of in-house capability in the long term (Quinn, 1992).

If a firm can manage to achieve knowledge acquisition while also maintaining a strong and supportive internal knowledge environment (i.e. strong internal absorptive capacity), it might be expected that the potential downsides of technology buy-in can be negated.

On the basis of this analysis, we hypothesize that:

H2a: *External technology purchase, via a 'crowding out' effect, tends to reduce future innovation performance (as measured by the internal commitment of R&D resources) of SMEs.*

H2b: *External technology purchase, when occurring alongside strong internal absorptive capacity, can improve the innovation performance (as measured by the internal commitment of R&D) of SMEs.*

Investment in absorptive capacity and innovation performance

The potentially positive significance of absorptive capacity for innovation has been supported by much empirical work (e.g., DeSanctis *et al.*, 2002; Tsai, 2001). Hendry *et al.* (2007) addressed SMEs' deficiencies in absorptive capacity as a key barrier to their innovation and growth. On the other hand, Stock *et al.* (2001) have reported an inverted-U relationship between absorptive capacity and the new product development success, finding that only up to a certain level did absorptive capacity contribute to the high levels of new product development. It generally takes time and other resources to transform incompatible organizational routines into capabilities embodying absorptive capacity. These conditions might bring out a mixed effect on innovation performance.

H3a: *Investment in absorptive capacity is positively related to innovative performance (as measured by the internal commitment of R&D).*

Versus

H3b: *Investment in absorptive capacity is negatively related to innovative performance (as measured by the internal commitment of R&D).*

Interaction between networking, absorptive capacity and innovation performance

Empirical studies have tended to suggest that complex interaction of knowledge search and acquisition activities tend to co-occur with internal absorptive capacity. For example, international networking has been found to be complementary to internal R&D only when the focal firms possessed a level of absorptive capacity (Vinding, 2006). Absorptive capacity might mitigate the potential threats resulting from some approaches to open innovation. It was shown that costs were likely to be lower with adequate knowledge accumulation and prior experience in the knowledge base (Teece *et al.*, 1997; Zander and Kogut, 1995; Zott, 2003). Therefore, we hypothesize a significant and positive mediating effect of absorptive capacity on the relationship between interorganizational networking and innovation performance.

This argument can be stated in the following hypothesis:

H4: *The interaction effect of networking and absorptive capacity on innovation performance (as measured by the internal commitment of R&D) will be greater than the direct effect of networking alone.*

Sample and Measures

Sample

The data for analysis are drawn from the Australian Bureau of Statistics' Business Longitudinal Survey. Through a process of careful screening, 292 Australian manufacturing SMEs were identified that were active innovators and had continuous operations during the period of the surveys (1994–1998).

Dependent variable

Since the surveys provide only limited information about innovation output, R&D intensity (*RDI*) has been employed as a proxy for innovation performance. This has been calculated as the proportion of expenditure on research and development activities within the firm divided by total expenditure on the development of new or changed products or processes in 1997–98. While actually being an input measure, R&D input as a proxy for output is commonly used in previous empirical studies on innovation (e.g., Veugelers, 1997; Mahmood and Lee, 2004) in spite of some noted limitations (Patel and Pavitt, 1993). However, in the absence of any direct measure of innovation output such as "the level of sales of different types of innovation" (Negassi, 2004), a proxy of this kind is often the most appropriate measure of innovation performance.

Independent variables

Interorganizational networking (*Networking*) is measured by responses to survey questions relating to whether the business has engaged in any formal networking with other businesses in 1997–98. It is a dummy variable taking the value of 1 when the business indicates that it has used networking, and 0 otherwise. Technology purchase (*Techbuyin*) is calculated dividing the total expenditure on the development of new products and processes by the value of patents, trademarks and licenses acquired in 1997–98.

We develop a proxy measure of absorptive capacity, following previous work. Much previous empirical research has emphasized the importance of training as a major driver for absorptive capacity (e.g., Daghfous, 2004; Graça *et al.*, 2005). Moreover, the role of training and education has been shown to be a central determinant of knowledge absorption which is in turn highly correlated with absorptive capacity (Knudsen *et al.*, 2001). Thus, investment in training is used to approximate the extent of the investment in absorptive capacity (*ACAP*). We develop this as a relative measure, training intensity, which is computed as the expenditure on training and education related to innovation divided by total expenditure on the development of new or changed products and processes.

Control variables

Firm size (*LogSize*) is a common control variable in previous innovation research. It is expressed as the logarithm of sales, which is also a form commonly employed in past studies (Veugelers, 1997; Vermeulen, 2004). To avoid possible simultaneously causal relation between sales and innovation performance, sales of goods and services in the previous year (1996–97) is used. To further systematically control for the effects of other factors on innovation performance, we include four other control variables: the value of exports (*Export*), again to avoid potential simultaneous problem, the value of exports in 1996–97 is used; the type of legal organization (*Legalstatus*) — a dummy variable with the value of 1 if the business is incorporated, and 0 otherwise after recoding the original responses (this is important as some firms in the survey are quite small, with only a limited degree of operation); whether the business contracted out its R&D (*Outsourcing*) — a dummy variable taking the value of 1 when contracting out and 0 otherwise; and whether the firm was previously R&D active (*RDactive*) — a dummy variable with the value of 1 if the business engaged in any R&D activities of the previous year (1996–97) and 0 otherwise.

Statistical Methods and Results

Analyzes and weighting

A hierarchical OLS regression model is employed for this study following three steps. First, all the control variables and the three basic independent variables (*ACAP*, *Techbuyin*, and *Networking*) are entered into the first model (Model 1) to test Hypotheses 1, 2a, 3a/3b. Next, the interaction of *Networking* and *ACAP* is entered to assess Hypothesis 4 (Model 2). Finally, the interaction of *Techbuyin* and *ACAP* is entered to examine Hypothesis 2b (Model 3). The possibility of multicollinearity was considered, though rejected as all of the Variance Inflation Factors are less than 2, thus far below the generally recommended threshold of 10 (Zatzick and Iverson, 2006).

An underlying assumption of OLS is that the relationship between dependent variable and independent variables is one-directional and not simultaneous (they are not simultaneously determined). We were aware that Colombo and Garrone (1996) had previously found a simultaneous relationship between firms' R&D intensity and their technological cooperation agreements. To avoid any possible simultaneity problem that might cause OLS estimators to be biased or inconsistent, the Pairwise Granger Causality Test is conducted between dependent variable and the three basic independent variables respectively (control variables have been chosen from the data of 1996–97 to avoid this problem). The null hypothesis of this test is that (a) the dependent variable does not granger cause the independent variable chosen for test and (b) all p values exceed 0.1 or 0.05. Our calculations show that the null hypothesis for causality test are supported for the independent variables in our sample for analysis. Therefore, simultaneity is not a problem in our study and OLS can be employed for model building and hypotheses testing.

Finally, the sample for analysis used in the analysis was not randomly selected since the original population was stratified depending on firm size, industry type, and innovation status by the Australian Bureau of Statistics in their data gathering. The ABS provided bootstrap-derived weights for their sample to ensure that the subsample is representative of Australian firms generally (Australian Bureau of Statistics, 2002). We used relevant weights provided by the BLS to ensure the sample does not over-represent observations from any particular stratum. The usage of such weights has been found to be useful in many similar studies (e.g. Mark and Graversen, 2004).

Results

Descriptive statistics and correlations for dependent variable, independent variables and control variables are presented in Table 1.

Table 2 displays the findings with regard to hypotheses tests. Model 1 reports the results of the hierarchical model which only includes control variables and three independent variables to be tested. The R^2 is 66.6% (adjusted R^2 is 66.5%) showing that approximately 66.5% of the variation in dependent variable can be explained by the variation in the independent variables and control variables based on our sample. It indicates a relatively high goodness of fit for our first model.

The coefficient for the variable *Networking* is positive and significant ($p < 0.01$), supporting Hypothesis 1. However, it is far less than 1, suggesting that innovation performance improves with the engagement in networking, but less than in direct proportion. With respect to Hypothesis 2a which states that technology purchase will have a negative effect on future innovation performance, the coefficient for the variable *Techbuyin* is negative (as we expected) and highly significant ($p < 0.001$). Given the coefficient of the variable investment in absorptive capacity (*ACAP*) is negative and strongly significant ($p < 0.001$), H3b rather than its alternative hypothesis is supported.

Of our control variables, the size of firm seems to influence the innovation performance. However, it is reported as a negative coefficient. There is a quite weak effect with regard to the value of exports despite its high significance ($p < 0.001$), implying it has a small but consistently significant impact on innovation performance. On the other hand, the type of legal organization, which is not a standard explanatory variable in innovation research, does exhibit a significantly positive effect, which shows that incorporated organizations exhibit comparatively better innovation performance than unincorporated firms in our sample.

In Model 2, we introduce the interaction variable through multiplying *Networking* by *ACAP* to assess those variables' combined effect on innovation. After this interaction variable is included, there is a noticeable and significant improvement of the model fit ($\Delta R^2 = 0.4\%$). Furthermore, *Networking* now is no loner significant ($p > 0.1$) which implies that with the role of this combined variable, inter-firm networking itself has limited importance in explaining the variance of innovation output. The coefficient of the interaction variable is significantly positive and much stronger than the coefficient of *Networking* alone in the first model. Therefore, Hypothesis 4, which proposed that investment in absorptive capacity may interact positively with interorganizational networking's impact on innovation, is supported.

Table 1. Means, standard deviations, and correlations.

Variable	Mean	S.D.	1	2	3	4	5	6	7	8
1. RDI	0.52	0.41								
2. ACAP	0.03	0.1	−0.22**							
3. Techbuyin	0.03	0.14	−0.23**	−0.02						
4. Networking	0.28	0.45	−0.06**	0.08**	0.11**					
5. LogSize	3.09	0.84	−0.06**	0.09**	0.12**	0.30**				
6. Export	563.56	2280.87	0.14**	−0.05**	−0.04*	0.16**	0.31**			
7. Legalstatus	0.72	0.45	−0.01	0.11**	0.02	0.17**	0.51**	0.08**		
8. Outsourcing	0.08	0.27	0.05**	0.003	0.06**	0.10**	−0.002	0.10**	−0.08**	
9. R&D active	0.72	0.45	0.79**	−0.09**	−0.20**	−0.08**	0.007	−0.10**	−0.08**	0.14**

**Correlation is significant at the 0.01 level (one-tailed).

*Correlation is significant at the 0.05 level (one-tailed).

Table 2. The results of hierarchical regression analysis.

Variable	Innovation performance (RDI)		
	Step 1 (Model 1)	Step 2 (Model 2)	Step 3 (Model 3)
(Constant)	0.16***	0.17***	0.17***
Absorptive Capacity (ACAP)	−0.61***	−0.84***	−0.85***
Technology Buy-in	−0.20***	−0.20***	−0.25***
Networking	0.03**	0.006	0.004
LogSize	−0.07***	−0.07***	−0.07***
Export	1.60*E*−05***	1.67*E*–05***	1.66*E*–05***
Legal Status	0.11***	0.12***	0.12***
Outsourcing	−0.09***	−0.08***	−0.08***
Prior R& D active	0.71***	0.72***	0.72***
ACAP × Networking		0.57***	0.58***
ACAP × Technology Buy-in			1.83***
ΔR^2	66.6%***	0.4%***	0.1%***
Total R^2	66.6%	67.0%	67.1%

†p < 0.10
*p < 0.05
**p < 0.01
***p < 0.001
Two-tailed tests

In Model 3, we further include the interaction of *Techbuyin* and *ACAP*. Although this interaction variable does not make a large contribution to model fit ($\Delta R^2 = 0.1\%$), the ΔR^2 is significant ($p < 0.05$). According to Aguinis *et al.* (2004, 2005), even small amount of improvement in explanatory fit can indicate meaningful and practical effects. The effect of this interaction variable reverses the directionality of *Techbuyin* from the first model, reversing it from negative to positive in its interactive form. Based on the result of Model 3, H2b, asserting a better effect for technology purchase with inclusion of investment in absorptive capacity, is also supported.

Discussion

In this study, our objective is to examine the effects of different practices of open strategy on innovation performance, with a special focus on the role of absorptive capacity in facilitating open innovation effectiveness.

The weak impact of networking

Despite the advantages presented by the previous theoretical research in networking, such as sharing risks or gaining access to complementary skills, the coefficient of *Networking* reveals a weak but significantly positive effect. This finding is consistent with previous studies that have tended to report a weak effect of R&D collaboration on innovation output (Brouwer and Kleinknecht, 1996; Negassi, 2004). The minimal role of networking shown by our study could probably be attributed to three aspects related to the characteristics of our sample.

First, our study seems to only measure those formal networking arrangements that are contractually agreed upon for knowledge exchange between engaged organizations. However, the roles of informal forms of knowledge spillovers resulted from labour mobility or interpersonal interaction between individuals are often difficult to capture (Agrawal and Henderson, 2002; Murray, 2002). Almeida *et al.* (2003)'s finding asserted that small firms accrue more benefits from informal knowledge flow mechanisms than their larger peers. That might possibly provide some further explanation to the results of our study that the formal networking does not show substantial beneficial returns for innovation among SMEs.

Second, when compared to other larger firms in the same industry, SMEs by definition have relatively fewer resources such as skilled employees, revenue, or assets which are critical to innovation (Alvarez and Barney, 2002). According to Rothwell and Dodgson (1991), an SME's ability to access external know-how is conditioned by its in-house employment of qualified technical specialists, scientists and engineers. The lack of any of these resources might cause the specific problems for SMEs in establishing external linkages.

Third, networks are very prevalent in knowledge-intensive industries, where due to questions of interoperability, standards and technological convergence, individual firms find it difficult to keep up with technologies from diverse areas (Gemünden *et al.*, 1996). Therefore, our observations in manufacturing industry are probably related to a lower lever of networking engagement.

The negative effect of technology purchase

The results of Model 1 also provide strong support to Hypothesis 2a asserting that external technology buy-in has a negative influence on innovation performance of SMEs. This is in line with most other similar empirical studies which have demonstrated a significantly negative effect of externally acquired technology.

For example, Lowe and Taylor's (1998) research found that the filling of existing technology gap as the main motive for technology acquisition usually implies a negative relationship between technology purchase and the current research capabilities of the firm. It further implies that firms often choose to buy in technologies externally rather than develop them in-house due to the inabilities of their in-house R&D or technological blocks that may exist to further research.

Additionally, a negative relationship might be associated with the absence of absorptive capacity. Absorptive capacity contains two essential elements: the ability to identify and access external opportunities, and the ability to assimilate the exogenous know-how and incorporate it into the internal knowledge base. The lack of the latter element might elongate the time required for a firm to truly benefit from innovation outputs, which will be commercialized more slowly in the market because of a lengthening process of learning (Chesbrough *et al.*, 2006). That will hamper the benefits of technology purchase and reduce current and short term returns from innovation. Therefore, although a commonly used driver of open innovation strategy, technology buy-in should be employed conditionally with adequate consideration of the required absorptive capacity as it may result in negative influence without presence of such capabilities.

The two-fold effect of absorptive capacity

The role of absorptive capacity in the innovation process has been widely recognized. However, its effect on current innovation performance is still controversial. The reader may recall that in Model 1, the dependent variable and the three independent variables (excluding control variables) are drawn from the data of the same year 1997–1998, thus only the current effect or short-term future effects can be detected. Although the one-period lagged measure for absorptive capacity can also be used, a much longer lag effect (which may be occurring in some instances) cannot be examined by our study.

The modeling results of our sample shows that a firm's absorptive capacity has a negative effect on current innovation performance, suggesting investment in absorptive capacity in the short term might not bring about direct economic returns from innovation. This outcome is in accordance with the Cohen and Levinthal's (1990) finding that absorptive capacity is "path-dependent" in nature. Our findings are also consistent with those of Knudsen *et al.* (2001), namely that a precondition for absorptive capacity's positive impact on innovation is sufficient levels of training undertaken within the firm coincident with externally-sourced supplementary knowledge.

It is also indicated that this lagged effect might be not necessarily restricted to one period or a predicted short term within a fast business cycle. It may take a number of years for innovation outcomes to eventuate from current investment in absorptive capacity (namely, the accumulation of a firm's stock of innovation-specific knowledge). According to Zahra and George (2002), continued explorations in a firm's knowledge search zone is required for building a more developed and refined capability to acquire and assimilate external knowledge.

On the other hand, as expected, the individual drivers of open innovation strategy such as networking or technology acquisition, combined with the role of absorptive capacity, do have a better impact on innovation. The positive effect of networking engagement is stronger when absorptive capacity is taken into account. Moreover, this capacity also plays an effective role in lessening the negative effect of technology buy-in. In fact, as we use R&D intensity (a form of internal commitment of R&D resources) as the proxy for R&D output, our hypotheses are essentially built to test the relationship between internal R&D effort and external R&D behaviors. One of the major assumptions of open innovation paradigm for openness is "complementary with internal R&D, instead of serving as a substitute" (Lichtenthaler, 2008). The interaction effects evidently observed in our model provide evidence for such an approach.

Open innovation, in its extant literature, has been axiomatically asserted as a kind of "comprehensive strategy" that, through the successful integration of externally-sourced knowledge and in-house capabilities, can form sustainable competitive advantage greater than an internally driven innovation strategy alone (Lichtenthaler, 2008). Our study supports this view, although only when absorptive capacity is present. In essence, external approaches to openness may stimulate rather than replace the internal R&D effort of the firm on the condition that internal R&D of the receiving firm has developed the proper capability to effectively absorb what they obtained externally. Under such a circumstance, open innovation strategy is expected to optimally benefit a firm's innovation activities and exhibit its advantage over the closed innovation pattern.

Limitations and future research

Inevitably, there are some limitations associated with our study and the results should be carefully interpreted with these limitations in mind. First, as mentioned earlier, one of the limitations of this study is using R&D input intensity as a proxy to measure innovative output. Although currently there is no standard measure of innovative output, R&D input is somewhat a flawed performance indicator since it does not take account of the commercialization process of R&D — a crucial part to the innovation success (Leiponen and Helfat, 2003). Nonetheless, within a reasonable sample it is assumable that firms engage in R&D to create future profits and across a sample, R&D expenditures will create R&D driven sales over the medium to long term. It could be argued that under the open innovation paradigm, this traditional measure seems no longer the most suitable proxy to capture the actual innovation process because now the external R&D activities are also likely to greatly contribute to ultimate innovative outcome rather than internal R&D investment alone.

We undertook our analysis with a different dependent variable which might be seen as more closely aligned with innovation effectiveness, namely the "total expenditure on the development of new products and processes" which is not confined to the internal R&D activities of the firm. The results from this analysis did not differ in any notable way from the results of this study. This value shares the weakness of our dependent variable, that it is in fact an input rather than an output measure of innovation effectiveness. A more direct output measure, such as the "fraction of the firm's turnover relating to innovations with different degrees of novelty" (Laursen and Salter, 2006) available in other databases, will be considered for future study, though was not available in this microdata. Second, the purpose of our quantitative research is to draw inferences from the sample to a larger population. However, generalizing from sample to population for our study is confined to Australian manufacturing SMEs. Thus, the situation in different contexts will be an area for future research as well.

Finally, the database generated from 1994 to 1998 is somewhat dated now. But after adequate consideration on the strengths of this database, it was decided the strengths definitely outweigh this weakness. In the future, similar research design can be applied to more recent dataset to examine other current issues associated with open innovation.

Conclusion

This paper attempted to fill some of gaps in previous open innovation studies through a quantitative study on a data sample of 292 small and medium-sized manufacturing firms in Australia. Despite the restricted research design in terms of sample size and variables involved, the results of this empirical study create some interesting findings that have received limited attention elsewhere. The findings of this study contribute to the understanding of open innovation in three main ways.

First, while there is a growing trend towards the adoption of an open innovation strategy, it is not a panacea, particularly for smaller businesses. With respect to the basic means of openness, we reported a weak impact of networking as well as a negative relationship between technology buy-in and innovation performance based on SMEs in our sample. It suggests that great care must be taken with the adoption of open strategies, particularly for SMEs, as they might detract from innovation output rather than contribute to it.

Second, the relationship between absorptive capacity (one of the vital components of the open model) and innovation performance should be interpreted as twofold. On one hand, investment in absorptive capacity in the short term might negatively affect the current or short-term performance. On the other hand, it plays an essential role in strengthening the positive effect (or moderating the negative effect) of other modes of openness such as networking and technology buy-in. In other words, an investment in absorptive capacity will ensure that firms are more likely to combine internal and external sources of innovation to build a comprehensive innovation strategy instead of benefiting from innovating exclusively.

Additionally, there are some particular challenges for the common proxy for innovation (using R&D input as a measure for output) in innovation research, as this seems no longer suitable under an open innovation paradigm. Some of the external activities resulted from openness (which may accrue costs not picked up in this proxy) might also drive innovation performance, rather than internal R&D investment alone. The challenge relating to the accurate choice of innovation proxy measures in an open innovation environment is a perplexing, but ultimately fundamentally interesting, issue for future research.

Acknowledgments

We thank the Australian Bureau of Statistics for the permission to use the Business Longitudinal Survey data.

References

Agrawal, A and R Henderson (2002). Putting patterns in context: Exploring knowledge transfer from MIT. *Management Science*, 48(1), 44–60.

Aguinis, H (2004). *Regression Analysis for Categorical Moderators*. New York: Guilford Press.

Aguinis, H, JC Beaty, RJ Boik and CA Pierce (2005). Effect size and power in assessing moderating effects of categorical variables using multiple regression: A 30-year review. *Journal of Applied Psychology*, 90, 94–107.

Ahuja, G (2000). The duality of collaboration: Inducements and opportunities in the formation of interfirm linkages. *Strategic Management Journal*, 21, 317–343.

Almeida, P, G Dokko and L Rosenkopf (2003). Startup size and the mechanisms of external learning: Increasing opportunity and decreasing ability. *Research Policy*, 32, 301–315.

Alvarez, SA and JB Barney (2002). Wasps and tarantulas: Alliances between entrepreneurial firms and large firms. Working Paper, Fisher College of Business, Ohio State University.

Arora, A and A Gambardella (1990). Complementarity and external linkages: The strategies of the large firms in biotechnology. *Journal of Industrial Economics*, 38, 361–379.

Australian Bureau of Statistics (12 June 2000). *ABS Business Longitudinal Survey: Confidentialised Unit Record File: 1994 to 1998.*

Bee, E (2003). Knowledge networks and technical invention in America's metropolitan areas: A paradigm for high-technology economic development. *Economic Development Quarterly*, 17(2), 115–131.

Brouwer, E and A Kleinknecht (1996). Firm size, small business presence and sales of innovative products: A micro-econometric analysis. *Small Business Economics*, 8, 189–201.

Carr, CS (1995). Does outsourcing (outside contracting) provide a strategic advantage? *InTech*, 42, 72–79.

Cassiman, B and R Veugelers (2002). Complementarity in the innovation strategy: Internal R&D, external technology acquisition, and cooperation in R&D. IESE Business School Working Paper No. 457.

Chaston, I (2000). Organisational competence: Does networking confer advantage for high growth entrepreneurial firms? *Journal of Research in Marketing and Entrepreneurship*, 2(1), 37–56.

Chesbrough, H (2003). *Open Innovation: The New Imperative for Creating and Profiting from Technology*. Boston, MA: Harvard Business School Press.

Chesbrough, H, W Vanhaverbeke and J West (eds.) (2006). *Open Innovation: Researching a New Paradigm*. Oxford: Oxford University Press.

Cohen, WM and DA Levinthal (1989). Innovation and learning: The two faces of R&D. *Economic Journal*, 99, 569–596.

Cohen, WM and DA Levinthal (1990). Absorptive capacity: A new perspective on learning and innovation. *Administrative Science Quarterly*, 35, 128–152.

Colombo, M and P Garrone (1996). Technological cooperative agreements and firm's R&D intensity: A note on causality relations. *Research Policy*, 25, 923–932.

Cooke, P (1996). The new wave of regional innovation networks: Analysis, characteristics and strategy. *Small Business Economics*, 8, 159–171.

Daghfous, A (2004). Absorptive capacity and the implementation of knowledge-intensive best practices. *SAM Advanced Management Journal*, 69(2), 21–27.

Dahlander, L and D Gann (2007). Appropriability, proximity, routines and innovation: How open is open innovation? Paper presented at the *Druid Summer Conference*.

DeSanctis, G, J Glass and I Ensing (2002). Organizational designs for R&D. *Academy of Management Executive*, 16(3), 55–66.

Dodgson, M, D Gann and A Salter (2006). The role of technology in the shift towards open innovation: The case of Procter & Gamble. *R&D Management*, 36, 333–346.

Drechsler, W and M Natter (2008). Open innovation — management trend with impact? An empirical investigation of antecedents, drivers and performance outcomes. Retrieved 19th January from http://ssrn.com/abstract=1129809, Working Paper Series.

Freeman, C (1991). Networks of innovators: A synthesis of research issues. *Research Policy*, 20, 499–514.

Gassmann, O (2006). Opening up the innovation process: Towards an agenda. *R&D Management*, 36, 223–228.

Gemünden, HG, T Ritter and P Heydebreck (1996). Network configuration and innovation success: An empirical analysis in German high-tech industries. *International Journal of Research in Marketing*, 13(5), 449–462.

Graça, M, M Easterby-Smith, E Antonacopoulou and J Ferdinand (2005). Absorptive capacity: A process perspective. Paper for *Academy of Management Meeting*. Hawaii, August 2005.

Grandori, A (1997). An organizational assessment of interfirm coordination modes. *Organization Studies*, 18, 897–925.

Grindley, PC and DJ Teece (1998). Managing intellectual capital: Licensing and cross-licensing in semiconductors and electronics. *IEEE Engineering Management Review*, 26(2), 92–111.

Hagedoorn, J and G Duysters (2002). External sources of innovative capabilities: The preference for strategic alliances or mergers and acquisitions. *Journal of Management Studies*, 39, 167–188.

Hendry, C, P Harborne and J Brown (2007). Innovation barriers to growth in high-technology SMEs: The role of absorptive capacity. Paper presented at the *37th EISB Conference*, University of Ljubljana, Slovenia, September.

Howells, J (1999). Research and technology outsourcing. *Technology Analysis & Strategic Management*, 11(1), 17–29.

Human, SE and KG Provan (1996). External resource exchange and perceptions of competitiveness within organizational networks: An organizational learning perspective. *Frontiers for Entrepreneurship Research*, 45(2), 327–669.

Jones, GK, A Lanctot and HJ Teegen (2001). Determinants and performance impacts of external technology acquisition. *Journal of Business Venturing*, 16, 255–283.

Katila, R and G Ahuja (2002). Something old, something new: A longitudinal study of search behaviour and new product introduction. *Academy of Management Journal*, 45(8), 1183–1194.

Knudsen, MP, B Dalum and G Villumsen (2001). Two faces of absorptive capacity creation: Access and utilisation of knowledge. Paper presented at the *Nelson and Winter Conference*, Denmark, June.

Kodama, M (2005a). How two Japanese high-tech companies achieved rapid innovation via strategic community networks. *Strategy and Leadership*, 33(6), 39–47.

Kodama, M (2005b). Technological innovation through networked strategic communities: A study on a high tech company in Japan. *SAM Advanced Management Journal*, 70(1), 21–35.

Lane, P, B Koka and S Pathak (2006). The reification of absorptive capacity: A critical review and rejuvenation of the concept. *Academy of Management Review*, 31(4), 833–863.

Laursen, K and A Salter (2006). Open for innovation: The role of openness in explaining innovation performance among U.K. manufacturing firms. *Strategic Management Journal*, 27, 131–150.

Leiponen, A and CE Helfat (2003). Innovation objectives, knowledge sources and the benefit of breadth. Paper presented at the *What Do We Know about Innovation Conference*. University of Sussex, Brighton, U.K, November.

Lichtenthaler, U (2006). Technology exploitation strategies in the context of open innovation. *International Journal of Technology Intelligence and Planning*, 2(1), 1–21.

Lichtenthaler, U (2008). Open innovation in practice: An analysis of strategic approaches to technology transactions. *IEEE Transactions on Engineering Management*, 55(1), 148–157.

Lowe, J and P Taylor (1998). R&D and technology purchase through licence agreements: Complementary strategies and complementary assets. *R&D Management*, 28(4), 263–278.

Macpherson, A (1997). A comparison of within-firm and external sources of product innovation. *Growth and Change*, 28, 289–308.

Mahmood, IP and C Lee (2004). Business groups: Entry barrier-innovation debate revisited. *Journal of Economic Behavior & Organization*, 54, 513–531.

Mark, M and EK Graversen (2004). Determinants of Danish firms' choice of R&D cooperation partners. Working paper by the Danish Center for Studies in Research and Research Policy.

McGee, J, H Thomas and M Pruett (1995). Strategic groups and the analysis of market structure and industry dynamics. *British Journal of Management*, 6(4), 257–270.

Monsted, M (1994). Strategic networks and social control: Dilemmas of network export strategies for small firms. *International Council for Small Business Conference*, Strasbourg, France.

Murray, F (2002). Innovation as co-evolution of science and technological networks: Exploring tissue engineering. *Research Policy*, 31, 1389–1403.

Negassi, S (2004). R&D co-operation and innovation: A microeconometric study on French firms. *Research Policy*, 33(3), 365–384.

Patel, P and K Pavitt (1993). Patterns of technological activity: Their measurement and interpretation. In Stoneman, P (ed.), *The Handbook of the Economics of Innovation and Technological Change*. Oxford: Blackwell Scientific Publication.

Pisano, GP (1990). The R&D boundaries of the firm: An empirical analysis. *Administrative Science Quarterly*, 35, 153–176.

Porter, ME and CHM Ketels (2003). UK competitiveness: Moving to the next stage. *DTI Economics Paper No 3.*

Powell, WW, KW Koput and L Smith-Doerr (1996). Inter-organizational collaboration and the locus of innovation: Networks of learning in biotechnology. *Administrative Science Quarterly*, 41(1), 116–145.

Quinn, JB (1992). *The Intelligent Enterprise*. New York: Free Press.

Rothwell, R and M Dodgson (1991). External linkages and innovation in small and medium-sized enterprises. *R&D Management*, 21(2), 125–137.

Sen, F and AH Rubenstein (1989). External technology and in-house R&D's facilitative role. *Journal of Product Innovation Management*, 6, 123–138.

Shaw, B (1998). Innovation and new product development in the UK medical equipment industry. *International Journal of Technology Management*, 15, 433–445.

Simard, C and J West (2006). Knowledge network and geographic locus of innovation. In *Open Innovation: Researching a New Paradigm*, W Vanhaverbeke and J West (eds.). Oxford: Oxford University Press.

Stock, GN, NP Greis and WA Fischer (2001). Absorptive capacity and new product development. *Journal of High Technology Management Research*, 12, 77–92.

Teece, DJ, G Pisano and A Shuen (1997). Dynamic capabilities and strategic management. *Strategic Management Journal*, 18, 509–533.

Tsai, W (2001). Knowledge transfer in intraorganizational networks: Effects of network position and absorptive capacity on business unit innovation and performance. *Academy of Management Journal*, 44(5), 996–1004.

Vanhaverbeke, W (2006). The inter-organizational context of open innovation. In *Open Innovation: Researching a New Paradigm*, W Vanhaverbeke and J West (eds.). Oxford: Oxford University Press.

Vermeulen, P (2004). Factor content, size, and export propensity at the firm level. *Economics Letters*, 82(2), 249–252.

Veugelers, R (1997). Internal R&D expenditures and external technology sourcing. *Research Policy*, 26(3), 303–316.

Vinding, AL (2006). Absorptive capacity and innovation performance: A human capital approach. *Economics of Innovation and New Technology*, 15, 507–517.

West, J and S Gallagher (2006). Challenges of open innovation: The paradox of firm investment in open-source software. *R&D Management*, 36, 319–331.

Zahra, SA and G George (2002). Absorptive capacity: A review, reconceptualization, and extension. *Academy of Management Review*, 27(2), 185–203.

Zander, U and B Kogut (1995). Knowledge and the speed of transfer and imitation of organizational capabilities. *Organization Science*, 6, 76–92.

Zatzick, CD and RD Iverson (2006). High-commitment management and workforce reduction: Competitive advantage or disadvantage? *Academy of Management Journal*, 49(5), 999–1015.

Zott, C (2003). Dynamic capabilities and the emergence of intra-industry differential firm performance: Insights from a simulation study. *Strategic Management Journal*, 24, 97–125.

Series on Technology Management

(Continuation of series card page)

Vol. 21 *Workbook for Opening Innovation: Bridging Networked Business, Intellectual Property and Contracting*
by Jaakko Paasi (VTT Technical Research Centre of Finland, Finland), Katri Valkokari (VTT Technical Research Centre of Finland, Finland), Henri Hytönen (VTT Technical Research Centre of Finland, Finland), Laura Huhtilainen (University of Eastern Finland, Finland) & Soili Nystén-Haarala (University of Eastern Finland, Finland)

Vol. 20 *Bazaar of Opportunities for New Business Development: Bridging Networked Innovation, Intellectual Property and Business*
by Jaakko Paasi (VTT Technical Research Centre of Finland, Finland), Katri Valkokari (VTT Technical Research Centre of Finland, Finland), Tuija Rantala (VTT Technical Research Centre of Finland, Finland), Soili Nystén-Haarala (University of Eastern Finland, Finland), Nari Lee (University of Eastern Finland, Finland) & Laura Huhtilainen (University of Eastern Finland, Finland)

Vol. 19 *From Knowledge Management to Strategic Competence: Assessing Technological, Market and Organisational Innovation (Third Edition)*
edited by Joe Tidd (University of Sussex, UK)

Vol. 18 *Perspectives on Supplier Innovation: Theories, Concepts and Empirical Insights on Open Innovation and the Integration of Suppliers*
edited by Alexander Brem (University of Erlangen-Nuremberg, Germany) & Joe Tidd (University of Sussex, UK)

Vol. 17 *Managing Process Innovation: From Idea Generation to Implementation*
by Thomas Lager (Grenoble Ecole de Management, France)

Vol. 16 *Perspectives on User Innovation*
edited by Stephen Flowers (University of Brighton, UK) & Flis Henwood (University of Brighton, UK)

Vol. 15 *Gaining Momentum: Managing the Diffusion of Innovations*
edited by Joe Tidd (SPRU, University of Sussex, UK)

Vol. 14 *Innovation and Strategy of Online Games*
by Jong H. Wi (Chung-Ang University, South Korea)

Vol. 13 *Building Innovation Capability in Organizations: An International Cross-Case Perspective*
by Milé Terziovski (University of Melbourne, Australia)

Vol. 12 *Project-Based Organization in the Knowledge-Based Society*
by Mitsuru Kodama (Nihon University, Japan)

Vol. 11 *Involving Customers in New Service Development*
edited by Bo Edvardsson (Karlstad University, Sweden),
Anders Gustafsson (Karlstad University, Sweden),
Per Kristensson (Karlstad University, Sweden),
Peter Magnusson (Karlstad University, Sweden) &
Jonas Matthing (Karlstad University, Sweden)

Vol. 10 *Open Source: A Multidisciplinary Approach*
by Moreno Muffatto (University of Padua, Italy)

Vol. 9 *Service Innovation: Organizational Responses to Technological Opportunities & Market Imperatives*
edited by Joe Tidd (University of Sussex, UK) &
Frank M. Hull (Fordham University, USA)

Vol. 8 *Digital Innovation: Innovation Processes in Virtual Clusters and Digital Regions*
edited by Giuseppina Passiante (University of Lecce, Italy),
Valerio Elia (University of Lecce, Italy) &
Tommaso Massari (University of Lecce, Italy)

Vol. 7 *Innovation Management in the Knowledge Economy*
edited by Ben Dankbaar (University of Nijmegen, The Netherlands)

Vol. 6 *Social Interaction and Organisational Change: Aston Perspectives on Innovation Networks*
edited by Oswald Jones (Aston University, UK),
Steve Conway (Aston University, UK) & Fred Steward (Aston University, UK)

Vol. 5 *R&D Strategy and Organisation: Managing Technical Change in Dynamic Contexts*
by Vittorio Chiesa (Università degli Studi di Milano-Bicocca, Milan, Italy)

Vol. 4 *Japanese Cost Management*
by Yasuhiro Monden (University of Tsukuba, Japan)

Vol. 3 *From Knowledge Management to Strategic Competence: Measuring Technological, Market and Organizational Innovation*
edited by Joe Tidd (SPRU, University of Sussex, UK)

Vol. 2 *The Knowledge Enterprise: Implementation of Intelligent Business Strategies*
by J. Friso den Hertog (MERIT, Maastricht University & Altuïtion bv, 's Hertogenbosch, The Netherlands) &
Edward Huizenga (Altuïtion bv, 's Hertogenbosch, The Netherlands)

Vol. 1 *Engines of Prosperity: Templates for the Information Age*
by Gerardo R. Ungson (University of Oregon, USA) &
John D. Trudel (The Trudel Group, USA)